Marine Science

Marine Science

Edited by **Theodore Roa**

R CALLISTO
REFERENCE

New York

Published by Callisto Reference,
106 Park Avenue, Suite 200,
New York, NY 10016, USA
www.callistoreference.com

Marine Science
Edited by Theodore Roa

International Standard Book Number: 978-1-63239-665-5 (Hardback)

Printed in the United States of America.

Contents

Preface   VII

Chapter 1   **Application of a Coupled Vegetation Competition and Groundwater Simulation Model to Study Effects of Sea Level Rise and Storm Surges on Coastal Vegetation**   1
Su Yean Teh, Michael Turtora, Donald L. DeAngelis, Jiang Jiang, Leonard Pearlstine, Thomas J. Smith III and Hock Lye Koh

Chapter 2   **Bitumen on Water: Charred Hay as a PFD (Petroleum Flotation Device)**   30
Nusrat Jahan, Jason Fawcett, Thomas L. King, Alexander M. McPherson, Katherine N. Robertson, Ulrike Werner-Zwanziger and Jason A. C. Clyburne

Chapter 3   **Viral and Bacterial Epibionts in Thermally-Stressed Corals**   46
Hanh Nguyen-Kim, Thierry Bouvier, Corinne Bouvier, Van Ngoc Bui, Huong Le-Lan and Yvan Bettarel

Chapter 4   **Domestication of Marine Fish Species: Update and Perspectives**   61
Fabrice Teletchea

Chapter 5   **Design Optimization for a Truncated Catenary Mooring System for Scale Model Test**   78
Climent Molins, Pau Trubat, Xavi Gironella and Alexis Campos

Chapter 6   **Human Genotoxic Study Carried Out Two Years after Oil Exposure during the Clean-up Activities Using Two Different Biomarkers**   98
Gloria Biern, Jesús Giraldo, Jan-Paul Zock, Gemma Monyarch, Ana Espinosa, Gema Rodríguez-Trigo, Federico Gómez, Francisco Pozo-Rodríguez, Joan-Albert Barberà and Carme Fuster

Chapter 7   **Effect of Vegetation on the Late Miocene Ocean Circulation**   113
Gerrit Lohmann, Martin Butzin and Torsten Bickert

Chapter 8   **A Novel Mooring Tether for Highly-Dynamic Offshore Applications; Mitigating Peak and Fatigue Loads via Selectable Axial Stiffness**   136
Tessa Gordelier, David Parish, Philipp R. Thies and Lars Johanning

Chapter 9   **Time Evolution of Man-Made Harbor Modifications in San Diego: Effects on Tsunamis**   160
Aggeliki Barberopoulou, Mark R. Legg and Edison Gica

Chapter 10   **Reducing Reliability Uncertainties for Marine Renewable Energy**   182
Sam D. Weller, Philipp R. Thies, Tessa Gordelier and Lars Johanning

Chapter 11   **Longer-Term Mental and Behavioral Health Effects of the Deepwater Horizon Gulf Oil Spill**   195
Tonya Cross Hansel, Howard J. Osofsky, Joy D. Osofsky and Anthony Speier

Chapter 12   **Marine Microphytobenthic Assemblage Shift along a Natural Shallow-Water CO$_2$ Gradient Subjected to Multiple Environmental Stressors**   207
Vivienne R. Johnson, Colin Brownlee, Marco Milazzo and Jason M. Hall-Spencer

Chapter 13   **Soil Organic Carbon in Mangrove Ecosystems with Different Vegetation and Sedimentological Conditions**   230
Naohiro Matsui, Wijarn Meepol and Jirasak Chukwamdee

Chapter 14   **Tsunamigenic Earthquakes at Along-dip Double Segmentation and Along-strike Single Segmentation near Japan**   251
Junji Koyama, Motohiro Tsuzuki and Kiyoshi Yomogida

Permissions

List of Contributors

Preface

The main aim of this book is to educate learners and enhance their research focus by presenting diverse topics covering this vast field. This is an advanced book which compiles significant studies by distinguished experts. This book addresses successive solutions to the challenges arising in the area of application, along with it; the book provides scope for future developments.

Marine science studies the fauna and flora present in water bodies and their interaction with others. It includes marine and coastal management, oceanography, marine biology and marine ecology. This book discusses a wide array of topics in the light of new information and findings conducted by experts across the world. Marine resources, marine chemistry, marine environment, marine biotechnology, etc. are some of the important topics explained in it. This book is compiled in such a manner, that it will provide in-depth knowledge about the theory and practice of marine science. It is a complete source of knowledge on the present status of this important field. It is a ripe text for marine biologists, ecologists, professionals, researchers and students involved with the study of marine science at various levels.

It was a great honour to edit this book, though there were challenges, as it involved a lot of communication and networking between me and the editorial team. However, the end result was this all-inclusive book covering diverse themes in the field.

Finally, it is important to acknowledge the efforts of the contributors for their excellent chapters, through which a wide variety of issues have been addressed. I would also like to thank my colleagues for their valuable feedback during the making of this book.

Editor

Application of a Coupled Vegetation Competition and Groundwater Simulation Model to Study Effects of Sea Level Rise and Storm Surges on Coastal Vegetation

Su Yean Teh [1,†], Michael Turtora [2,†], Donald L. DeAngelis [3,*], Jiang Jiang [4], Leonard Pearlstine [5], Thomas J. Smith III [6] and Hock Lye Koh [7]

[1] School of Mathematical Sciences, Universiti Sains Malaysia, Penang 11800, Malaysia;
E-Mail: syteh@usm.my

[2] U.S. Geological Survey, Caribbean-Florida Water Science Center, 4446 Pet Lane, Suite #108, Lutz, FL 33559-630, USA; E-Mail: mturtora@usgs.gov

[3] U.S. Geological Survey, Southeast Ecological Science Center, Coral Gables, FL 33124, USA

[4] Jiang Jiang, Forestry College of Nanjing Forestry University, Key Laboratory of soil and water conservation and Ecological Restoration, Nanjing Forestry University, Nanjing 210037, China; E-Mail: ecologyjiang@gmail.com

[5] Leonard Pearlstine, Everglades National Park, South Florida Natural Resources Center, 950 N Krome Ave, Homestead, FL 33030, USA; E-Mail: Leonard_Pearlstine@nps.gov

[6] U.S. Geological Survey, 600 Fourth Street South, St. Petersburg, FL 33701, USA;
E-Mail: tom_j_smith@usgs.gov

[7] Hock Lye Koh, Sunway University Business School, Jalan Universiti, Bandar Sunway, Selangor 47500, Malaysia; E-Mail: hocklyek@sunway.edu.my

[†] These authors contributed equally to this work.

[*] Author to whom correspondence should be addressed; E-Mail: don_deangelis@usgs.gov

Academic Editor: Rick Luettich

Abstract: Global climate change poses challenges to areas such as low-lying coastal zones, where sea level rise (SLR) and storm-surge overwash events can have long-term effects on vegetation and on soil and groundwater salinities, posing risks of habitat loss critical to native species. An early warning system is urgently needed to predict and prepare for the consequences of these climate-related impacts on both the short-term dynamics of salinity in the soil and groundwater and the long-term effects on vegetation. For this purpose, the U.S. Geological Survey's spatially explicit model of vegetation community dynamics along coastal salinity gradients (MANHAM) is integrated into the USGS groundwater model (SUTRA) to create a coupled hydrology–salinity–vegetation model, MANTRA. In MANTRA, the uptake of water by plants is modeled as a fluid mass sink term. Groundwater salinity, water saturation and vegetation biomass determine the water available for plant transpiration. Formulations and assumptions used in the coupled model are presented. MANTRA is calibrated with salinity data and vegetation pattern for a coastal area of Florida Everglades vulnerable to storm surges. A possible regime shift at that site is investigated by simulating the vegetation responses to climate variability and disturbances, including SLR and storm surges based on empirical information.

Keywords: coupled hydrology–vegetation model; salinity; coastal Everglades; hardwood hammock; mangroves; vadose zone; groundwater

1. Introduction

Sea Level Rise (SLR) is one of the most significant predicted consequences of global climate change and has the potential for severe effects on the vegetation of low-lying coastal areas and islands [1]. Rising sea level will also mean higher storm surges [2], even if the frequencies do not change. Mean SLR will have a gradual effect on shoreline retreat and subsequent loss of ecosystem area in these locations. However, large-scale marine water intrusion through storm surges may affect large areas on a short time scale, including the inundation of whole low-lying islands. The immediate effect will be on the freshwater lenses that sit on top of saline groundwater in these areas. Such effects on available fresh water may have negative consequences for the ecological and human populations of coastal areas and, particularly, islands, as they depend critically on fresh water stored in the lenses [3–5]. Longer-term consequences may involve large-scale vegetation regime changes, which can pose risks to conservation and restoration efforts in coastal national parks and preserves. For example, in southern Florida, USA, the beneficial effects of increased freshwater flow resulting from the Comprehensive Everglades Restoration Plan (CERP) [6] may be compromised in some places by increased saltwater intrusion and salinity overwash events.

In tropical and subtropical coastal areas, increase in salinity of the vadose zone (unsaturated soil zone) induced by storm surges might reduce or eradicate the salinity-intolerant (glycophytic) species and promote rapid landward migration of salinity-tolerant (halophytic) species such as mangroves. Inland expansion of mangroves at the expense of glycophytic vegetation has been noted in coastal ecosystems, e.g., see [7,8]. The effect of a disturbance may cause a rapid shift in the transition zone between vegetation types, also called the ecotone [9]. While many ecotones between floristic types are broad and diffuse, some are remarkably narrow. An example of the sharpening of ecotones in coastal areas involves halophytic vegetation (mangrove vegetation in tropical and sub-tropical regions) and glycophytic vegetation (including tropical hardwood trees forming "hammocks", and freshwater marsh) in southern Florida coastal areas [10]. Typically, these are not interspersed. Stability of the ecotone is promoted by self-reinforcing positive feedback as follows [11]. During the dry season, plant transpiration can lead to infiltration by highly brackish underlying ground water into the vadose zone. Hardwood hammock trees reduce transpiration when the salinity in the vadose zone increases. This limits the salinization of the vadose zone. Meanwhile, the transpiration of mangroves can continue even at relatively high salinities, sustaining salt-water infiltration. Thus, through self-reinforcing positive feedback, each type of vegetation has a tendency to promote the salinity condition that is favorable to itself in competition.

However, both SLR over decades and the acute effects of large storm surges may upset that stability. If a large enough pulse of salinity from overwash remains in the soil for a long period, it may overwhelm the feedback maintaining the favorable conditions for glycophytic vegetation, such that the ecotone can no longer be maintained; that is, a regime shift could occur. Large areas of glycophytic vegetation could be replaced by halophytic vegetation, which may lead to permanent salinization of the vadose zone. For example, Baldwin and Mendelssohn [12] studied the effects of salinity and inundation coupled with clipping of aboveground vegetation on two adjoining plant communities, *Spartina patens* and *Sagittaria lancifolia*. The study reported that that the levels of flooding and salinity at the time of disturbance determined the potential shift of vegetation to a salt tolerant species. Large storm surges created by Hurricanes Katrina and Rita (2005) affected the coastal areas of Louisiana, USA. Subsequently, both freshwater and brackish communities exhibited changes in vegetation [13]. It was noted that in the central region of their study area, marsh composition changed to a more saline classification, and high mean salinities exceeded mean pore-water salinity levels tolerated by the previous dominant species.

Hindcasting the effects of previous storm surges and on coastal vegetation and forecasting the future effects of both storm surges and SLR requires modeling. In particular, forecasting these effects on the halophyte-glycophyte ecotone requires that the competition of these two vegetation types be modeled along with hydrology and salinity dynamics. To investigate the dynamics of the ecotone between halophytic and glycophytic vegetation, including the possibility of a regime shift, a spatially explicit computer simulation model, MANHAM (MANgrove HAMmock model), has been developed [14].

MANHAM simulates the competition of hardwood hammock trees and mangrove trees on a grid of spatial cells; each a few square meters in area. Vegetation of each type may be present in a given cell, and growth and competition are modeled on this local scale. Dispersal of propagules of each species is also modeled. In cells where vadose zone salinity is low, hammock trees grow faster and outcompete mangroves, but in higher salinity cells, hammock tree growth is slowed and mangroves can outcompete the hammock trees.

MANHAM also models water flow and salinity in the vadose zone, which depends on precipitation, tides, evaporation, plant transpiration, and groundwater infiltration. The dynamics of hydrology and salinity are modeled on a time resolution of less than a day, whereas the dynamics of vegetation is modeled on monthly time steps. The key mechanism in the model is the self-reinforcing positive feedback relationship between each vegetation type and vadose zone salinity described above. In each spatial cell these feedbacks help maintain dominance of the current vegetation type in the cell, hammock or mangrove, against invasion by the other type. However, a large external impact on the vadose zone salinity of a cell dominated by hammock trees, such as from storm surge overwash, could lead to decline in hardwood growth and favor growth of mangrove propagules, or seedlings, into trees. Through positive feedback between the growing mangrove vegetation and the vadose zone salinity, the cell could eventually shift to mangrove domination. Shifts from mangrove to hardwood hammock are also possible if there is a sufficient external forcing of freshwater. Mathematical details of MANHAM are described in Appendix 1.

MANHAM has examined the impact of SLR on southern Florida coastal forests [15], showing that buttonwood forest (*Conocarpus erectus*), could be squeezed out by red mangrove (*Rhizophora mangle*). Simulations have also indicated that a significant one-day storm surge event could feasibly initiate a vegetation shift from hardwood trees to mangroves in areas initially dominated by the former. Mangroves in the model were able to take over large areas when storm surges saturated the vadose zone with over 0.015 kg/kg salinity, if the salinity was not quickly washed out of the vadose zone by precipitation, and if a sufficient density of mangrove propagules (seedlings, by which mangroves are spread) were present. It was observed that such a shift might not be conspicuous at first, but would be inevitable once a tipping point had been passed. These findings are relevant to many coastal areas, including the coastal Everglades. They have motivated us to apply the modeling to forecasting future changes in coastal vegetation and developing plans to meet the challenges of these changes.

MANHAM simulates the vadose zone as a uniform compartment and does not model underlying groundwater dynamics [14] and assumes that ground water is a constant boundary condition. These assumptions are certainly violated in real systems. For example, in low-lying coastal areas and small atoll islands, tides cause diurnal fluctuations in groundwater, and storm surges can cause major changes in groundwater that can last for years. A deficiency caused by these assumptions in MANHAM is that it does not consider the freshwater lens, which in coastal areas and atolls typically overlies deeper water salinity levels of the neighboring seawater. This freshwater lens is an important constituent of the water balance for the overlying vegetation through transpiration and plays a key role on the salinity balance as well. Because of these deficiencies we developed MANTRA (MANhamsuTRA), which builds in more detailed hydrological and salinity dynamics. Below we describe the USGS's Saturated–Unsaturated TRAnsport (SUTRA) groundwater model and its coupling with MANHAM to form MANTRA. We apply MANTRA to a coastal area of Everglades National Park (ENP), southern Florida, and demonstrate how it can be used to project the effects of both storm surges and SLR on coastal vegetation.

2. Methods

2.1. MANTRA Model

To overcome the limitation due to MANHAM's lack of a freshwater lens, MANHAM has been integrated with an established groundwater hydrology and salinity model, the United States Geological Survey (USGS)'s Saturated–Unsaturated TRAnsport (SUTRA) groundwater model [16,17]. The fluid pressure and salinity gradients in the transition zone between glycophytic and halophytic vegetation associated with the freshwater lens are quantified by the variable density flow simulated by SUTRA. SUTRA also simulates the unsaturated zone of the soil, and so can substitute for the hydrodynamics and salinity dynamics of the soil and groundwater. However, SUTRA does not include vegetation competition dynamics. By combining MANHAM with SUTRA, forming MANTRA (MANhamsuTRA), we provide an integrated model that simulates the possible effects of gradual SLR, as well as both short- and long-term effects of a single or a sequence of overwash events on a coastal area or small island, containing zones of glycophytic and halophytic vegetation.

MANTRA input data are (1) vegetation type and (2) groundwater conditions (fluid pressure and salinity) and it simulates the changes in vegetation type biomass over time subject to the groundwater conditions. Because MANTRA is an extension of SUTRA, it also delivers the output of SUTRA, including fluid pressure and solute concentration. The primary variable upon which the groundwater model SUTRA is based is fluid pressure, which varies spatially and temporally. Variations in fluid density and fluid pressure differences drive flow of groundwater, which is a fundamental mechanism upon which the solute transport model is based. MANTRA employs spatial discretization called mesh by quadrilateral finite elements. In a cross sectional form, the elements are organized in rows and columns with each element having four nodal points. Nodal points (or nodes) are shared by the elements adjoining the node. A cell is centered on a node, not an element. Cell boundaries are half way between opposite sides of an element as shown in Figure 1. Further details of MANTRA are described in Appendix 2.

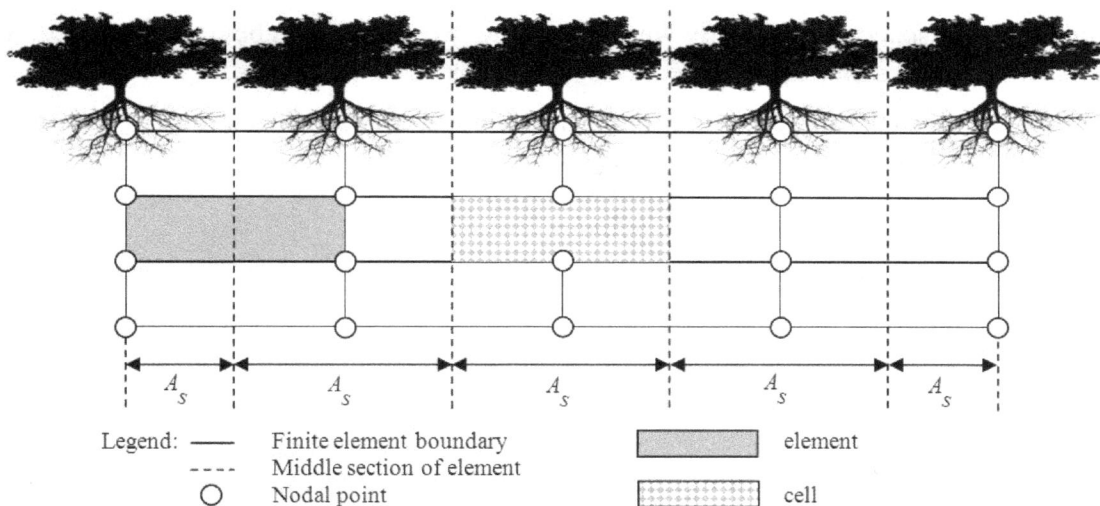

Figure 1. Schematic sketch of a hypothetical simulation case for illustration purposes.

Here we describe an important aspect of how the hydrology of SUTRA and the vegetation dynamics of MANHAM interact, which is through water uptake by plants. The fluid mass source/sink term Q [M/s] (where M = mass and s = second) in SUTRA, which accounts for external addition/subtraction of fluid including pure water mass plus the mass of any solute dissolved in the source fluid, can be used to characterize the uptake of water by plants (Q_p) [18–20]. This term in a mass balance equation is used to represent the addition (source) or extraction (sink) from the mass balance system. As a function of salinity, the total water uptake $R = f(C)$ [L/s] (where L here is the dimension of vertical distance or depth) by plants is determined by the salinity concentration C [M_s/M_f] (where M_s = mass of solute and M_f = mass of fluid) calculated by SUTRA. The salinity C is derived from the solute mass balance equation that includes processes such as fluid flow and diffusion. Then, assuming a closed canopy so that transpiration can be assumed constant for the vegetation types and evaporation ignored, the fluid mass per unit time (Q_p) required by the plants in a certain horizontal cell for transpiration can be estimated by:

$$Q_p = R \cdot A_s \cdot \rho \ [M/s] \tag{1}$$

Here, A_s represents the surface along the depth dimension $[L^2]$ and ρ is the fluid density of fresh water $[M/L^3]$. For cross-sectional model, the width of each cell is assumed to be 1.0 m. Thus, A_s depends on the length or horizontal grid size (Figure 1).

It may happen that the fluid mass required by the plant for transpiration is more than what is available (unsaturated flow). Hence, there should be a relation between the fluid mass required by transpiration and the fluid mass available. It is assumed that actual fluid mass being subtracted from a cell due to transpiration depends on the saturation S_w and porosity ε in the cell, leading to the following relation:

$$Q_{IN} = -Q_p \cdot \varepsilon \cdot S_w \ [M/s] \tag{2}$$

where, Q_{IN} = total mass sink (due to plant transpiration) $[M/s]$; ε = porosity $[V_v/V]$; S_w = water saturation $[V_w/V_v]$ with V = total volume, V_v = volume of voids, V_w = volume of water. Suppose porosity ε is kept constant at 1.0. When the void space is completely filled with fluid and is said to be saturated, that is $S_w = 1$, the actual water uptake by plants, Q_{IN}, will be equivalent to the amount of water required by the plants, Q_p. When the void space is only partly water filled and is referred to as being unsaturated, that is $S_w < 1$, the actual water uptake by plant will decrease as a factor of the saturation, S_w.

To simulate fluid inflow to the soil pores due to seawater inundation during a storm surge, Equation (3) below [21] was used to specify the pressure at surface nodes inundated by seawater.

$$p = \rho_{sea} \cdot g \cdot h \tag{3}$$

Here, p = pressure at top layer nodes $[M/(Ls^2)]$, ρ_{sea} = fluid density of seawater $[M/L^3]$, g = gravity $[L/s^2]$, h = inundation depth $[L]$.

2.2. Study Transect

To evaluate MANTRA's effectiveness in a coastal Everglades setting, we chose a site and a set of scenarios related to past and possible future events. MANTRA was used on a coastal site that has been exposed to storm surges. The main objective of MANTRA is to project possible future changes in hardwood hammocks in southern Florida under conditions of gradually rising sea level and/or major storm surges. As an example, we apply MANTRA to a specific hardwood hammock along the southwestern coast of ENP, bordering Florida Bay (25°12′24.13″ N, 80°55′47.48″ W). This hammock has been described by Saha *et al.* [22], where it is referred to as the Coot Bay Hammock (see aerial view in Figure 2). The hammock is in the middle of a low ridge with north-south orientation. Along a 370 m transect from west to east across the hammock (see Figures 2 and 3), vegetation changes from black mangroves (*Avicennia germinans*) at the low elevation (0.4–0.5 m above mean sea level) western end to mixed halophyte coastal prairie (*Batis/Salicornia*) (0.5–0.7 m) with individual buttonwoods (*Conocarpus erectus*) (0.7–0.8 m), to mixed-hammock, and hardwood hammock communities (0.9–1.5 m) at the peak in elevation in the middle of the ridge. As one continues farther along the transect, the associations occur in reverse order, except for a change from white to black mangroves. The transitions are relatively sharp. The geology is marl mixed with peat to a depth of 2–3 m over karst bedrock. Table 1 shows assumed ranges of conductivities. The site has a tropical climate with an average of 1570 mm annual precipitation, nearly 60% of which is from June through September. Mean January and

July temperatures are, respectively, 22 °C and 30 °C. Hourly salinity data of water at 0.5 m below land surface were available at two locations (blue dots near transect on Figure 2) and monthly point measurements of salinity and water depth at five locations (other blue dots on Figure 2) from in February 2011, through July 2012. Coot Bay Hammock is an ideal study transect because hardwood hammocks occur at the highest, and mangroves and coastal prairies occur at the lowest end of the gradient in elevation.

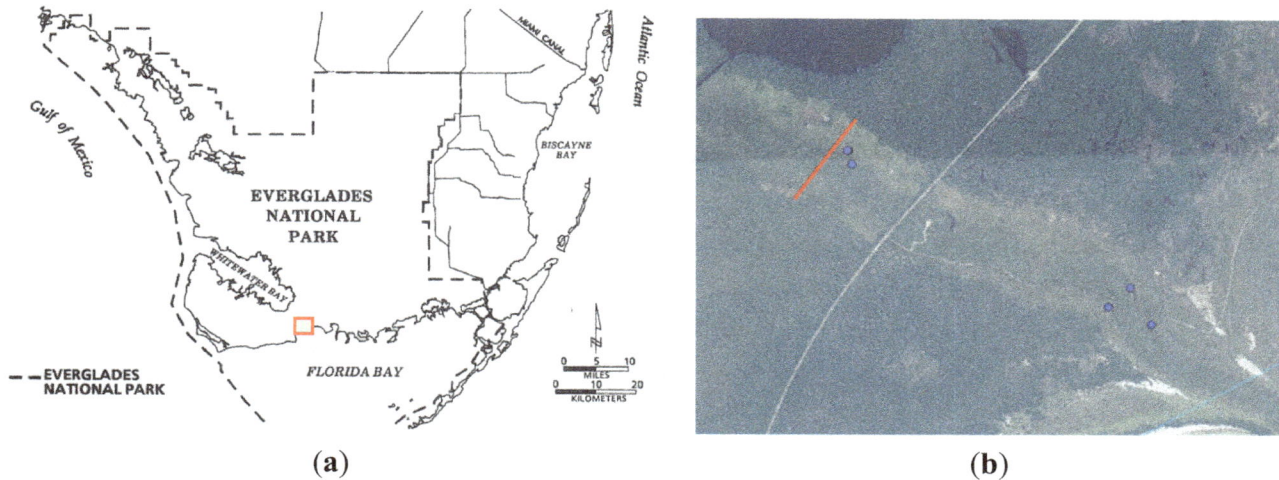

(a)   (b)

Figure 2. (**a**, **Left**) Southern Florida with site shown in red box (from Light and Dineen 1994 [23]). (**b**, **Right**) Aerial view of Coot Bay Hammock, with 370 m transect, shown in red. Blue dots indicate well sites. l. (Figure 2a adapted from reference [23] with copyright permission).

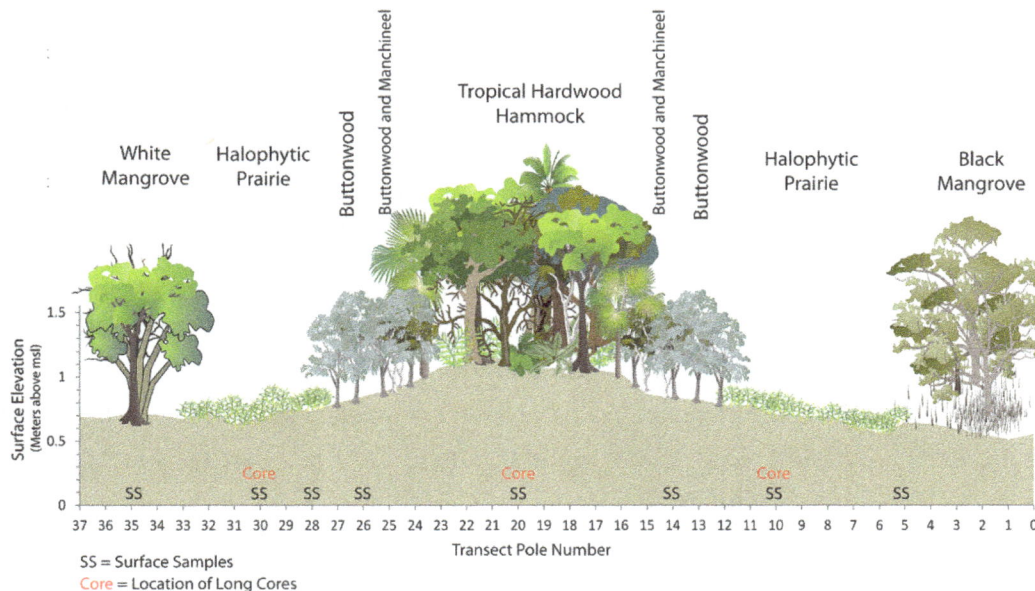

Figure 3. West-to-east transect of about 370 m across the Coot Bay Hammock showing the sharp gradations between vegetation types. Transect poles mark 10 m distances, and locations where surface samples and core samples were taken. Figure courtesy of Brandon Gamble, National Park Service. Graphic symbols courtesy of the Integration and Application Network, University of Maryland Center for Environmental Science.

Table 1. Stratigraphy and hydraulic conductivities used for the Coot Bay Hammock. Parameter values are assumed from other sources.

Layer Name	Lower Boundary (m)	Hydraulic Conductivity (Qualitative)	(m/Day)
Marl plus peat	~2–3	low	<0.3
karst	~6	high	>1000
sand	~9	medium	10 to 100
noflow	~9		

Saha *et al.* [22] noted an increase in salinity over the last decade (data from hydrological station maintained by ENP). Saha *et al.* [22] suggest that SLR can induce a rising water table, which will cause a shrinking of the vadose zone and an increase in salinity in the bottom portion of the freshwater lens, subsequently increasing brackishness of plant-available water. For these reasons, the Coot Bay Hammock was selected as a first site to test MANTRA. The 370 m transect was modeled as two-dimensional, with the horizontal dimension along the transect, and the vertical axis for depth.

Storm surge/SLR would most likely occur from the lower right of Figure 2, right panel. This happened in November 2005 due to Hurricane Wilma [24]. A storm surge could also enter the area through a canal system. The park road through ENP to the town of Flamingo on Florida Bay (visible in Figure 2) cuts across the southern tip of the tropical hardwood forest that extends to the NW. The ridge extends further SE with vegetation that was identified in 1981 as "collapsed hammock" [22] and is now referred to as a transitional buttonwood hardwood hammock.

The Coot Bay Hammock area is vulnerable to both wind damage and storm surges from hurricanes. Hurricanes have been important in shaping the vegetation of the region. The "Labor Day" hurricane of 1935 killed many buttonwoods, while Hurricane Donna (1960) produced a storm surge of 4 m in this area, causing 90% mortality of trees on lower ground, and 25%–50% mortality buttonwoods and hardwood hammock trees on coastal hammocks such as the Coot Bay Hammock. These hurricanes were factors that "produced new vegetation mosaics of white, black and red mangroves and buttonwoods" [25]. Buttonwoods have not recovered in some areas they once dominated. While fires have been suggested as a possible factor in their loss, Olmstead and Loope [25] discount this, which leaves open the possibility that a storm surge induced regime shift was the cause. Two hurricanes that more recently have affected the Coot Bay Hammock area were Hurricane Andrew (1992) and Hurricane Wilma (2005). The storm surge from Andrew measured 1.2–1.5 m at Flamingo, near Coot Bay Hammock, while that from Wilma measured 0.7 m at Black Forest, also near Coot Bay Hammock, see [26]. Hurricane Andrew caused very little structural damage to trees that far south, while Wilma caused moderate though not severe damage.

2.3. Model Simulations

A 2-D model of the transect across the Coot Bay Hammock was developed using MANTRA. For simplicity, buttonwoods were aggregated with the hardwood hammock, and hardwood hammock and mangroves were modeled as competing vegetation types. Coastal hammocks, tree islands, and buttonwood forests of Florida Bay experience tidal amplitude of only ~15 cm [27], so tides were ignored in the model. Model scenarios are described below.

2.3.1. Scenario 1: Existing Conditions

The model was first applied to the existing conditions of the Coot Bay Hammock. The aim was to calibrate the model to produce results that are consistent with the observed data; that is, with the observed sharp boundary and with virtually no mangroves in areas dominated by hardwood hammocks and *vice versa*.

2.3.2. Scenarios 2 and 3: Storm Surges

Preliminary MANTRA simulations (not shown here) showed that neither of the storm surges plus the associated light damage to the hammock from Hurricanes Andrew and Wilma would be sufficient to cause a regime shift from hardwood hammock to mangroves. However, sites such as Coot Bay Hammock have been struck by larger hurricane disturbances in the more distant past, and will be in the future. Therefore, to project the effects of greater disturbances, two storm surge scenarios were applied that used inundation depths consistent with those caused by Hurricane Andrew, *i.e.*, about 1 m, but which had some additional factors that would amplify the effects of the surge. Scenario 2 assumes that the storm inflicted heavy damage to the hardwood hammock trees, reducing their living biomass to below the level of mangrove seedlings. Scenario 3 consisted of a storm surge in which only moderate damage was done to the hardwood hammock vegetation, reducing the initial vegetation by one-half. However, the storm surge was followed immediately by a severe four-year drought in which mean precipitation was reduced by half. Although a four-year drought starting immediately following a hurricane would be unusual, periodic droughts are part of the climate of southern Florida, and a drought of this magnitude is plausible.

2.3.3. Scenario 4: Gradual SLR

A final scenario (Scenario 4) of gradual SLR was applied over a period of 150 years without storm surge events.

3. Results

3.1. Scenario 1: Existing Conditions

The results for the distribution of mangroves and hardwood hammock trees for the calibrated MANTRA are shown in Figure 4. Consistent with the vegetation distribution observed at the Coot Bay Hammock, our results show the hardwood hammocks occupy the slightly elevated ridge, where a freshwater lens of about 0.5 m is maintained by precipitation and feedback from hardwood hammocks, whereas the mangroves occupy the lower elevated areas on either side of the hardwood hammock area.

3.2. Scenarios 2 and 3: Storm Surges

Next, storm surge Scenario 2 was applied using MANTRA. The results of the simulation are shown in Figure 5. The process of the positive feedback involving increased mangrove invasion causing increased soil salinity, allowed the mangroves to take over in about 16 years. Scenario 3 consisted of a storm surge in which only moderate damage was done to the hardwood hammock vegetation, but was followed by a severe four-year drought. The results are shown in Figure 6, where again the mangroves took over the entire site within about 16 years. Again, the positive feedback loop of mangrove invasion and increasing salinity drove the transition.

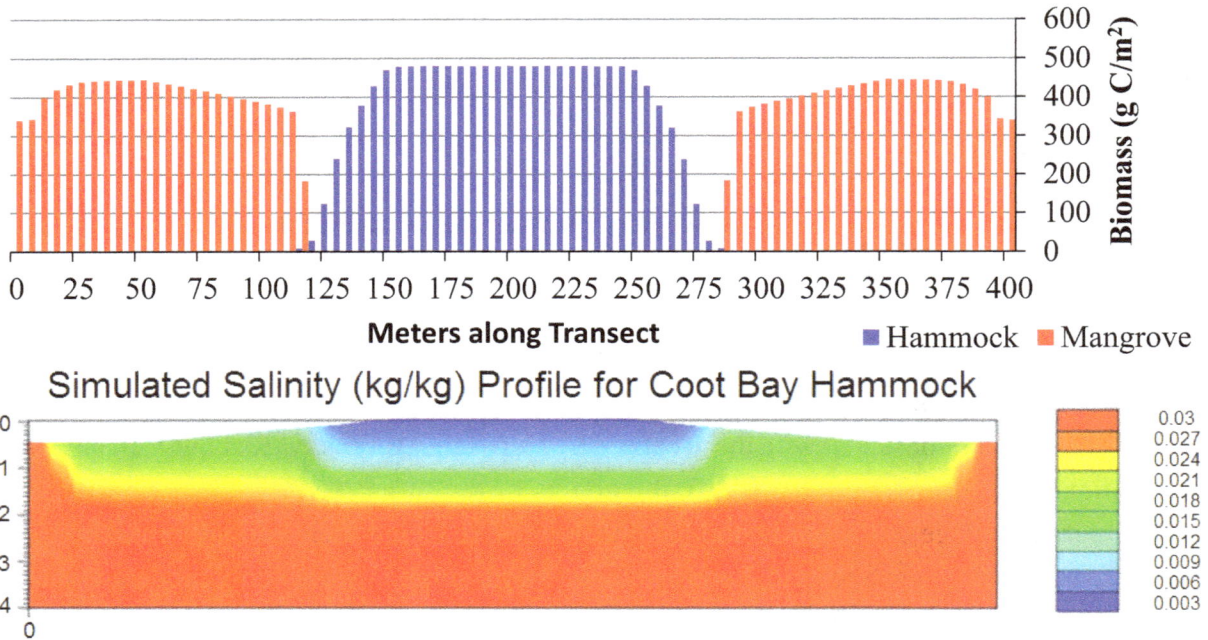

Figure 4. (**Top**) Simulated distribution of mangroves (red) and hardwood hammock (blue) trees along a 370-meter transect (rounded to 400 m in the model) across the Coot Bay Hammock. (**Bottom**) Simulated salinity profile with ground depth in meters and salinity in kg/kg.

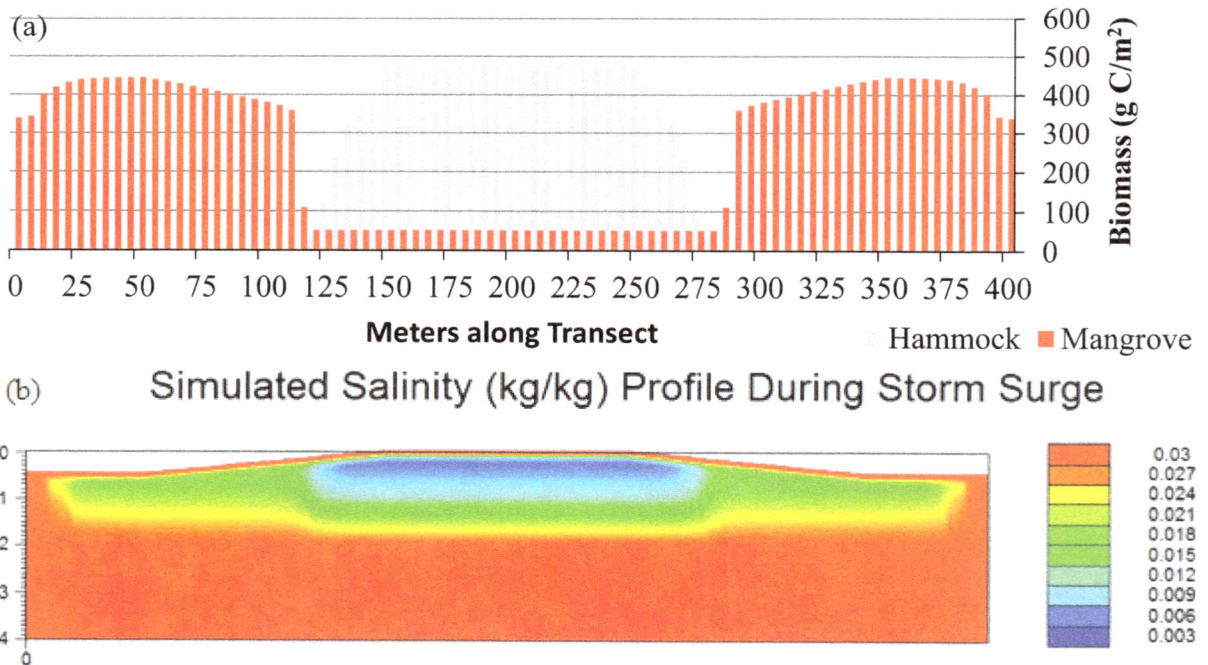

Figure 5. *Cont.*

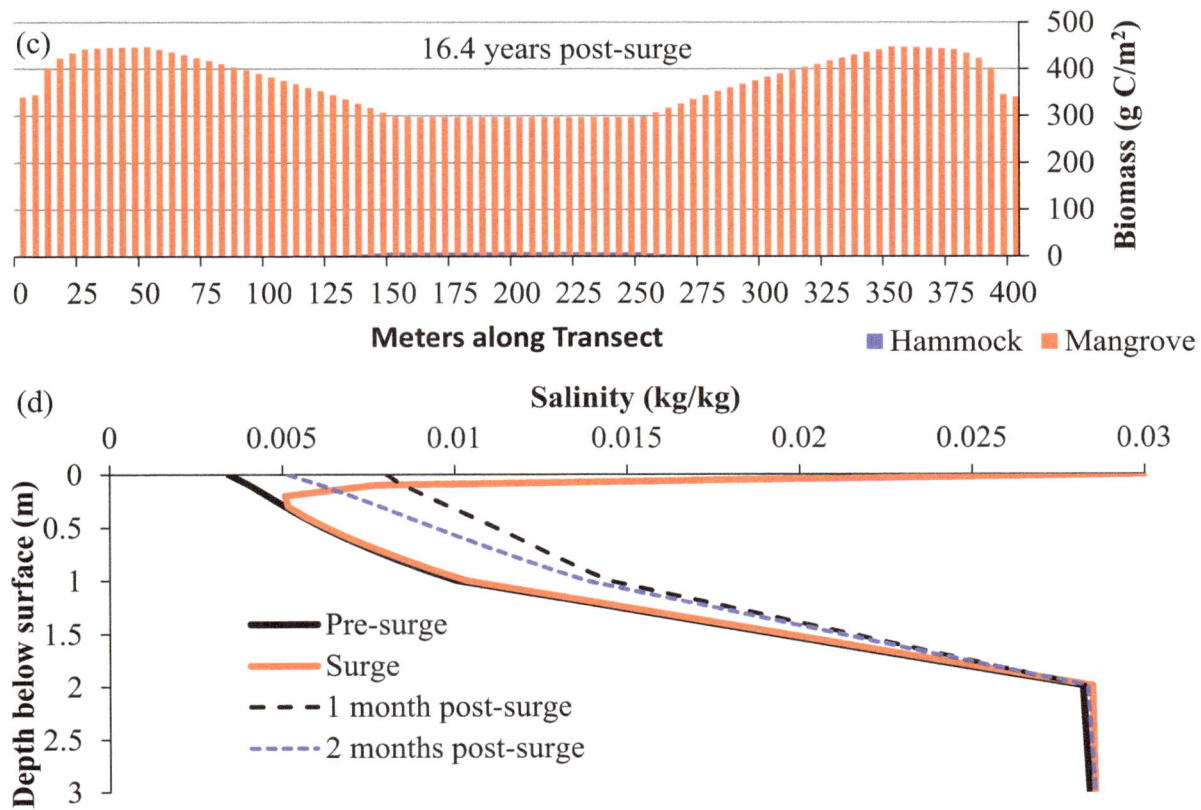

Figure 5. (**a**, **Top**) Depiction of initial conditions following storm surge, with almost complete elimination of hardwood hammock living biomass through knockdown of trees. Faded blue colors indicate destruction of initial trees. (**b**, **Second from top**) Simulated salinity profile during the storm surge. (**c**, **Third from top**) Simulated distribution of mangrove (red) vegetation 16.4 years (6000 days) after the storm surge. (**d**, **Bottom**) Sequence of salinity profiles starting before the surge until two months after, as predicted by MANTRA, with the particular precipitation pattern.

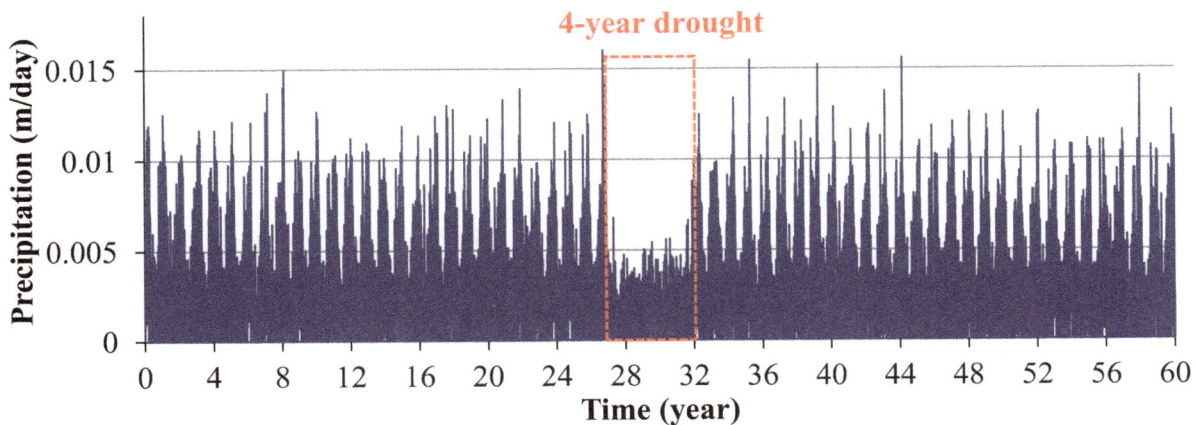

Figure 6. *Cont.*

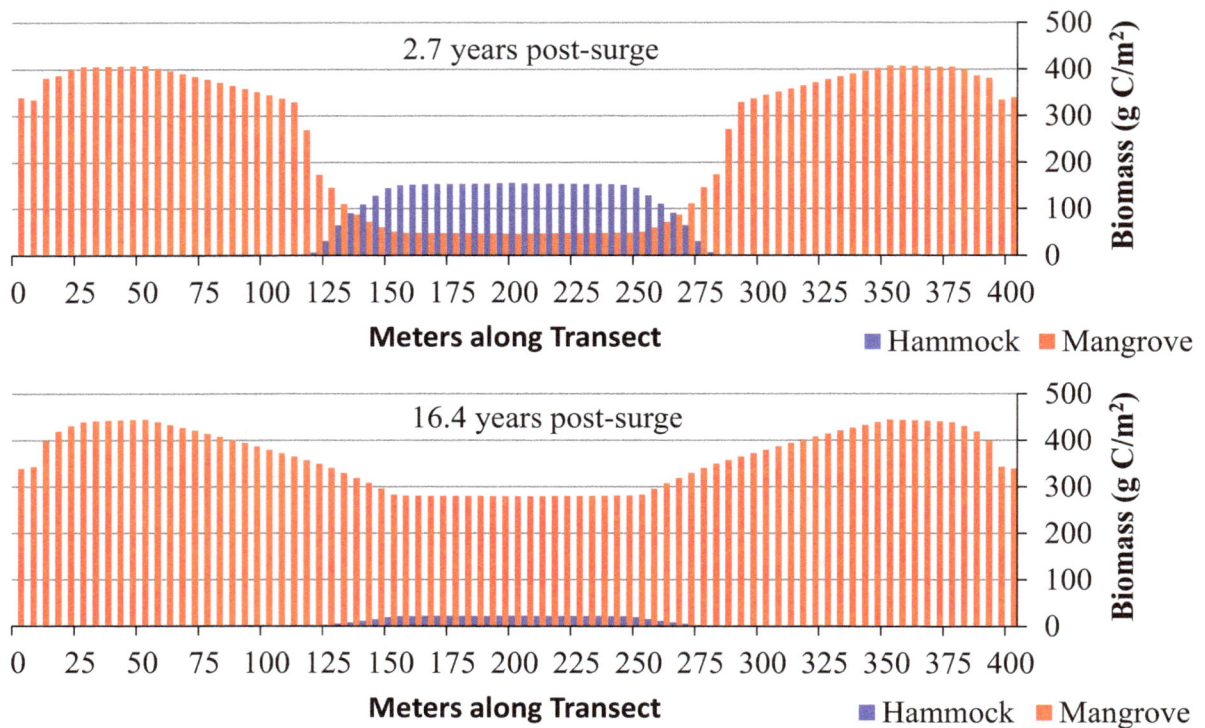

Figure 6. (**Top**) Depiction of precipitation over time used in the scenario, with four-year drought. (**Middle**) Simulated distribution of mangrove (red) and hardwood hammock (blue) vegetation 2.7 years (1000 days) after the storm surge, which occurs at 27.3 years into the simulation. (**Bottom**) Simulated distribution of mangrove (red) vegetation 16.4 years (6000 days) after the storm surge.

3.3. Scenario 4: Gradual SLR

Figure 7 illustrates Scenario 4, of SLR effect on the vegetation distribution at the Coot Bay Hammock transect. We consider here only the effects of sea level rise and ignore any possible effects of major storm surges. These simulation results indicate the mangroves will encroach into the areas of hardwood hammock, confining the freshwater vegetation to a smaller area. Hardwood hammock would persist on the elevated ridge. Further simulations (results not shown) indicate that the hardwood hammocks appear to persist at the elevated ridge unless the sea level rises to a level where the ridge is frequently inundated; e.g., every 20 years or so, with seawater.

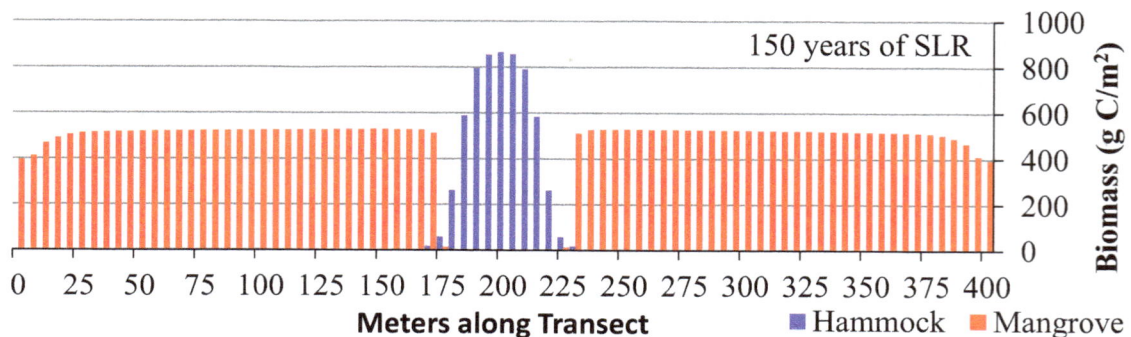

Figure 7. *Cont.*

Simulated Salinity (kg/kg) Profile - 150 years of SLR

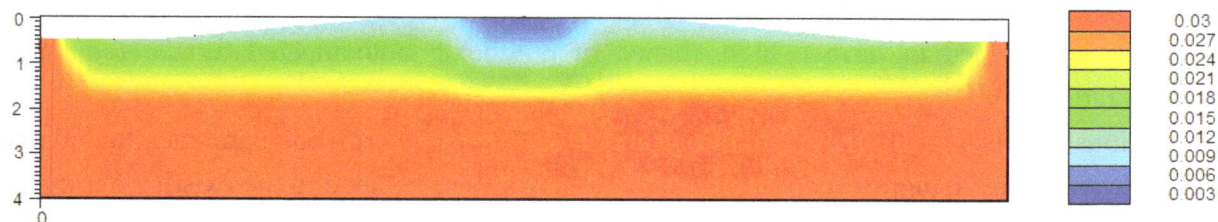

Figure 7. (Top) Simulated distribution of mangroves (red) and hardwood hammock (blue) trees along a 400-meter transect across the Coot Bay Hammock after 150 years subject to plausible SLR scenario (3 mm/year). **(Bottom)** Simulated salinity profile.

4. Discussion

Application of MANTRA to the Coot Bay Hammock transect provides insights on the potential vulnerability of the vegetation to storm surges and SLR. Simulations (not shown) that assumed storm surges of about 1 m, but little damage to trees and no prolonged drought did not produce regime shifts of the Coot Bay Hammock to halophytic vegetation. This is consistent with observations in the field showing no recent signs of the start of a shift. The scenarios that we presented that included damage or prolonged drought indicated that a shift might occur under those circumstances. Simulation results for Scenario 2 indicate that mangroves might be able to take over the slightly elevated ridge previously dominated by hardwood hammocks after a storm surge if the surge inflicts heavy damage to the hardwood hammock trees. It is possible that the effects of earlier hurricanes, Hurricane Donna (1960) in particular, may have led to permanent changes over parts of the Coot Bay hammock area. That hurricane produced a 4 m storm surge in the area and destroyed many buttonwood trees, which have not returned. This could be explained by an event like Scenario 2. Precipitation is the source for groundwater lens recharge. Scenario 3, a moderate storm surge followed by a prolonged period of drought, could also cause the shift from hardwood hammocks to mangroves, even though the hardwood hammocks were not badly damaged by the surge.

MANTRA improves greatly over MANHAM on the resolution with which hydrology and salinity are simulated, particularly along the vertical axis. The previous results of MANHAM suggested the possibility of a regime shift from a storm surge [14], but it was cautioned that that it was only a hypothetical result based on simple assumptions of water budget for hydrological dynamics. SUTRA is a highly detailed hydrologic model that has been tested in many contexts and can reliably predict hydrology and solute dynamics if provided good parameter values. Therefore, in MANTRA, the vadose zone is now connected seamlessly with the ground water, rather than the latter being treated as a boundary condition. A freshwater lens emerges naturally above the groundwater in simulations. Importantly, this more realistic treatment of hydrology and salinity dynamics does not change the emergence of sharp boundaries between glycophytic and halophytic vegetation, which was observed in MANHAM [11,14]. This gives us confidence that the self-reinforcing positive feedbacks hypothesized to be acting between each vegetation type and its local soil environment are a reasonable explanation for the sharp ecotone observed, and that these feedbacks may provide resilience to storm surge disturbances that are not too strong. MANTRA shows, like MANHAM, that a storm surge can cause a regime shift, but it is more conservative, as it shows salinity washing out faster unless there is a drought. So the results of MANTRA show that wind damage to the freshwater vegetation must be severe enough that the mangrove seedlings washed in have a high chance of not being outcompeted by the remaining freshwater vegetation.

Our simulations underscore that three conditions are necessary for a hardwood hammock to undergo a regime shift leading to a mangrove community; sufficiently severe damage to the existing hammock to open a gap to allow growth of invading seedlings, a large input of salinity persisting for a long enough period of time to favor growth of mangrove seedlings in competition remaining freshwater vegetation, and an input of enough mangrove seedlings to allow mangroves to be present in sufficient number to influence the future soil salinity. Surveys of hurricane damage to Everglades hardwood hammocks from Hurricane Andrew (1992) provide an estimate of what a major hurricane can inflict. Studies show heavy damage to 85% of all stems >2 cm diameter and loss of almost all leaves [28,29]. As pointed out in another study [30], the damage provided opportunities for invasive seedlings, which negatively influenced regrowth of the native vegetation. If such damage were inflicted on a coastal hammock, the damage would likely be accompanied by storm surge overwash. Hydrodynamic simulations have been performed of the effects of hurricanes on southern Florida [31]. In particular, those authors performed a hindcast of the "Great Miami Hurricane" of 18 September 1926, including the subsequent meteorological conditions. Simulations of surface water and groundwater salinity in an on shore area showed that salinity levels above 5 g/kg could remain in the soil for close to three years (until July 1929 in their Figure 10), and longer in the upper layer of groundwater. Of course, persistence of soil salinity conditions will vary from location to location with geology and freshwater influxes. However, even short-term exposure to salinity would kill much salinity-intolerant vegetation [32], and the persistence of the high salinity levels for two or three years shown in the modeling of [31] is more than sufficient in MANTRA to favor mangrove seedlings over freshwater vegetation regrowth. Input of mangrove seedlings by storm surges has rarely been studied, but on the basis of one study [33], Jiang *et al.* [34] estimated that up to 2000 propagules per ha could be input from nearby mangrove forest. It has also been suggested that a strategy of mangroves is to constantly produce a large number of seedlings [35] that can be spread by high tides, wind, or animals to provide a "sit-and-wait" seedling bank. Mangrove seedlings and small plants are commonly observed in nearby freshwater areas (*personal observation*).

MANTRA was developed as a tool to study the potential impact of SLR and storm surges on competing halophytic and glycophytic vegetation and, in particular, to investigate the hypothesis that a large input of salinity to a community such as hardwood hammock could result in a regime shift to a halophytic community, such as mangroves. The scenario simulations of MANTRA indicate the feasibility of such shifts, but it can be asked what the evidence is for occurrences in the past. Solid evidence for past regime shifts in southern Florida may be lacking, but that may reflect that up until now there have been few studies focused on vegetation changes following storm surges. A regime shift of vegetation would also take at least a decade to be noticed, and so might appear to be ordinary gradual change rather than an irreversible transition.

Nonetheless, there are some additional examples where regime shifts may be inferred to have occurred in southern Florida. A report on Cape Sable at the southwestern tip of Florida documented the shift in much of this region from freshwater to marine marsh [36]. This shift started to occur suddenly in the 1930s and shows no signs of returning to its original state, so it appears to be an irreversible change, perhaps a regime shift. It is possible that the Labor Day Hurricane of 1935, which produced a nearly four-meter storm surge over Cape Sable, was at least a partial cause of this shift. As another case, Ross *et al.* (2009) [37], studying vegetation changes in the Florida Keys, noted that "Once sea level reaches a critical level, the transition from a landscape characterized by mesophytic upland forests and freshwater wetlands to one dominated by mangroves can occur suddenly, following a single storm-surge event. We document such a trajectory, unfolding today in the Florida Keys. With sea level projected to rise substantially during the next century, ex-situ actions may be needed to conserve individual species of special concern".

In the Introduction, we noted general evidence from outside of southern Florida that salinity input to a glycophytic community could lead to apparent long-term vegetation shifts [12,13]. In addition, the role of overwash salinity pulses in causing long-term effects on vegetation has been noted in islands of the southern Pacific, where widespread sea flooding by storm surges around the coastlines of South Pacific atolls is a serious hazard during tropical cyclones. For example, in 2005, tropical cyclone Percy inundated the three atolls of Tokelau. The high surge allowed waves to sweep across the low-lying atoll islands. It also inundated the Pukapuka Atoll in the Northern Cook Islands. The immediate effect was on the freshwater lenses that sit on top of saline ground water in these areas [38,39]. Both [38] and [40] stressed that recovery from an overwash event may be prolonged, depending on the amount of seawater that accumulates in the central depression of the atolls. The reason that we extended our original MANHAM model to be combined with SUTRA in MANTRA was to be able to simulate the changes in the freshwater lens and groundwater from storm surges, which could have a long-lasting effect on vegetation.

4.1. Relevance of MANTRA for Management

A goal of the CERP for restoration of the Everglades is to bring additional fresh water south into ENP to restore freshwater habitats. It has been challenging, however, to deliver historic quantities of fresh water sufficient to improve conditions all the way to the coast and Florida Bay, and climate change and SLR will complicate this further [41]. Coastal hardwood communities provide unique habitat for a high diversity of species from plants to mammals [42,43] and protect fresh marsh communities behind them from storm surges. Mangrove migration inland along the west coast, at the expense of hardwood communities and freshwater marsh, has often kept pace with current rates of SLR, although the coastal forests are further stressed by the historic and current reductions of fresh water flow to the Everglades. Along the southern coast, including the study transect of this paper, there is an elevation dip inland of the mangrove/buttonwood/hardwood zone that isolates these hardwood forest communities from simple migration to higher elevations inland, e.g. see [15]. Projections of increasing rates of SLR heighten concerns for maintaining a fresh water hydrologic head that slows salt-water intrusion and allows coastal hardwood communities to have critical time to adapt to changing conditions. MANTRA provides information on how SLR and storm surges may affect vulnerable hardwood hammocks.

4.2. Future Plans

MANTRA development and application is documented in this paper to provide a start in developing a robust model for projecting the effects of overwash and climate change events on groundwater salinity as well as potential changes in vegetation composition. In this paper, the halophytic plant species at Coot

Bay Hammock are generally grouped together as mangroves. However, in fact, there were other halophytic plants that have different dynamics than mangroves along the Coot Bay Hammock transect. Therefore, a better representation of the plant community at Coot Bay can be obtained by including more vegetation types in the model. MANTRA will be improved by revising the plant root network horizontally and vertically, which allows water uptake farther from the main stem and deeper into the ground. Future uses of MANTRA will be extended to a three-dimensional environment.

Precipitation is a major source of groundwater lens recharge. Hence, changes in precipitation pattern will affect groundwater lens recharge and vegetation distribution. Precipitation interception by plant foliage is not modeled explicitly in this model but it should be noted that rainfall interception by plant foliage is an important component in hydrological studies. Zinke [44] reported that interception loss is commonly 10% to 20% in hardwoods. A simulation study assuming this common interception loss will not change the conclusion of this paper.

Based upon the findings of a recent study [45], MANTRA may be revised by using SUTRA-MS to simulate oxygen isotope transport in addition to salt transport, because ^{18}O may be an early indicator of salinity stress on trees. MANTRA shall be applied to study sites along the Waccamaw River, South Carolina, USA. Potential applications to study sites in Malaysia and Mekong River, Vietnam, where floodwaters brought devastating damage to crops like paddy, are also planned. Transitioning to the three-dimensional version of SUTRA will be necessary for some of these projects.

5. Conclusions

The object of this paper was to describe a new model, MANTRA, that combines hydrologic and salinity dynamics based on USGS's SUTRA model, with MANHAM, which models two vegetation types, glycophytic and halophytic, having different transpiration properties with respect to soil salinity. The purpose of the model is to accurately describe effects of SLR and storm surges on the ecotone between these types, given that positive feedbacks between vegetation and soil salinity are important components of this system. The application to Coot Bay Hammock shows consistency with historical data showing that the last major hurricanes, Andrew and Wilma, did not cause a major change (regime shift) of the ecotone, but indicates that larger disturbances, which cause substantial damage to existing vegetation, might have such an effect.

Acknowledgments

S.Y.T. was supported in part by the USGS's Across Trophic Level System Simulation program. Financial support by Grants 305/PMATHS/613418 and 203/PMATHS/6730101 to S.Y.T. and H.L.K. is gratefully acknowledged. M.T. and D.L.D. were supported in part by the USGS's Natural Resources Preservation Project. J.J. was supported in part by National Basic Research Program of China (No. 31200534). Use of trade or product names does not imply endorsement by the U.S. Government. We greatly appreciate the many suggestions and edits of two reviewers for the journal and a USGS reviewer.

Appendix

Appendix 1. Description of MANHAM Model

MANHAM is spatially explicit model of two competing vegetation types, mangroves and hardwood hammocks (though it is adaptable to other competing vegetation types). The vegetation types compete for light and have different tolerances of salinity. The basic assumptions are similar to those in Sternberg *et al.* [11], in which the formation of a sharp boundary between the vegetation types was modeled. Both vegetation types use water from the vadose zone, which overlies a saturated zone of brackish groundwater. Hammock species are assumed to be better competitors in low salinity areas, but cannot grow well under high salinity, where they are out-competed by mangroves. If enough water is withdrawn from the vadose layer by plant water uptake or evaporation, groundwater will infiltrate by capillary action into the vadose layer and increase its salinity. On the other hand, if precipitation exceeds evaporation plus the transpiration of water, then salinity in the vadose layer is percolated towards the underlying ocean water layer and salinity decreases [46]. High vadose zone salinity that develops during Florida's dry season is considered to be the major determinant of vegetation distribution in Florida in this model.

The main mechanism in the model is based on the feedback relationship between the two vegetation types and vadose zone salinity mentioned above. For example, consider a microsite and assume the vadose zone has a particular average salinity during the dry season, which is not sufficiently high to decrease the complete domination of hammock species in an area. During the dry season, as freshwater hammock species continue to transpire water from the vadose layer, ocean water tends to infiltrate and to increase the salinity of the groundwater of the microsite. Because freshwater plants are sensitive to salinity [47], they decrease their transpiration rates, reducing further infiltration of ocean water. In this way the salinity of the vadose layer may be stabilized at low concentrations that are not lethal to freshwater plants. Conversely, consider the alternate equilibrium state where mangroves dominate an area. As mangroves transpire the water in the vadose layer, underlying groundwater with ocean water salinity infiltrates upwards into the vadose zone, but unlike the hammock species, mangroves will continue to transpire and continue to increase the salinity of the vadose layer to levels which would not be tolerated by freshwater hammock species. Thus there will be a tendency for one or the other vegetation type to stabilize itself in a given area, by reinforcing salinity conditions favorable to itself.

This mechanism may explain observations at the landscape level. Our model conceptualizes the landscape as a grid of microsites, or spatial cells, and assumes each grid cell is occupied by a closed canopy of a small number of plants, which can include both mangrove and hammock individuals. Each cell, whether currently dominated by mangrove or hammock species, always contains at least some small fraction of the other type, which can act as 'seeds' for growth under more favorable conditions. Each cell is exposed to precipitation, soil evaporation, tidal deposition of saline water (depending on the cell's elevation in the landscape) and transpiration, which produce vertical fluxes of water in a cell and either increase or decrease the salinity of the vadose zone of that cell. (Evaporation of intercepted water is not considered in this version of the mode). The evapotranspiration depends on the fractions of each of the vegetation types in the cell. The vadose layer of the cell is also assumed coupled to neighboring cells through lateral movement of salinity. The strongest mechanism for this transport may be water uptake by the roots of plants in the neighboring cells, which redistributes water and salinity between cells. Thus there is some tendency for adjacent cells to approach over time the same vadose zone salinity, allowing the possibility for each vegetation type to spread horizontally from one cell to dominate adjacent cells.

We hypothesize that a model landscape of mangrove and hardwood hammock trees, initially randomly mixed and then subjected to these abiotic factors, will self-organize into a pattern similar to those observed in nature, having strong aggregation into areas of either solid hammock or mangrove vegetation (vegetation clumping), such that there can be rapid changes between the vegetation types along gradual clines in microtopography. However, we also hypothesize that a large enough disturbance can change this pattern. For example, a storm surge that deposits a large amount of saline water across the landscape may cause hammock trees to slow their growth sufficiently to be outcompeted by mangroves over a sufficiently long time period to allow mangroves take over. Thus a large area may "switch" vegetation type quickly.

This model was implemented quantitatively as a two-dimensional grid of square spatial cells, where the sides of each cell were assumed to be in the range of a few to several meters.

Hydrology and salinity: The salinity in a given spatial cell is determined first of all by the difference between the precipitation, P, which brings in fresh water to the top of the vadose zone, and the evaporation, E, and plant uptake of water, R. This difference is called the infiltration rate, I_{NF};

$$I_{NF} = E + R - P \text{ (mm day}^{-1}) \tag{A1.1}$$

and the dynamics of salinity in the vadose zone are given by the equations

$$nz \frac{dS_V}{dt} = I_{NF} S_{wt} \text{ for } I_{NF} > 0 \tag{A1.2}$$

$$\varepsilon z \frac{dS_V}{dt} = I_{NF} S_V \text{ for } I_{NF} < 0 \tag{A1.3}$$

where z (mm) is the depth of the vadose zone of a given cell, ε is the porosity, and S_V and S_{wt} are the salinities of the pore water in the vadose zone and of the underlying saline groundwater, respectively. Positive values of infiltration (A1.2) occur when precipitation is less than the water demanded by evaporation and transpiration; then water from the underlying saline groundwater infiltrates upward into the vadose zone. Note that when $I_{NF} > 0$, salt is deposited in the vadose zone by evapotranspiring water, so concentrations can build up to high levels. Negative values occur when precipitation exceeds evaporation and transpiration demands; then water percolates downward into the underlying groundwater table.

The assumption of a groundwater table with fixed salinity is a useful first approximation, but it is also possible that the salinity dynamics of at least the surface layer of groundwater is more complex. The effect of an upper layer of groundwater that is affected both by precipitation and flow of groundwater from higher elevations is examined in an appendix (see on-line Appendix 2), while here we restrict ourselves to examining the model with the simpler assumption.

Evaporation was assumed to be small compared with transpiration and was neglected, since we are assuming a dense canopy in each cell, which inhibits evaporation. Moreira, *et al.* [48] and Harwood, *et al.* [49] both observed that in forests transpiration dominates as the vapor generator compared to evaporation. R_{TOTAL} depends on the transpiration and gross productivity of each vegetation type in the spatial cell (see A1.13 below). The maximum possible water uptake rate by freshwater hammocks is assumed to be 2.6 mm/d. This value is based on previous studies indicating that transpiration in tropical forests lies within this range [38]. Uptake of water as a function of salinity by the hardwood hammock $R_1(S_v)$ and mangrove species $R_2(S_v)$ is given by the respective empirical relations (Figure 2):

$$R_1(S_v) = 2.6\left(1 - \frac{S_v}{3.14 + S_v}\right) \text{ mm day}^{-1} \tag{A1.4a}$$

$$R_2(S_v) = 4.4\left(\frac{100 - S_v}{15 + 100 - S_v}\right) \text{ mm day}^{-1} \tag{A1.4b}$$

in which hammocks reduce their transpiration by ½ when the salinity of the pore water is 3.14 ppt, while mangrove transpiration is not reduced by ½ until the salinity of the pore water is 85 ppt.

In addition to the above hydrologic processes, tidal effects were imposed on all spatial cells at elevations low enough to be affected. The effect of tides on the salinity of spatial cells was calculated as follows. On each day a single high tide was assumed. The height of the tide above the surface of each spatial cell was generated as a function of the mean and a randomly generated variation within the observed limits of tidal flux of the empirical data, so that the number of spatial cells covered by the tide on a given day varied in number. The amount of salt contained in the volume of water above the cell, assumed to have a salinity of 30 ppt, was allowed to mix homogeneously with the vadose zone below. In all model simulations precipitation and effects of tides were prescribed on a daily basis. Means and standard deviations of daily precipitation (NOAA, National Weather Services Forecast Office, Florida) and daily tidal height (NOAA, Tide & Current Historic data base, Key West Station) for each month were derived from 162 and 5 years of empirical data respectively. Daily values were determined using a normal random number generator, with values truncated at zero.

We assume there is also horizontal diffusion of salinity between cells. We used a diffusion constant of $D = 0.0005$, which is about seven times the theoretical value used by Passioura, et al. [50] and ten times the laboratory values of [51]. However, we assume that classical diffusion is not the only process causing mixing of solute between cells. The extension of roots across cell boundaries can contribute to the mixing among cells, and we believe our value is reasonable.

Vegetation dynamics: A given cell could be occupied by the two types of vegetation simultaneously, and, in fact, even in cells dominated by one type, small amounts of the other type tended to persist. The biomasses of each species in a given spatial cell were explicitly modeled, as well as the mechanism of competitive dominance of the hammock vegetation over mangrove vegetation under very low soil salinity conditions. We used an approach similar to that of Herbert, et al. [52] for competition between species of different functional types, with a slight difference. This is explained in detail in Teh et al. [14] and Herbert, et al. [52]. Herbert et al. [52] assumed that the plant types differed in their abilities to compete for light and nutrients. Here we assumed that the plants differed only in their ability to compete for light, with the additional assumption that the hardwood species were superior in low salinity. We assumed that in any particular microsite the equations for the different vegetation types (hardwood hammock and mangrove in this case) were

$$\frac{dB_{Ci}}{dt} = U_{Cvi} - M_{Cvi} - L_{Cvi} \quad (i = 1,2)$$

(A1.5)

where B_{Ci} is carbon in plant biomass (gCm^{-2}), U_{Cvi} is gross productivity (gCm^{-2}day^{-1}),

$$U_{Cvi} = \frac{Q(S_v)g_{Ci}w_{Ci}I(1-e^{-k_lS_{CT}})}{w_{C1} + w_{C2}} \quad (i = 1,2)$$

(A1.6)

where w_{C1} and w_{C2} incorporate competition for light, depending on how much of the canopy of a spatial cell each occupies (see Herbert *et al.* [52] for details);

$$w_{Ci} = \frac{2(1-e^{-k_lS_{Ci}})}{(1+e^{-k_lS_{Ci}})} \prod_{j=1}^{2} \left(\frac{f_{Cj}e^{-k_lS_{Ci}} + f_{Ci}}{f_{Ci} + f_{Cj}} \right) \quad (i = 1,2)$$

(A1.7)

where the $f_{Ci} = 0.58/c_{ii}$ parameters are measures of canopy dominance (e.g., the relative degree to which trees of one species shade another due to height differentials). S_{Ci} (m^2 m^{-2}) is the leaf area index of species i in a spatial cell,

$$S_{Ci} = b_{Ci}B_{Ai} \quad (i = 1,2)$$

(A1.8)

B_{Ai} (gCm^{-2}) is active tissue carbon of each plant species,

$$B_{Ai} = \frac{c_{ii}B_{Amax,i}B_{Ci}}{B_{Amax,i} + c_{1i}B_{C1} + c_{2i}B_{C2}}$$

(A1.9)

where the parameters c_{ij} are allometric parameters governing the amount of energy allocated to active tissue (leaves). Note that the biomass of species j can affect the allocation of biomass of species i. We assumed the effect of salinity on productivity of species i, $Q_i(S_v)$, occurs through its effect on the water uptake rate, normalized by the maximum possible rate;

$$Q_i(S_v) = \frac{R_i(S_v)}{R_i(0)} \quad (i = 1,2)$$

(A1.10)

M_{Cvi} (g C m^{-2} day^{-1}) is the respiration of each plant;

$$M_{Cvi} = m_{Ai}B_{Ai} + m_{wi}(B_{Ci} - B_{Ai}) \quad (i = 1,2)$$

(A1.11)

where the rates differ between dead and living matter. L_{Cvi} (g C m^{-2} day^{-1}) is litterfall of each plant,

$$L_{Cvi} = l_{Ai}B_{Ai} + l_{wi}(B_{Ci} - B_{Ai}) \quad (i = 1,2)$$

(A1.12)

Total evapotranspiration from a cell is linearly related to the evapotranspiration of each species, multiplied by its fraction of the primary production in that spatial cell;

$$R_{TOTAL} = \frac{U_{Cv1}}{U_{Cv1} + U_{Cv2}}R_1 + \frac{U_{Cv2}}{U_{Cv1} + U_{Cv2}}R_2$$

(A1.13)

We do not have parameter values related to light competition for these vegetation types, but made estimates that allowed hardwood hammock vegetation to dominate for salinities below 7 ppt.

To improve the fit of the model to the boundary between hardwood hammocks and mangroves of the Coot Bay Hammock, some moderate modifications were made to the original MANHAM formulation and the parameter value. Notably, we varied the light extinction coefficients, k_i, for the two species, as well as the parameters and equation for leaf area index of each species i in a spatial cell. We gave the hardwood hammock an advantage in light use shading out of the mangroves, which would allow hardwood hammock to outcompete the mangroves at lower salinity areas. The equation for leaf area index, Sc_i, of species i in a spatial cell is revised from

$$Sc_i = bc_i B_{Ai} \qquad \text{(A1.14a)}$$

(where bc_i is the leaf area index per unit active tissue and B_{Ai} is the active tissue carbon, see Equation (8) in Teh *et al.* [14]) to Equation (A1.14b) below to include the plant water uptake effort $Q_i(C)$, following Herbert *et al.* [52].

$$Sc_i = bc_i B_{Ai} \cdot Q_i(C) \qquad \text{(1)}$$

where C = solute concentration [M_s/M_f]. The changes to (A1.14a) would cause the hardwood hammocks to diminish in biomass quickly at higher salinities, as the water uptake effort $Q_i(S_v)$ is less than halved when the salinity is higher than 3.14 ppt. The revised formulation (A1.14b) alone results in a situation where the mangroves outcompete the hardwood hammocks at the elevated ridge, which also is not realistic. However, the increase of light extinction factor k_i for hardwood hammocks ensures that hardwood hammocks outcompete the mangroves at lower salinities. These revisions allow the mangrove and hardwood hammocks to evolve, in a more realistic manner, into the distribution observed at Rowdy Bend.

Appendix 2. MANTRA Version 1 Manual: August 1, 2014

This report briefly describes the coupled hydrology-salinity-vegetation model, MANTRA, for analysis of the feedback effect of competing glycophytes (hardwood hammocks) and halophytes (mangroves) on the groundwater flow and salinity regime. MANTRA can be further revised to include more competing plants [15] by incorporating the dynamics of these plants in the MANHAM module. MANTRA was developed by integrating the USGS spatially explicit models of vegetation community dynamics along coastal salinity gradients (MANHAM) into the USGS groundwater models (SUTRA). Table A2.1 lists the files needed to run MANTRA. The MANHAM [11,14] module is incorporated into SUTRA [16] in sutra_2_2.f by accounting for the exchange of fluid and solute mass between the models. A storm surge event is incorporated into usubs_2_2.f as time-dependent specified pressure.

In MANTRA, the total amount of water required for plant transpiration is subtracted from the SUTRA cells covered by plants. The fluid mass source/sink term Q_{IN}, originally available in SUTRA, is used to characterize the uptake of water by plant [18–20]. This source/sink term accounts for external addition/subtraction of fluid including pure water mass plus the mass of any solute dissolved in the source fluid. This fluid uptake by plant reduces the fluid mass in the cells, which in turn increases the solute (salt) concentration. Plant growth in MANHAM depends on the actual fluid available in the cells for transpiration and the solute concentration. Units are presented in square brackets [ ] with M being the unit of mass, L the unit of length and s the unit of time in seconds. The relations are formulated for simulations of a 2D cross-sectional fishnet domain with saturated-unsaturated, variable-density (pressure) with single species (solute) transport. It is assumed at present that the plants will only withdraw water from the uppermost layer of cells/nodes and that the plant roots cover the entire surface area of the cells (Figure A2.1). Currently, the vegetation and groundwater modules operate at the same time step. This is expected to slow down the computation, particularly when there are large numbers of element. As the hydrological processes and vegetation dynamics operate at different time scale, a method similar to those employed in the SEHM model by Jiang, *et al.* [53] can be employed to optimize the computation time.

Table A2.1. List of files needed to run MANTRA.

Filename	Remarks
SUTRA.FIL	These are the files that are also needed to run the original version of SUTRA.
SUTRA.inp	These files contain the input parameters for the groundwater flow and solute
SUTRA.ics	transport simulation. These files can be created by using ArgusONE.
MANTRA.exe	This is the executable file of MANTRA. This executable file is built from fmods_2_2.f, sutra_2_2.f, ssubs_2_2.f, usubs_2_2.f. Modifications have been made to incorporate the MANHAM module into sutra_2_2.f. A storm surge event is incorporated into usub_2_2.f as time-dependent specified pressure.
MANHAM.DAT	This is the input file for the MANHAM module in MANTRA. This file contains the input parameters related to the vegetation.

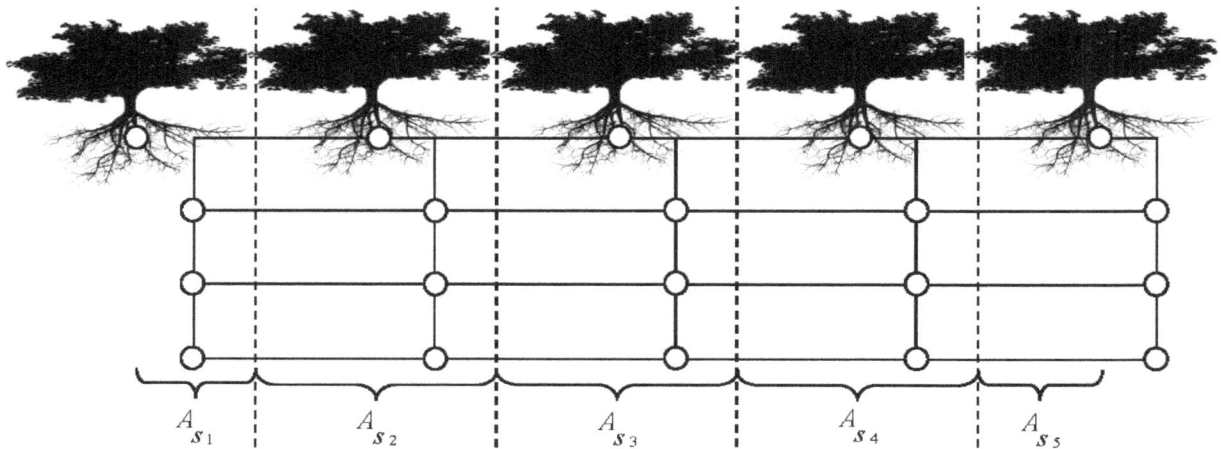

Figure A2.1. Schematic sketch of a simulation case. A_{s_i} = Surface area for node i at uppermost layer of cells.

The uptake of water as a function of salinity for hardwood hammock (R_1) and mangrove (R_2) are estimated by the empirical relations Equations (A2.1) and (A2.2).

$$R_1(C) = R_{max,1}\left(1 - \frac{C}{S_{half,1} + C}\right) \quad \text{[L/s]} \tag{A2.1}$$

$$R_2(C) = R_{max,2}\left(\frac{0.1-C}{S_{half,2}+0.1-C}\right) \quad \text{[L/s]} \tag{A2.2}$$

with C = solute concentration [M_s/M_f] calculated by SUTRA. Here, Fluid mass per unit time (Q_p) required by the plants in a certain cell is then estimated by (A2.3).

$$Q_p = (R_1 + R_2) \cdot A_s \cdot \rho \quad \text{[M/s]} \tag{A2.3}$$

Here, Q_p = fluid mass per unit time extracted in a certain cell for plant transpiration [M/s]; A_s = cell surface area [L^2]; ρ = fluid density [M/L^3]. For cross-sectional model, the width of each cell is assumed to be 1.0. Thus, the cell surface area depends mainly on the length or horizontal grid size (Figure A2.1). The actual fluid mass being subtracted from a cell due to evapotranspiration depends on the saturation S_w and porosity ε in the cell, leading to Equation (A2.4).

$$Q_{IN} = -Q_p \cdot \varepsilon \cdot S_w \quad \text{[M/s]} \tag{A2.4}$$

Here, Q_{IN} = total mass sink (due to plant transpiration) [M/s]; ε = porosity [V_v/V]; S_w = water saturation [V_w/V_v] with V = total volume, V_v = volume of voids, V_w = volume of water. Suppose porosity ε is kept constant at 1.0. When the void space is completely filled with fluid and is said to be saturated, that is $S_w = 1$, the actual water uptake by plant will be equivalent to the amount of water required by the plants Q_p. When the void space is only partly water filled and is referred to as being unsaturated, that is $S_w < 1$, the actual water uptake by plant will decrease as a factor of the saturation S_w. Similar to transpiration, precipitation is implemented as a source term in MANTRA. The precipitation rate is varied stochastically on daily basis. Daily values are determined using a normal random number generator, with values truncated at zero.

Since the storm surge event is incorporated into usubs_2_2.f as time-dependent specified pressure, time-dependent specified pressure for the surface nodes should be indicated in the SUTRA input file (sutra.inp) so that the usubs_2_2.f routine will be called for implementation. The storm surge event is simulated by allowing the surface nodes to be inundated with seawater of certain depth and salinity. This form of inundation will change the hydrostatic pressure at the surface nodes. Equations (A2.5) and (A2.6) [21] are respectively used to specify the pressure and concentration at surface nodes inundated by seawater during a storm surge event.

$$p = \rho_{sea} \cdot g \cdot (h_{surge} - y) \quad \text{[M/(L·s}^2\text{)]} \tag{A2.5}$$

$$C = S_{sea} \quad \text{[M}_s/\text{M}_f\text{]} \tag{A2.6}$$

Here, p = pressure at top layer nodes [$M/(L\cdot s^2)$], ρ_{sea} = fluid density of seawater [M/L^3], h_{surge} = inundation depth [L], y = node height [L], and S_{sea} = seawater salinity [M_s/M_f]. Figure A2.2 shows an example of input file (MANHAM.DAT) for MANHAM module in MANTRA for the example case of Rowdy Bend. Table A2.2 summarizes the list of parameters in MANHAM.DAT with their description, type, value and unit.

Table A2.2. List of parameters in MANHAM.DAT with their description, type, value and unit.

Variable Name				
Input File	Eqn	Description	Type	Value	Unit
SUTRA Node Control					
NSNODE	–	Number of sets of surface nodes	Integer, I10	1	–
NFIRST	–	First surface node number in SUTRA domain	Integer, I10	1	–
NLAST	–	Last surface node number in SUTRA domain	Integer, I10	1601	–
NDIFF	–	Interval between surface node numbers	Integer, I10	20	–
Storm Surge Control					
IT_Surge	–	Iteration time of a storm surge event	Integer, I10	60,000	–
SDepth	h_{surge}	Storm surge inundation depth in relation to the height of SUTRA computational domain	Real, F10.2	10.5	m
SSalinity	S_{sea}	Storm surge water salinity	Real, F10.3	0.030	kg/kg
Plant Initial Condition (IC) Control					
NSPEC	–	Number of plant species	Integer, I10	2	–
NRAND	–	Seed for random number generator	Integer, I10	1234	–
BEGINHAM	–	Ratio of cells dominated by hardwood hammock	Real, F10.2	0.50	–
BC0	–	Total initial biomass in a cell	Real, F10.2	20,000.00	g C m^{-2}
PERSPEC	–	Ratio of plant *i* in a cell	Real, F10.2	0.50	–

Table A2.2. *Cont.*

Variable Name Input File	Eqn	Type	Description	Value		Unit
Light Parameters						
SI	I	Real, F10.3	Solar irradiance	0.010		GJ m^{-2}day^{-1}
EKI	k_i	Real, F10.4	Light extinction factor	0.600	0.400	
GC(NSPEC)	g_{Ci}	Real, F10.1	Light-use efficiency	520.0	380.0	g C GJ^{-1}
BAMAX(NSPEC)	$B_{Amax,i}$	Real, F10.1	Maximum value attainable by B_{Ai}	350.0	350.0	g C m^{-2}
Plant Parameters						
bc(NSPEC)	b_{ci}	Real, F10.4	Leaf area per unit carbon	0.0355	0.0170	m^2/g C
RA(NSPEC)	r_{Ai}	Real, F10.4	Active tissue respiration rate	4.0000	4.0000	year^{-1}
RW(NSPEC)	r_{Wi}	Real, F10.4	Woody tissue respiration rate	0.0296	0.0296	year^{-1}
RMA(NSPEC)	m_{Ai}	Real, F10.4	Active tissue litter loss rate	1.7000	1.7000	year^{-1}
RMW(NSPEC)	m_{Wi}	Real, F10.4	Woody tissue litter loss rate	0.0148	0.0148	year^{-1}
RFWMAX(NSPEC)	$R_{max,i}$	Real, F10.4	Maximum water uptake for plant NSPEC	0.0026	0.0088	mm day^{-1}
SATK(NSPEC)	$S_{half,i}$	Real, F10.4	Half saturation constant for maximum water uptake for plant NSPEC	3.1400	15.0000	ppt
C(NSPEC, NSPEC)	c_{ii}	Real, F10.4	Parameters for plant allometry	0.1000	0.1000	–
				0.5000	0.5000	
Environmental Parameters						
VPRE(12)	M_{pre}	Real, F10.4	Means precipitation rate for twelve months	1.590	1.360 ...	mm day^{-1}
VPRESD(12)	SD_{pre}	Real, F10.4	Standard deviations of precipitation rate	1.870	1.200 ...	mm day^{-1}

```
INPUT FILE FOR MANHAM

----+----+----+----+----+----+----+----+----+----+----+----+----+----+----+
  SUTRA Node Control
----+----+----+----+----+----+----+----+----+----+----+----+----+----+----+
 NSNODE              1
 NFIRST              1
 NLAST            1601
 NDIFF              20


----+----+----+----+----+----+----+----+----+----+----+----+----+----+----+
  Storm Surge Control
----+----+----+----+----+----+----+----+----+----+----+----+----+----+----+
 IT_Surge        60000
 SDepth           10.5
 SSalinity        0.030


----+----+----+----+----+----+----+----+----+----+----+----+----+----+----+
  Plant IC Control
----+----+----+----+----+----+----+----+----+----+----+----+----+----+----+
 NSPEC               2
 NRAND            1234
 BEGINHAM         0.50
 BC0          20000.00
 PERSPEC          0.50


----+----+----+----+----+----+----+----+----+----+----+----+----+----+----+
  Light Parameters
----+----+----+----+----+----+----+----+----+----+----+----+----+----+----+
 SI               0.010
 EKI              0.600      0.400
 GC(NSPEC)        520.0      380.0
 BAMAX(NSPEC)     350.0      350.0


----+----+----+----+----+----+----+----+----+----+----+----+----+----+----+
  Plant Parameters
----+----+----+----+----+----+----+----+----+----+----+----+----+----+----+
 bc(NSPEC)        0.0355     0.0170
 RA(NSPEC)        4.0000     4.0000
 RW(NSPEC)        0.0296     0.0296
 RMA(NSPEC)       1.7000     1.7000
 RMW(NSPEC)       0.0148     0.0148
 RFWMAX(NSPEC)    0.0026     0.0088
 SATK(NSPEC)      3.1400    15.0000
 C(NSPEC,NSPEC)   0.1000     0.1000
                  0.5000     0.5000    ( C11,C21,C12,C22)


----+----+----+----+----+----+----+----+----+----+----+----+----+----+----+
  Environmental Parameters
----+----+----+----+----+----+----+----+----+----+----+----+----+----+----
+----+----+----+----+----+----+----+----+----+----+----+----+
 VPRE(12)          1.590     1.360     1.320     1.500     2.730     3.820
                   3.090     4.050     5.480     4.640     2.020     1.570
 VPRESD(12)        1.870     1.200     1.290     1.800     2.220     2.710
                   1.970     2.130     2.710     3.470     2.840     1.520
```

Figure A2.2. Example input file (MANHAM.DAT) for MANHAM module in MANTRA for the example case of Rowdy Bend.

References

1. Nicholls, R.J.; Cazenave, A. Sea-level rise and its impact on coastal zones. *Science* **2010**, *328*, 1517–1520.

2. Gornitz, V. *Rising Seas: Past, Present, Future*; Columbia University Press: New York, NY, USA, 2013; p. 344.

3. Anderson, W.P., Jr. Aquifer salinization from storm overwash. *J. Coast. Res.* **2002**, *18*, 413–420.

4. Anderson, W.P., Jr.; Lauer, R.M. The role of overwash in the evolution of mixing zone morphology within barrier islands. *Hydrogeol. J.* **2008**, *16*, 1483–1495.

5. Terry, J.P.; Falkland, A.C. Responses of atoll freshwater lenses to storm-surge overwash in the Northern Cook Islands. *Hydrogeol. J.* **2010**, *18*, 749–759.

6. Sklar, F.H.; Chimney, M.J.; Newman, S.; McCormick, P.; Gawlik, D.; Miao, S.; McVoy, C.; Said, W.; Newman, J.; Coronado, C. The ecological-societal underpinnings of Everglades restoration. *Front. Ecol. Environ.* **2005**, *3*, 161–169.

7. Ross, M.S.; O'Brien, J.J.; Flynn, L.J. Ecological site classification of Florida Keys terrestrial habitats. *Biotropica* **1992**, *24*, 488–502.

8. Smith, T.J.I.; Foster, A.M.; Tiling-Range, G.; Jones, J.W. Dynamics of mangrove–marsh ecotones in subtropical coastal wetlands: fire, sea-level rise, and water levels. *Fire Ecol.* **2013**, *9*, 66–77.

9. Gosz, J.R. Ecotone hierarchies. *Ecol. Appl.* **1993**, 3, 369–376.

10. Snyder, J.R.; Herndon, A.; Robertson, W.B.J. South Florida rockland. In *Ecosystems of Florida*; Myers, R.L., Ewel, J.J., Eds.; The University of Central Florida Press: Orlando, FL, USA, 1990; pp. 230–279.

11. Sternberg, L.D.L.; Teh, S.Y.; Ewe, S.M.L.; Miralles-Wilhelm, F.; DeAngelis, D.L. Competition between hardwood hammocks and mangroves. *Ecosystems* **2007**, *10*, 648–660.

12. Baldwin, A.H.; Mendelssohn, I.A. Effects of salinity and water level on coastal marshes: An experimental test of disturbance as a catalyst for vegetation change. *Aquatic Botany* **1998**, *61*, 255–268.

13. Steyer, G.D.; Cretini, K.F.; Piazza, S.; Sharp, L.A.; Snedden, G.A.; Sapkota, S. *Hurricane Influences on Vegetation Community Change in Coastal Louisiana*; U.S. Geological Survey: Reston, VA, USA, 2010.

14. Teh, S.Y.; DeAngelis, D.L.; Sternberg, L.D.L.; Miralles-Wilhelm, F.R.; Smith, T.J.I.; Koh, H.L. A simulation model for projecting changes in salinity concentrations and species dominance in the coastal margin habitats of the Everglades. *Ecol. Model.* **2008**, *213*, 245–256.

15. Saha, A.; Saha, S.; Sadle, J.; Jiang, J.; Ross, M.; Price, R.; Sternberg, L.; Wendelberger, K. Sea level rise and South Florida coastal forests. *Clim. Chang.* **2011**, *107*, 81–108.

16. Voss, C.I.; Provost, A.M. *SUTRA, A Model for Saturated-Unsaturated Variable-Density Ground-Water Flow with Solute or Energy Transport*; U.S. Geological Survey Water-Resources Investigations Report 02–4231; U.S. Geological Survey: Reston, VA, USA, 2010; p. 291.

17. Voss, C. USGS SUTRA code—History, practical use, and application in Hawaii. In *Seawater Intrusion in Coastal Aquifers—Concepts, Methods and Practices*; Practices, J., Bear, A., Cheng, H.D., Sorek, S., Ouazar, D., Herrera, I., Eds.; Kluwer Acdemic Publishers: Dordrecht, Netherlands 1999; pp. 249–313.

18. Vrugt, J.; Wijk, M.V.; Hopmans, J.W.; Šimunek, J. One-, two-, and three-dimensional root water uptake functions for transient modeling. *Water Resour. Res.* **2001**, *37*, 2457–2470.

19. Zhu, Y.; Ren, L.; Skaggs, T.H.; Lü, H.; Yu, Z.; Wu, Y.; Fang, X. Simulation of Populus euphratica root uptake of groundwater in an arid woodland of the Ejina Basin, China. *Hydrol. Process.* **2009**, *23*, 2460–2469.

20. Tian, W.; Li, X.; Wang, X.-S.; Hu, B. Coupling a groundwater model with a land surface model to improve water and energy cycle simulation. *Hydrol. Earth Syst. Sci. Discuss.* **2012**, *9*, 1163–1205.

21. Kooi, H.; Groen, J.; Leijnse, A. Modes of seawater intrusion during transgressions. *Water Resour. Res.* **2000**, *36*, 3581–3589.

22. Saha, A.; Moses, C.; Price, R.; Engel, V.; Smith, T., III; Anderson, G. A hydrological budget (2002–2008) for a large subtropical wetland ecosystem indicates marine groundwater discharge accompanies diminished freshwater flow. *Estuaries Coasts* **2012**, *35*, 459–474.

23. Light, S.S.; Dineen, J.W. Water control in the Everglades: A historical perspective. In *Everglades: The Ecosystem and its Restoration*, Davis, S.M., Ogden, J.C., Eds.; St. Lucie Press: Florida, FL, USA, 1994; pp. 47–84.

24. Sadle, J. Everglades National Park. Homestead, Florida, FL, USA. Personal communication, 2014.

25. Olmstead, I.C.; Loope, L.L. Vegetation along a Microtopographic Gradient in the Estuarine Region of Everglades National Park, Florida, USA; South Florida Research Center: Florida, FL, USA, 1981; p. 41.

26. Smith, T.; Anderson, G.H.; Tiling, G. *Science and the Storms: The USGS Response to the Hurricanes of 2005*; Farris, G.S., Smith, G.J., Crane, M.P., Demas, C.R., Robbins, L.L., Lavoie, D.L., Eds.; U. S. Geological Survey Circular 1306: Reston, VA, USA, 2007; pp. 169–174.

27. Wanless, H.R.; Parkinson, R.W.; Tedesco, L.P. Sea level control on stability of Everglades wetlands. In *Everglades: The Ecosystem and Its Restoration*; St. Lucie Press: Delray Beach, FL, USA, 1994, pp. 199–223.

28. Armentano, T.V.; Doren, R.F.; Platt, W.J.; Mullins, T. Effects of Hurricane Andrew on coastal and interior forests of southern Florida: Overview and synthesis. *J. Coast. Res.* **1995**, *21*, 111–144.

29. Slater, H.H.; Platt, W.J.; Baker, D.B.; Johnson, H.A. Effects of Hurricane Andrew on damage and mortality of trees in subtropical hardwood hammocks of Long Pine Key, Everglades National Park, Florida, USA. *J. Coast. Res.* **1995**, *21*, 197–207.

30. Horvitz, C.C.; Pascarella, J.B.; McMann; S.; Freedman, A.; Hofstetter, R.H. Functional roles of invasive non-indigenous plants in hurricane-affected subtropical hardwood forests. *Ecol. Appl.* **1998**, *8*, 947–974.

31. Swain, E.D.; Krohn, D.; Langtimm, C.A. Numerical computation of hurricane effects on historic coastal hydrology in southern Florida. *Ecol. Process.* **2015**, *4*, 4, doi:10.1186/s13717-014-0028-3.

32. Hook, D.D.; Buford, M.A.; Williams, T.M. Impact of Hurricane Hugo on the South Carolina coastal plain forest. *J. Coast. Res.* **1991**, *8*, 291–300.

33. Rathcke, B.J.; Landry, C.L. Dispersal and recruitment of white mangrove on San Salvador Island, Bahamas after Hurricane Floyd. In Proceedings of the Ninth Symposium on the Natural History of the Bahamas, San Salvador, Bahamas, 14–18 June 2001.

34. Jiang, J.; DeAngelis, D.L.; Anderson, G.H.; Smith, T.J., III. Analysis and simulation of propagule dispersal and salinity intrusion from storm surge on the movement of a marsh-mangrove ecotone in South Florida. *Estuar. Coasts* **2014**, *37*, 24–35.

35. López-Hoffman, L.; Ackerly, D.D.; Anten, N.P.R.; Denoyer, J.L.; Martinez-Ramos, M. Gap-dependence in mangrove life-history strategies: A consideration of the entire life cycle and patch dynamics. *J. Ecol.* **2007**, *95*, 1222–1233.

36. Wanless, H.R.; Brigitte, M.V. *Coastal Landscape and Channel Evolution Affecting Critical Habitats at Cape Sable*; Final Report to ENP; Everglades National Park: Homestead, FL, USA, 2005.

37. Ross, M.S.; O'Brien J.J.; Ford, R.G.; Zhang, K.; Morkill, A. Disturbance and the rising tide: The challenge of biodiversity management for low island ecosystems. *Front. Ecol. Environ.* **2009**, *9*, 471–478.

38. Cabral, O.M.; McWilliam, A.; Roberts, J. In-canopy microclimate of Amazonian forest and estimates of transpiration. In *Amazon Deforestation and Climate*; Gash, J.; Nobre, C.; Roberts, J.; Victoria, R., Eds; Wiley Press: Chichester, UK, 1996; pp. 207–220.

39. White, I.; Falkland, T. Management of freshwater lenses on small Pacific islands. *Hydrogeol. J.* **2010**, *18*, 227–246.

40. Chui, T.F.; Terry, J.P. Modeling fresh water lens damage and recovery on atolls after storm-wave washover. *Ground Water* **2012**, *50*, 412–420, doi:10.1111/j.1745–6584.2011.00860.x.

41. Pearlstine, L.G.; Pearlstine, E.V.; Aumen, N.G. A review of the ecological consequences and management implications of climate change for the Everglades. *J. Am. Benthol. Soc.* **2010**, *29*, 1510–1526.

42. Odum, W.E.; McIvor, C.C.; Smith, T.J., III. *The Ecology of the Mangroves of South Florida: A Community Profile*; Bureau of Land Management Fish and Wildlife Service: Washington, DC, USA, 1982.

43. Meshaka, W.; Loftus, W.F.; Steiner, T. The herpetofauna of Everglades National Park. *Fla. Sci.* **2000**, *63*, 84–103.

44. Zinke, PJ. Forest interception study in the United States. In *Forest Hydrology*; Sopper, W.E., Lull, H.W., Eds.; Pergamon: Oxford, UK, 1967; pp. 137–161.

45. Zhai, L.; Jiang, J.; DeAngelis, D.L.; Sternberg, L.S.L. Prediction of plant vulnerability to salinity increase in a coastal ecosystem by stable isotopic composition of plant stem water: A model study. In review.

46. Swain, E.D.; Wolfert, M.A.; Bales, J.D.; Goodwin, C.R. Two-dimensional hydrodynamic simulation of surface-water flow and transport to Florida bay through the Southern Inland and Coastal Systems (SICS); U.S. Geological Survey Water-Resources Investigations Report 03-4287, U.S. Geological Survey: Reston, VA, USA, 2003.

47. Munns, R. Comparative physiology of salt and water stress. *Plant Cell Environ.* **2002**, *25*, 239–250.

48. Moreira, M.; Sternberg, L.D.L.; Martinelli, L.; Victoria, R.; Barbosa, E.; Bonates, L.; Nepstad, D. Contribution of transpiration to forest ambient vapour based on isotopic measurements. *Glob. Change Biol.* **1997**, *3*, 439–450.

49. Harwood, K.; Gillon, J.; Roberts, A.; Griffiths, H. Determinants of isotopic coupling of CO2 and water vapour within a Quercus petraea forest canopy. *Oecologia* **1999**, *119*, 109–119.

50. Passioura, J.B.; Ball, M.C.; Knight, J.H. Mangroves may salinize the soil and in so doing limit their transpiration rate. *Funct. Ecol.* **1992**, *6*, 476–481.

51. Hollins, S.E.; Ridd, P.V.; Read, W.W. Measurement of the diffusion coefficient for salt in salt flat and mangrove soils. *Wetl. Ecol. Manag.* **2000**, *8*, 257–262.

52. Herbert, D.A.; Rastetter, E.B.; Gough, L.; Shaver, G.R. Species diversity across nutrient gradients: an analysis of resource competition in model ecosystems. *Ecosystems* **2004**, *7*, 296–310.

53. Jiang, J.; DeAngelis, D.; Smith, T.J.I.; Teh, S.Y.; Koh, H.L. Spatial pattern formation of coastal vegetation in response to external gradients and positive feedbacks affecting soil porewater salinity: a model study. *Landsc. Ecol.* **2012**, *27*, 109–119.

Bitumen on Water: Charred Hay as a PFD (Petroleum Flotation Device)

Nusrat Jahan [1], Jason Fawcett [1], Thomas L. King [2], Alexander M. McPherson [1], Katherine N. Robertson [1], Ulrike Werner-Zwanziger [3] and Jason A. C. Clyburne [1,*]

[1] Atlantic Centre for Green Chemistry, Departments of Chemistry and Environmental Science, Saint Mary's University, Halifax, NS B3H 3C3, Canada;
E-Mails: drnusrat2001@hotmail.com (N.J.); jafawcett@gmail.com (J.F.);
mcpherson.alexm@gmail.com (A.M.M.); Katherine.Robertson@smu.ca (K.N.R.)

[2] Centre for Offshore Oil, Gas and Energy Research, Bedford Institute of Oceanography, Dartmouth, NS B2Y 4A2, Canada; E-Mail: Tom.King@dfo-mpo.gc.ca

[3] Department of Chemistry and Institute for Research in Materials, Dalhousie University, Halifax, NS B3H 4J3, Canada; E-Mail: ulrike.wernerzwanziger@gmail.com

* Author to whom correspondence should be addressed; E-Mail: Jason.Clyburne@smu.ca

Academic Editors: Merv Fingas and Tony Clare

Abstract: Global demand for petroleum keeps increasing while traditional supplies decline. One alternative to the use of conventional crude oils is the utilization of Canadian bitumen. Raw bitumen is a dense, viscous, semi-liquid that is diluted with lighter crude oil to permit its transport through pipelines to terminals where it can then be shipped to global markets. When spilled, it naturally weathers to its original form and becomes dense enough to sink in aquatic systems. This severely limits oil spill recovery and remediation options. Here we report on the application of charred hay as a method for modifying the surface behavior of bitumen in aquatic environments. Waste or surplus hay is abundant in North America. Its surface can easily be modified through charring and/or chemical treatment. We have characterized the modified and charred hay using solid-state NMR, contact angle measurements and infrared spectroscopy. Tests of these materials to treat spilled bitumen in model aquatic systems have been undertaken. Our results indicate that bitumen spills on water will retain their buoyancy for longer periods after treatment with charred hay, or charred hay coated with calcium oxide, improving recovery options.

Keywords: petroleum; bitumen; dilbit; crude oil; asphaltene; remediation; spill; recovery; hay; flotation

---

1. Introduction

The development of the Canadian oil sands in northern Alberta has become a significant contributor to the Canadian economy. It could also become a significant contributor of oil to the world economy. Bituminous sands in Canada have been assessed to hold approximately 43% of the total global bitumen deposits, which represents approximately 26.9 billion m^3 or about 169.3 billion barrels of crude bitumen [1]. The locations of these Canadian bitumen deposits are far from both ocean and refinery access. Using current techniques crude bitumen can be refined to approximately 20% by weight of petroleum coke. This is of little value while landlocked but it could be valuable, even though high in sulfur, in "leading edge" environmental applications such as a source of activated carbon to reduce the toxins content in oil sands tailings [2]. In order to produce and transport bitumen profitably, ocean access for shipment is essential. Possible transportation routes might include: (1) railway to refinery or ocean port; (2) pipeline to railway or refinery or ocean port; or (3) truck to railway or refinery or ocean port.

The nature of these transportation methods, and the frequency and volume of product being transported, increases the risk of accidental spills or pipeline leaks. A review of the scientific literature indicates that there are limited options available to treat diluted bitumen (dilbit) spills. An article published by the Royal Society of Canada in 2010 assessed the environmental and health impacts of Canada's oil sands industry. This report outlines and summarizes a number of areas of concern regarding the environmental impacts of the oil sands industry, and suggests some of the necessary reclamation and monitoring practices necessary to mitigate these impacts [3]. A later 2012 publication in the journal, Environmental Science and Technology, discussed a number of shortcomings and oversights in the 2010 assessment [4]. However, both reports neglect to mention the need for new oil spill treatment technologies, given that bitumen and dilbit, under the correct circumstances, will sink. Conventional technologies, such as dispersants, will be rendered less effective in the event of a major dilbit spill either on site (e.g., land based) or during transport (e.g., entering aquatic systems).

The physical properties and composition of crude bitumen make it a very challenging material to manipulate. Bitumen is a heavy, viscous, semi-solid form of petroleum and is composed of a complex mixture of materials. At 15 °C, the complex viscosity of Athabasca bitumen has been reported to be 1.75×10^7 mPa·s [5]. The dynamic viscosity is reported to range from 1.9×10^4 to greater than 7.0×10^5 mPa·s at the same temperature [6], compared to conventional heavy crude, such as heavy fuel oil, HFO 6303, which has a reported viscosity of 2.28×10^4 mPa·s under the same conditions [6]. The same reference [6] reports the density of Athabasca bitumen as being 1.006 to 1.016 g cm^{-1}. A 2011 report in the Journal of Chemical and Engineering Data [5] details "Saturates, Aromatics, Resins and Asphaltenes" (SARA) analyses and Mass Fraction results which have been used to characterize the composition of Athabasca bitumen. These findings are outlined in Table 1 and notably include an asphaltene component of 18.6% ± 1.86% by weight in the bitumen studied.

Table 1. Composition analyses of Athabasca bitumen (adapted from Bazyleva *et al.* [5], with permission from © 2011 American Chemical Society).

Elemental Composition	Weight %
Carbon	83.2 ± 0.9
Hydrogen	9.7 ± 0.4
Nitrogen	0.4 ± 0.2
Sulphur	5.3 ± 0.2
Oxygen	1.7 ± 0.3
SARA Analysis	**Weight %**
Saturates	16.1 ± 2.1
Aromatics	48.5 ± 2.3
Resins	16.8 ± 1.2
Asphaltene (C_5)	18.6 ± 1.8

A series of publications spanning the years 2010 to 2012 by Murray R. Gray *et al.*, have tackled the arduous challenge of characterizing the structures of various bitumen fractions [7–9]. The most significant component of bitumen, the one that differentiates it from conventional crude oil, is the abundant asphaltene fraction. Asphaltenes are the heaviest fraction of crude bitumen, and consist mostly of polycyclic-aromatic rings complexed with metals including nickel and vanadium.

Asphaltenes are problematic for bitumen processing for a number of reasons, arising mainly by their tendency towards aggregation. Aggregation occurs because of various acid-base interactions, hydrogen bonding and the formation of metal-containing coordination complexes. This aggregation results in the drastically higher viscosity observed for crude bitumen as compared to crude oil. This, in turn, gives rise to the observed difficulties in pumping and processing bitumen. In the case of an ocean or fresh water-based bitumen spill aggregation will more than likely result in the clumping and sinking of the spilled materials. Understanding the aggregation behavior of asphaltenes in bituminous oils is essential to developing methods and materials for spill treatment/recovery.

A recent report from the Federal Government of Canada assesses the spill behavior and fate of two diluted bitumen (dilbit) samples under different weathering conditions. The dilbit products selected were those most frequently transported in Canada. Preliminary laboratory investigations showed that the dilbit products remained buoyant under natural ocean-simulated weathering conditions (0–15 °C) except when mixed with fine to moderately sized sediments [10]. One gap in this investigation was that only two samples of dilbit (e.g., Cold Lake Blend and Access Western Blend) were tested, and they were not compared to a base sample of crude bitumen. Furthermore, the products were studied only in sea water conditions. It must be remembered that there is also significant risk of spills occurring in fresh or brackish waters.

To extend the initial results to such waters, King *et al.*, (2014) have investigated dilbit weathering, through meso-scale (e.g., wave tank) studies, under natural conditions. One of the same dilbit products (e.g., Access Western Blend) was shown to weather enough, without interaction with sediments, such that its density exceeded that of fresh and brackish waters [11,12]. The authors concluded that this product would initially float on aquatic systems, but that after 6 days of natural attenuation, the product would sink in aquatic systems. A very recent paper by Stevens *et al.*, offers proof that oil weathering can result in its sinking [13]. The authors have developed an evaporation/sinking (EVAPOSINK) model that can be used to predict such behavior.

The potential for diluted bitumen products to sink when spilled is problematic from both environmental and industrial perspectives. Sunken oil is more difficult to find and track, and there are no known spill countermeasures to treat submerged dilbit. Preliminary findings have shown dispersants to be ineffective in the treatment of a diluted bitumen spill [10]. Submerged oil could potentially cause significant and persistent loss of potable water, ecosystems (e.g., rivers and lakes, marine systems, *etc.*) and aquatic life. Further investigation into the spill behavior of crude bitumen in aquatic systems is essential for the development of a cheap and effective countermeasure for spill impact mitigation and recovery. There is a definite need to identify a material capable of reducing the bulk density of the bitumen to keep it floating on the aquatic surface for as long as possible. This would prolong the window of opportunity available during the flotation phase to treat the spill by either mechanical means, such as booming or skimming the surface, or through *in situ* combustion.

Our preliminary investigations led us to hay, a cheap and abundant material with a large surface area. We felt that it might be a suitable material to adsorb bitumen and act as a flotation device. Attempts were made to modify the surface of the green hay so that it would also act as a natural dispersant. The hay was first immersed and coated with the organic-based surfactant, "Zep", a limonene-based household degreasing product. This surface modification was unsuccessful; the surfactant did not result in a modification of the surface properties of the hay. When this preliminary treatment failed, charring the hay and/or coating it with calcium oxide were investigated as means of surface modification. It was anticipated that charring the hay surface would render it more hydrophobic by removing surface OH groups and exposing the carbon backbone, while addition of CaO could possibly generate an *in situ* surfactant, improving the dispersant properties of the system. The results of the investigation are reported herein.

2. Experimental Section

2.1. Chemicals, Oils and Oil Spill Treating Agents

Athabasca bitumen was provided by the Centre for Oil Sands Innovation, Edmonton, Alberta and was used as received. Timothy hay (*Phleum pretense*) was purchased at Walmart, as supplied by Pestell Pet Products of Ontario, Canada. The composition of the hay is listed as follows: crude protein (min. 7.5%), crude fat (min. 2.0%), crude fibre (max. 35%), moisture (max. 12.0%), and calcium (min. 0.25%–max. 0.60%). "Instant Ocean" Sea Salt is distributed by United Pet Group Inc. of Cincinnati, OH, USA. It was prepared as directed on the packaging. "Zep" Heavy Duty Citrus Degreaser, with the active ingredients, d-limonene and monoethanolamine, was obtained from the Home Depot (Zep Superior Solutions, Atlanta, GA, USA). Reagent grade nitric acid, ACS reagent grade dichloromethane ≥99.5%, PCR reagent grade chloroform ≥99% and reagent grade calcium oxide and potassium bromide were purchased from Sigma-Aldrich Canada (Oakville, ON, Canada) and used as obtained.

2.2. Experimental Design

Simulated bitumen slicks were prepared in 250 mL glass beakers. A measured volume of 100 mL of either deionized water or Instant Ocean solution was added to each beaker. Bitumen slicks were generated by applying a known mass (1.8 g) of crude bitumen to the surface for both fresh water and Instant Ocean (artificially created salt water) samples. Samples were left at room temperature (23 °C) and were stirred for 2–3 min every 12 h. Samples were also periodically photographed to record bitumen aggregation and subsequent sinking of the product over time. At the end of the observation period, samples of the experimental solutions were collected. These were analyzed for trace metals, total petroleum hydrocarbon content and density.

Hay samples were cut into lengths of approximately 1.0 cm, sufficiently short to fit into the experimental beakers. The cut hay (all from a single source) was mixed to randomize its distribution before use, but no other attempts were made to homogenize the hay in the samples and replicate measurements were not performed for the charring process itself. The surface properties of the straw were then altered as follows: (1) the hay was charred to remove hydrogen and oxygen from its surface; and/or (2) the hay was coated with calcium oxide for, potentially, *in situ* surfactant formation. Addition of bitumen to the CaO-treated hay could possibly result in the deprotonation of the carboxylic acids, which would generate an *in situ* surfactant.

(1) The charred hay was prepared by placing the clippings in a sealed Schlenk flask and then placing the flask under vacuum. A propane torch was carefully applied to the bottom of the flask as it was mixed to endure uniform heating. Heating was performed at 10 min intervals, and the flask allowed to cool between heating cycles. Depending on the experiment, heating was continued for approximately 30 or 60 min total. During the hay charring process, a clean solvent trap was inserted into the Schlenk line and liquid nitrogen was used to condense the evolved gases. The condensate was washed from the trap using acetone, which was subsequently removed by evaporation. Preliminary experiments have been carried out to analyze the condensate for its principle components using gas chromatography coupled with mass spectrometry (GC/MS).

(2) Calcium oxide coated hay samples were prepared using the following procedure. A supersaturated solution of calcium carbonate (5 g) was prepared by adding just enough deionized water to make a paste. Then 2.5 g of uncharred hay clippings were mixed and coated with the paste and the mixture was left for 24 h at room temperature. A portion of the original mixture (the CaO-coated, uncharred hay sample) was then transferred to a Schlenk flask and charred under vacuum (see above) to produce the CaO-coated, charred hay samples. Heating was continued until the surface of the hay turned dark brown-black.

Buoyancy and bitumen adsorption of the charred hay samples (30 or 60 min) were evaluated by preparing sample slicks, containing approximately 2.0–2.3 g of bitumen in 100 mL of solution, as outlined above. The slicks were treated by adding 1.0 g of charred hay. Samples were shaken daily, and observed and photographed as outlined in the procedure above. A final set of buoyancy experiments examined the effectiveness of charred straw relative to CaO-coated charred straw. Instant Ocean solution (350 mL) was added to 125×65 mm^2 glass dishes to which were also added 3 g of bitumen and either 2 g of charred hay or 5 g of CaO-coated charred hay. The bitumen and straw were well mixed and then the dishes were placed on an orbital shaker operating at 65 rpm at room temperature. Once again, samples were observed and photographed periodically as outlined above.

2.3. Sample Analyses

2.3.1. Density

The densities of the deionized water and the Instant Ocean solution were measured by accurately determining the mass and volume of a specified quantity of each solution at room temperature. The density of Athabasca bitumen has been reported to be 1.006 to 1.016 g cm^{-1} [6].

2.3.2. Contact Angle Measurements

The differences in the *potential* strength of adsorption to the altered hay surfaces were evaluated via contact angle measurements of water droplets on flat surfaces of both the charred and the uncharred hay. Contact angle measurements were performed using a First Ten Angstroms (FTA) 135 Drop Shape Analyzer and FTA-32 Video software (Portsmouth, VA, USA).

2.3.3. Infrared Spectroscopy

Infrared spectra were recorded on a Bruker Vertex 70 Infrared Spectrometer (Bruker Optics Ltd., Milton, ON, Canada), with samples prepared as KBr pellets. Data processing was completed using OPUS 6.0 software (Bruker Optics Ltd., Milton, ON, Canada).

2.3.4. Nuclear Magnetic Resonance (NMR) Spectroscopy

The solid state ^{13}C cross polarization (CP)/magic angle spinning (MAS) NMR spectrum of raw hay was compared to those of two different samples of charred hay, one charred for 30 min and the other charred for 60 min. These NMR experiments were carried out in the NMR-3 Facility of Dalhousie University on a Bruker Avance DSX NMR spectrometer with a 9.4 Tesla magnet (400.24 MHz ^{1}H and 100.64 MHz ^{13}C Larmor frequencies) using a probe head for rotors of 4 mm diameter (Billerica, MA, USA). The parameters for the ^{13}C CP/MAS experiments with TPPM proton decoupling were optimized on glycine, whose carbonyl resonance also served as an external, secondary chemical shift standard at 176.06 ppm. For the final ^{13}C CP/MAS NMR spectra 1200 scans were acquired with 13.5 kHz sample spinning, 2.6 ms cross-polarization times and 3 s repetition times, as determined from the ^{1}H spin lattice relaxation times, T_1. Additional spectra, taken at 5.0 kHz sample spinning and also with a ^{13}C CP/MAS sequence followed by TOSS (TOtal Sideband Suppression), showed that there is no significant overlap between spinning sidebands and center bands.

2.3.5. Gas Chromatography with Flame Ionization Detection

Residual oil in the samples was analyzed using the method outlined by Cole *et al.* [12]. Briefly, the method is a modified version of EPA 3500C, where the sample container is used as the extraction vessel. Dichloromethane (DCM) was added to the sample bottle containing dispersed oil in solution. The sample was placed on a Wheaton R$_2$P roller (VWR International Ltd., Mississauga, ON, Canada) for 18 h. The roller had been modified to accommodate 3 inch diameter PVC pipe into each roller slot, so that sample containers of different sizes could be used. Once extraction was complete, the samples were removed and the DCM recovered. The recovered DCM was placed in a pre-weighed 15 mL centrifuge tube and the solvent volume reduced under a nitrogen evaporator to 1.0 mL. The extracts were analysed by gas chromatography using flame ionization detection. The original bitumen product was used to prepare calibration standards that were then used to generate a calibration curve from which oil concentrations in the extracts could be calculated. A mean percent recovery of 90.8 ± 4.6% was calculated from all oils spiked into water. The method detection limit was <0.5 mg/L. The method of extraction and analyses has been validated against the US EPA 3510C and provides better extraction efficiency for oils. The GC-FID method (EPA 8015B) is a standard US EPA method for analysing oils. The method has been published as supplementary material in an article in Environmental Engineering Science in 2015 [14].

2.3.6. Inductively Coupled Plasma Mass Spectrometry (ICP-MS)

All samples for ICP-MS analysis were freshly prepared. The required components for each sample, bitumen, green straw, charred straw, CaO-coated green straw, or CaO-coated charred straw, were placed into either deionized water or Instant Ocean solution. All straw-containing samples included 100 mL of solution (deionized water or Instant Ocean), 1 g of bitumen, and 0.5 g of straw (green or charred, with or without a CaO coating). The non-straw samples contained 100 mL of solution, 2 g of bitumen and in half of the samples added CaO (0.5 g). They were all left in the refrigerator for 48 h. The sample solutions were filtered through a 0.45 μm pore size (GHPP, Pall Gelman Acrodisc, purchased from Sigma-Aldrich Canada, Oakville, ON, Canada) syringe filter and acidified using 10% nitric acid to a pH of less than 2, prior to ICP-MS analyses for dissolved metals. Inductively Coupled Plasma Mass Spectrometry (ICP-MS) was performed at the Saint Mary's University Center for Environmental Analysis and Remediation (CEAR) on a VG PQ ExCell instrument (Thermo Elemental, Winsford, UK) by Patricia Granados.

3. Results and Discussion

3.1. Density

The densities of deionized water and the Instant Ocean solution were measured at 23 °C and determined to be 0.980 and 1.004 g/mL, respectively. Both are lower than a literature value reported for bitumen of 1.006 to 1.016 g/cm^3 [6].

3.2. Characterization of Charred versus Uncharred Hay

Preliminary experiments have been carried out on the condensate collected during the hay charring process. GC/MS analysis of the hay condensate showed that it contained many compounds, with three significant contributors being vanillin lactoside, 2,6-dimethoxyphenol (syringol) and 5-hydroxymethyl-2-furaldehyde. All of these are known components of combustion extracts [15].

3.2.1. Contact Angle Measurements

Contact angle measurements were made to characterize the impact of modification of the hay surface through the charring process on the bulk properties of the material. The expectation was that charring the hay would liberate hydrogen and oxygen from its surface, increasing C=C bond formation, and result in an overall increase in the hydrophobicity. Contact angle measurements for deionized water on charred and raw hay surfaces showed a significant increase in the contact angle after charring (e.g., from approximately 64°, Figure 1 (left) to approximately 126°, Figure 1 (right) and therefore, a definite change in the hydrophobic properties of the surfaces.

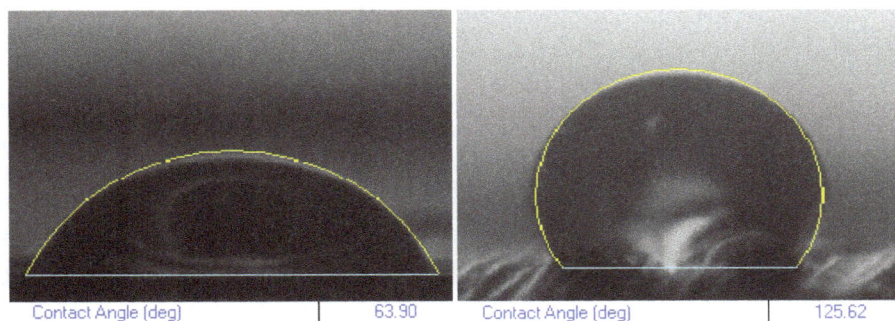

Figure 1. Contact angle measurements, before (**left**) and after (**right**) charring of the hay surface.

If the contact angle of water is less than 30°, a surface is designated as hydrophilic; wetting of the surface is favourable, and the water will spread over a large area. On a hydrophobic surface, water forms into distinct droplets. As the hydrophobicity increases, the contact angle of the droplets with the surface increases. Surfaces with contact angles greater than 90° are designated as hydrophobic [16]. The measured change in contact angle for our samples supports the conclusion that chemical modifications resulting from charring have produced a more hydrophobic surface. Variations in the observed contact angles were noted after subsequent measurements, however, samples consistently revealed larger contact angles, and thus increased hydrophobicity, after charring.

3.2.2. Infrared Spectroscopy

Infrared spectra were recorded for samples of both raw and charred hay. Transmittance in the 1600–1700 cm^{-1} (wavenumber) region was observed to be reduced after the hay was charred (boxed region in Figure 2). This has been attributed to the removal of absorbed water on charring, with a concomitant decrease in the observed OH bending signal. Charring should also expose the carbon backbone and thereby alter the hydrophobic properties of the hay surface. In this regard it is important to note the changes in the O–CH$_3$ methyl stretching region [17] at 2850–2815 cm^{-1} (* in Figure 2), and in the small peak at 809 cm^{-1} (C–C–O and C–O–C deformations) which disappears completely on charring. Other peaks also change in relative intensity, and all of this suggests that actual chemical modification has occurred. There is also a small possibility that these differences could have arisen from differences in the original hay samples themselves since the experiment was not performed on replicate samples. Although the spectra may appear only slightly different overall, the effect on the surface properties supports the evidence from the contact angle measurements that chemical modification of the surface has been achieved.

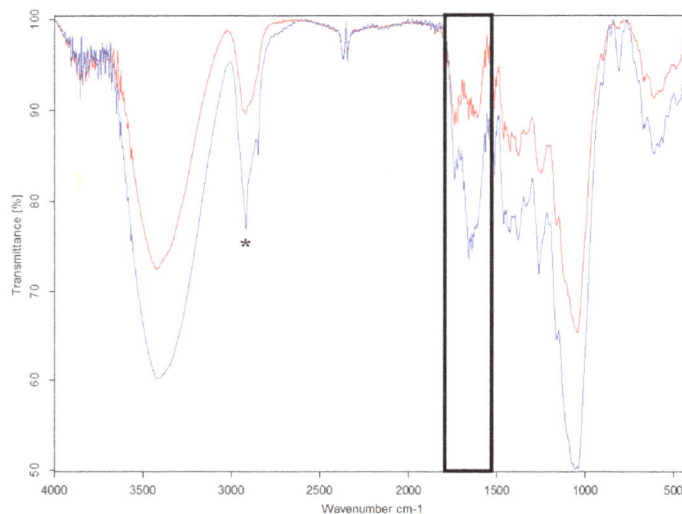

Figure 2. Overlay of infrared spectra comparing a raw hay sample (blue/bottom) to a charred hay sample (red/top). The boxed and starred areas show regions where the two spectra are distinctly different (see text).

3.2.3. ^{13}C CP/MAS Nuclear Magnetic Resonance Spectroscopy

All of the ^{13}C CP/MAS NMR spectra shown in Figure 2 exhibit typical cellulose signatures, with the alcoholic carbons between 50 and 90 ppm and the acetyl groups around 105 ppm. In addition, they show resonances for aliphatic groups between about 50 and 0 ppm. On the high chemical shift side, signals of carboxyl groups (around 173 ppm) and of unsaturated carbon groups, between about 110 and 155 ppm, of aromatic and possibly aliphatic origins are detected. In particular, the region between 140 and 155 ppm, corresponds to aromatic carbons bridging to other carbons or hetero-nuclei, such as oxygen. Comparison between the three samples shows that, relative to the largest peak, the intensities of the unsaturated (aromatic) region, the carboxyl groups and some aliphatic groups increase with increased charring time (indicated by *) (see Figure 3).

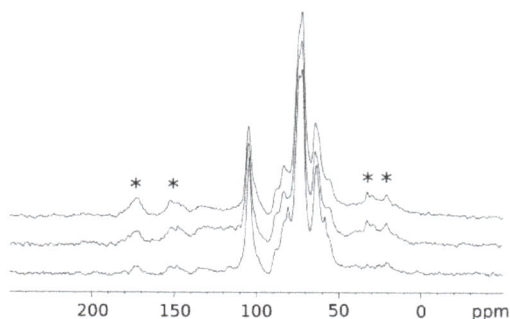

Figure 3. ^{13}C CP/MAS NMR spectra of raw hay (bottom) and charred hay samples (30 min middle, 60 min top) showing the relative intensity increases in the carboxyl, aryl and alkyl regions (isotropic center bands indicated by *) with charring time.

3.3. Application of Treated Hay to Bitumen Samples in Aqueous Environments

3.3.1. Flotation Analysis

A sample of bitumen in tap water with no additives was found to aggregate in the beaker and subsequently sink after a period of 4 days (96 h) at room temperature (23 °C). In the initial stages of the experiment, there was an even distribution of the bitumen on the surface of both the water solution and the Instant Ocean solution (Figure 4A,B, respectively). At 4 days, the bitumen sample in deionized water was observed to aggregate into a ball and sink to the bottom of the beaker on stirring (Figure 5A). As expected, the sample of bitumen in Instant Ocean solution remained afloat longer than the water sample because of the greater density of the Instant Ocean solution. As time passed the Instant Ocean solution became gradually more yellow in color and the bitumen slick increased in diameter (Figure 5B). It showed a tendency to aggregate and sink on stirring but it would remain suspended in the solution (never sinking to the bottom of the beaker) before rising again and dispersing across the surface of the solution. When the experiment was terminated after 121 h, the bitumen was still floating on the Instant Ocean solution.

Figure 4. Bitumen samples in deionized water (**A**) and Instant Ocean solution (**B**) at room temperature (23 °C) and time = 0 h.

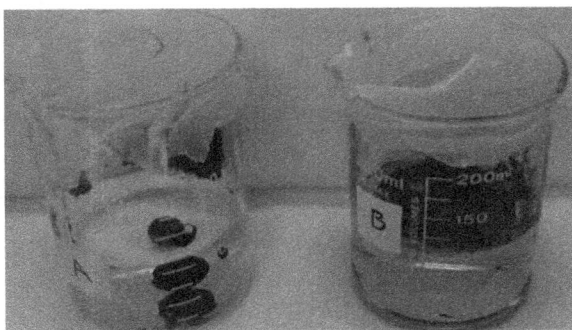

Figure 5. Bitumen samples in deionized water (**A**) and Instant Ocean solution (**B**) at room temperature (23 °C) and time = 96 h.

In solution, bitumen did not interact to any great extent with green, uncharred hay. Instead, the hay just became waterlogged and sank while the bitumen floated on the surface of the solution. However, samples of bitumen treated with charred hay as a flotation additive were buoyant on the water surface for up to 186 h. The application of hay charred for 60 min to the bitumen slicks is shown in Figure 6 (at time = 0 h) and in Figure 7 (at time = 186 h, where both samples have sunk). Samples containing the hay charred for 60 min remained buoyant for equally long (Instant Ocean) or longer (water) than the samples containing hay charred for only 30 min. In fact, the bitumen and charred hay sample (30 min) sank in water after only 72 h. Comparing Figure 7 to Figure 6, dramatic color changes can be seen in the solutions (both water and Instant Ocean). These changes can be attributed to surface interactions between the charred hay and the hydrophobic fractions of the bitumen resulting in dissociation of some of the more polar fractions of the bitumen sample into the solutions over time. For the same reason, all of these samples are more highly colored than the samples of bitumen alone (Figure 5).

Figure 6. Bitumen samples with charred hay (60 min) added in deionized water (**left**) and Instant Ocean solution (**right**) at room temperature (23 °C) and time = 0 h.

Figure 7. Bitumen samples with charred hay (60 min) added in deionized water (**right**) and Instant Ocean solution (**left**) at room temperature (23 °C) and time = 186 h.

Samples of bitumen treated with charred hay (Figure 8) or CaO-coated charred hay (Figure 9), a flotation *and* dispersant additive, were buoyant for more than 408 h (17 days) at which point the experiment was terminated. In both samples, it is clear that the bitumen adsorbed to the surface of the hay undergoes dispersion. However, this dispersion appears to be greater when the charred straw has been treated with calcium oxide. It was also noted that the Instant Ocean solution did not yellow as much over time when CaO-coated hay was used as the dispersant. This may be the result of the increased dispersion or it may indicate a reduction in the fractional dissolution of the polar compounds with time for the coated sample. The latter idea is supported by the leached hydrocarbon analyses presented in the following section.

Figure 8. A bitumen sample in Instant Ocean treated with charred hay at room temperature (23 °C) and time = 288 h.

Figure 9. A bitumen sample in Instant ocean treated with CaO-coated charred hay sample at room temperature (23 °C) and time = 336 h.

3.3.2. Residual Oil Concentration

The solutions remaining in the beakers (both Instant Ocean and deionized water) at the end of the floatation experiments were analyzed for leached hydrocarbons. The liquid phase was separated from the hay and bitumen residues prior to measurement. As can be seen in Table 2, all samples containing bitumen showed leaching of hydrocarbons into solution, and this leaching was generally greater in Instant Ocean than in fresh water (deionized). Bitumen samples treated with raw hay in both Instant Ocean solution and deionized water showed slightly more leaching of petroleum hydrocarbons than any of the other treatments. However, this relative amount is statistically insignificant with p-values > 0.05 (2-factor ANOVA). Bitumen samples treated with only CaO showed the smallest quantity of leached petroleum hydrocarbons in both Instant Ocean solution and deionized water. We speculate that the strong base, CaO, either saponifies or deprotonates acidic species in the bitumen thus generating an *in situ* surfactant that better bonds the hydrocarbons to the charred straw.

Samples of bitumen treated with hay did show ppm levels of hydrocarbons (e.g., aliphatic, and parental and alkylated polycyclic aromatic hydrocarbons), entering the water phase at a slow rate over several days due to dispersion. The aromatics that enter the water column would contribute to toxicity in organisms exposed to the contaminated water. However, in an open marine environment, where there are no boundaries, these chemicals would spread over a greater spatial area and be exposed to natural dilution depending on sea states and environmental conditions. The natural dilution of these hydrocarbons would reduce their environmental impacts.

Table 2. Total Petroleum Hydrocarbons (TPH) in water samples collected from water column after treatment.

Sample	TPH (mg/L)	
	Instant Ocean Solution	Deionized Water
Standard (water)	<1.0	<1.0
Bitumen	47	28
Bitumen + raw hay	60	52
Bitumen + charred hay	54	21
Bitumen + CaO	18	18
Bitumen + CaO-coated raw hay	19	21
Bitumen + CaO-coated charred hay	52	29

3.3.3. Inductively Coupled Plasma—Mass Spectroscopy

ICP-MS experiments were performed under a variety of conditions in order to assess, if possible, any observable trends resulting from metal ion interaction (leached from the bitumen) with the biomaterial (hay). Reliable results were obtained from the fresh water samples only. Results of experiments performed in the Instant Ocean solutions were complicated due to interference from its high concentration of salts with the metal ions being studied. All samples were tested for V, Cr, Mn, Co, and Cu.

Analysis of the deionized water used in sample preparation showed no (or only trace levels of) V, Mn and Co. Levels of Cr measured 20 ppb and Cu 90 ppb. Analysis of the Instant Ocean solution showed that the product itself contained no Co or Cr, though Cr was present in the solution at the same level found in the water used to prepare it. Instant Ocean was also found to contain V 60 ppb, Mn 30 ppb and Cu 230 ppb (partially from the water).

Addition of bitumen to deionized water resulted in no change to the levels of V, Cr or Co measured. Mn was found to leach into the water, the concentration increasing from a trace (2 ppb) to 20 ppb. The opposite effect was observed for the levels of Cu in solution. Bitumen appears to absorb copper from the solution as the levels decreased from 90 to 30 ppb for Cu. The results when bitumen was added to Instant Ocean were similar. No change in the concentration was observed for V, Cr, Mn or Co. In the case of Mn there already was a concentration of 30 ppb in the solution which may have prevented more from leaching in from the bitumen. The bitumen in Instant Ocean also appeared to absorb some of the copper from solution.

The addition of straw (no CaO) to solutions of bitumen in deionized water had little observable effect on metal ion concentrations. V, Cr and Co levels were totally unaffected. The addition of green straw slightly increased the levels of Mn and Cu in the water, while charred straw had no effect. This may be because the hay itself contains both Cu and Mn which may leach more easily from the green straw. The addition of straw (no CaO) to bitumen in Instant Ocean did not affect the measured concentrations of Cr, Co or Cu. The level of V increased, while the level of Mn increased appreciably upon the addition of the straw, and in both cases the impact of the charred hay was greater than that of green hay.

The effect of a CaO-coating on the straw samples was assessed by comparing the green and charred straw numbers to the corresponding CaO-coated green and charred straw results. For the deionized water solutions the results, whether green or charred straw was used, were very consistent. The V, Cr and Co levels remained relatively constant while the Mn concentrations increased and the Cu concentrations decreased upon using straws coated with CaO. For the Instant Ocean solutions the results were a bit more scattered. However, overall the Cr, Co and Cu concentrations remained relatively constant while the V levels increased and the Mn levels decreased with the introduction of the CaO coating on the straws.

4. Conclusions

This work has shown that charred hay (and in particular CaO-coated charred hay) is an effective substrate for adsorption and flotation of bitumen. With further testing and fine-tuning it could become a valuable tool in the treatment of bitumen and dilbit spills (dilbit tests are ongoing) in aqueous environments. We have shown that its use should prolong the window of opportunity for skimming or *in situ* combustion of spilled oil by increasing the time the bitumen remains afloat. One clear advantage of charred hay is the strong interactions that form between its modified surface and the bitumen, limiting any washing out effect that might occur via wave action or weathering. Hay is also cheap, biodegradable and easily produced in bulk quantities. Additionally, charred hay demonstrates potential for prophylactic treatment on shorelines to prevent bitumen/dilbit from adhering to and contaminating shoreline materials such as sand, rocks and sensitive habitat.

Acknowledgments

We thank the Natural Sciences and Engineering Research Council of Canada (through the Discovery Grants Program to JACC). JACC acknowledges generous support from the Canada Research Chairs Program, the Canadian Foundation for Innovation and the Nova Scotia Research and Innovation Trust Fund. We are grateful to NMR-3 (Dalhousie University) for NMR data acquisition and Patricia Granados (CEAR) for helpful discussions of the ICP-MS results. We would also like to acknowledge the Dalhousie University research group of J. Dahn for their assistance with contact angle measurements.

Author Contributions

Nusrat Jahan: Lab Experimentation;

Jason Fawcett: Lab Experimentation;

Thomas King: Residual Oil Analysis, Report Writing/Editing;

Alex McPherson: Experiment Design, Lab Experimentation, Report Writing/Editing;

Katherine N. Robertson: Data Analysis and Manuscript Preparation;

Ulrike Werner-Zwanziger: NMR Data Acquisition, NMR Analysis and NMR Writing/Editing; Jason Clyburne: Funding, Intellectual Property, Experiment Design and Report Writing/Editing.

Conflicts of Interest

The authors declare no conflict of interest.

References

1. *ST98-2011: Alberta's Energy Reserves 2010 and Supply/Demand Outlook 2011–2020*; Energy Resources Conservation Board: Calgary, AB, Canada, 2011.

2. Zubot, W.; MacKinnon, M.D.; Chelme-Ayala, P.; Smith, D.W.; El-Din, G.M. Petroleum coke adsorption as a water management option for oil sands process-affected water. *Sci. Total Environ.* **2012**, *427–428*, 364–372.

3. Gosselin, P.; Hrudey, S.E.; Naeth, M.A.; Plourde, A.; Therrien, R.; van Der Kraak, G.; Xu, Z. *The Royal Society of Canada Expert Panel: Environmental and Health Impacts of Canada's Oil Sands Industry*; The Royal Society of Canada: Ottawa, ON, Canada, 2011.

4. Timoney, K. Environmental and health impacts of Canada's bitumen industry: In search of answers. *Environ. Sci. Technol.* **2012**, *46*, 2496–2497.

5. Bazyleva, A.; Fulem, M.; Becerra, M.; Zhao, B.; Shaw, J.M. Phase behavior of Athabasca bitumen. *J. Chem. Eng. Data* **2011**, *56*, 3242–3253.

6. Environmental Science and Technology Division. Environment Canada Oil Spill Properties Database. Available online: http://www.etc-cte.ec.gc.ca/databases/OilProperties/ (accessed on 22 May 2014).

7. Borton, D., II; Pinkston, D.S.; Hurt, M.R.; Tan, X.; Azyat, K.; Scherer, A.; Tykwinski, R.; Gray, M.; Qian, K.; Kenttämaa, H.I. Molecular structures of asphaltenes based on the dissociation reactions of their ions in mass spectrometry. *Energy Fuels* **2010**, *24*, 5548–5559.

8. Gray, M.R.; Tykwinski, R.R.; Stryker, J.M.; Tan, X. Supramolecular assembly model for aggregation of petroleum asphaltenes. *Energy Fuels* **2011**, *25*, 3125–3134.

9. Rueda-Velásquez, R.I.; Freund, H.; Qian, K.; Olmstead, W.N.; Gray, M.R. Characterization of asphaltene building blocks by cracking under favorable hydrogenation conditions. *Energy Fuels* **2013**, *27*, 1817–1829.

10. *2013 Federal Government Technical Report: Properties, Composition and Marine Spill Behaviour, Fate and Transport of Two Diluted Bitumen Products from the Canadian Oil Sands*; Cat. No.: En84-96/2013E-PDF; Government of Canada: Ottawa, ON, Canada, 2013; pp. 1–85.

11. King, T.L.; Robinson, B.; Boufadel, M.; Lee, K. Flume tank studies to elucidate the fate and behavior of diluted bitumen spilled at sea. *Mar. Pollut. Bull.* **2014**, *83*, 32–37.

12. Cole, M.G.; King, T.L.; Lee, K. *Analytical Technique for Extracting Hydrocarbons from Water Using Sample Container as Extraction Vessel in Combination with Roller Apparatus*; Canadian Technical Report of Fisheries and Aquatic Sciences #2733; Fisheries and Oceans Canada: Québec, QC, Canada, 2007; pp. 1–12.

13. Stevens, C.C.; Thibodeaux, L.J.; Overton, E.B.; Valsaraj, K.T.; Nandakumar, K.; Rao, A.; Walker, N.D. Sea surface oil slick light component vaporization and heavy residue sinking: binary mixture theory and experimental proof of concept. *Environ. Eng. Sci.* **2015**, *32*, 694–702.

14. King, T.; Robinson, B.; McIntyre, C.; Toole, P.; Ryan, S.; Saleh, F.; Boufadel, M.; Lee, K. Fate of surface spills of Cold Lake blend diluted bitumen treated with dispersant and mineral fines in a wave tank. *Environ. Eng. Sci.* **2015**, *32*, 250–261.

15. Baldwin, I.T.; Staszak-Kozinski, L.; Davidson, R. Up in smoke: I. Smoke-derived germination cues for postfire annual, Nicotiana attenuata Torr. Ex. Watson. *J. Chem. Ecol.* **1994**, *20*, 2345–2371.

16. Arkles, B. Hydrophobicity, hydrophilicity and silanes. *Paint Coat. Ind. Mag.* **2006**, *22*, 114–123.

17. Coates, J. *Interpretation of Infrared Spectra: A Practical Approach. Encyclopedia of Analytical Chemistry*; Meyers, R.A., Ed.; John Wiley & Sons Ltd.: Chichester, UK, 2000; pp. 10815–10837.

Viral and Bacterial Epibionts in Thermally-Stressed Corals

Hanh Nguyen-Kim [1,2], **Thierry Bouvier** [1], **Corinne Bouvier** [1], **Van Ngoc Bui** [3], **Huong Le-Lan** [2] **and Yvan Bettarel** [1,†,*]

[1] Institute of Research for Development (IRD), National Center for Scientific Research (CNRS), UMR MARBEC, Montpellier 34095 cedex, France; E-Mails: nguyenkimhanh84@gmail.com (H.N.-K.); tbouvier@univ-montp2.fr (T.B.); cbouvier@univ-montp2.fr (C.B.)

[2] Institute of Oceanography (IO), Vietnam Academy of Science and Technology (VAST), Nha Trang 650000, Vietnam; E-Mail: lelanhuongio@gmail.com

[3] Institute of Biotechnology (IBT), Vietnam Academy of Science and Technology (VAST), Hanoi, 100000, Vietnam; E-Mail: bui@ibt.ac.vn

[†] Present address: IRD—Van Phuc Diplomatic Compound, Bldg 2G, Appt 202, 298 Kim Ma, Ba Dinh, Hanoi 100000, Vietnam.

[*] Author to whom correspondence should be addressed; E-Mail: yvan.bettarel@ird.fr

Academic Editor: Jose Victor Lopez

Abstract: The periodic rise in seawater temperature is one of the main environmental determinants of coral bleaching. However, the direct incidence of these episodic thermal anomalies on coral-associated microbiota and their subsequent effects on coral health are still not completely understood. In this study, we investigated the dynamics of three main microbial communities of the coral holobiont (e.g., *Symbiodinium*, bacteria and viruses), during an experimental thermal stress (+4 °C) conducted on the scleractinian *Fungia repanda*. The heat-treatment induced coral bleaching after 11 days and resulted in a final elevation of *ca.* 9, 130 and 250-fold in the abundance of mucosal viruses, bacteria, and *Symbiodinium*, respectively. On the contrary, the proportion of actively respiring bacterial cells declined by 95% in heat-stressed corals. The community composition of epibiotic bacteria in healthy corals also greatly differed from bleached ones, which also exhibited much higher production rates of viral epibionts. Overall, our results suggest that the shift in temperature induced a series of microbial changes, including the expulsion and transfer of *Symbiodinium* cells from the coral polyps to the mucus, the collapse of the physiological state of the native bacterial associates, a substantial alteration in their community structure, and accompanied by the development of a cortege of highly active virulent phages. Finally, this study provides new insights into the environmentally-driven microbial and viral processes responsible for the dislocation of the coral holobiont.

Keywords: coral bleaching; thermal stress; bacteria; viruses; holobiont; Vietnam

---

1. Introduction

Coral bleaching is a widespread phenomenon in tropical waters that has caused the massive decline of coral cover surface over the last decades [1]. This event typically occurs after the expulsion of the symbiotic dinoflagellates (*i.e.*, *Symbiodinium* sp.) by the coral animal (the polyp). The disruption of this symbiotic relationship is generally caused by environmental stresses, the most common of which being the elevation of seawater temperature [2–4], and, to a lesser extent, changes in solar radiation [5] and in salinity [6], ocean acidification [7], presence of contaminants [8], and eutrophication [9].

Recently, it has been shown that prokaryotes could be also involved in coral bleaching, although the underlying mechanisms still remain unclear and controversial. Coral-associated bacteria through a long history of selection and co-evolution with their host act as a nutrition supplier for corals [10,11]. They also represent a natural barrier against pathogen colonization due to their ability to synthesize antimicrobial compounds, such as peptides and antibiotics [12–15]. Any alteration in their physiological state, metabolic capacities and/or community structure may then directly affect their ecological functions and in turn impact coral health. For example, the access of the surrounding pathogens to coral surface may be facilitated by a weaker line of bacterial defense. This has been conceptualized in the coral probiotic hypothesis [16,17]. Evidence came from *in situ* study and laboratory experiments, which revealed the presence of pathogens in bleached corals, namely *Vibrio shiloi* and *Vibrio coralliilyticus* in the scleractinian *Oculina patagonica* and *Pocillopora damicornis*, respectively [18–21]. However, these findings faced controversial debates since other studies showed no involvement of pathogens in coral bleaching [22,23]. Nonetheless, several studies have reported remarkable shifts in the entire coral bacteriome during bleaching events in both natural and experimental conditions [24–27]. However, until now, no studies have clarified whether these shifts are a cause or a consequence of bleaching and whether these changes are also driven by biotic or abiotic factors.

Among the biological sources of bacterial control in marine habitats, viruses are certainly one of the most prominent [28,29]. Recently, viruses have been also recognized for their large abundance in coral mucus [30–33] and their potential ability to control both bacterial symbionts and pathogens [34–36], and also to an unknown extent the zooxanthellae *Symbiodinium* [26,37–39]. By using a complex combination of lytic and lysogenic strategies, viruses have been hypothesized to be capable of either protecting corals from pathogen colonization and viral surinfection, or also conversely hastening their decline, especially during times of adverse conditions (elevated temperature, for instance) [40]. However more data are needed to validate such presumptions. For example, knowledge about viral occurrence in bleached corals still remains limited, and finally little is known about their effective control of epibiotic bacterial communities.

To tentatively clarify the dynamics of mucosal viruses and their bacterial hosts during a bleaching event, we conducted a thermal stress experiment over 11 days using individuals of the free-living coral *Fungia repanda*, collected in the Nha Trang Bay, Viet Nam. In this study, we targeted the following objectives: (1) to compare the concentrations of the three main communities of the coral holobiont (viruses, bacteria, and *Symbodinium*) between healthy and bleached corals; (2) to track potential shifts in the physiological activity and community composition of coral-associated bacterial communities during the thermal stress; (3) and to estimate the viral lytic pressure on such associates.

2. Materials and Methods

2.1. Experimental Design

The experiment was conducted from the 9th to the 19th of October in 2012 at the Institute of Oceanography of Nha Trang (Nha Trang, Viet Nam). In this study, individuals of the plate coral *Fungia repanda* were collected in the Nha Trang Bay. This free-living coral is easy to handle and typically produces large quantities of mucus (>10 mL/individual/5 min).

Prior to the experiment, all the corals were kept in tanks filled with sand-filtrated seawater collected at corals' site of origin for 10 days to allow for acclimatization. The experimental design consisted of two 40-L aquaria with triplicate individuals of *F. repanda* corals of similar sizes (15–17 cm in diameter). At the end of the acclimatization period, the temperature was gradually increased (over 3 days) to 32 °C in the heat-treatment tank by using an immersion thermostat (LAUDA Ecoline Staredition E200), while the water temperature was kept at ambient temperature (28 °C) in the control aquarium. Water was renewed once a day in both aquaria with seawater previously adjusted to the different tank temperatures. During the experiment, both tanks received aeration to maintain air-saturated conditions. Lighting was provided by a fluorescent lamp (VHO: General Electric, 175 watt), with irradiance of 200 μmoL photons $m^{-2} \cdot s^{-1}$ (on a 12 h:12 h light:dark cycle). Water temperature, salinity, oxygen concentration, and light intensity were monitored twice a day. HOBO Pendant loggers (Onset, Massachusetts, MA, USA) were used to log light intensity and temperature in the control and heat treatment aquaria. After 11 days under this thermal stress, all the corals were sampled using each set of measurements described below.

Thirty milliliters of water, and 5–7 mL of mucus were collected from each coral at the beginning (T_0) and at the end (after 11 days) of incubation (T_{end}). Mucus was collected by using the desiccation method described in details elsewhere [32,41]. Briefly, the corals were taken out of the water and exposed to air for 1 to 3 min. This stress caused the mucus to be secreted, forming long gel-like threads dripping from the coral surface. The first 30 s of mucus production was discarded to prevent contamination and dilution by seawater. Mucus was then distributed in cryotubes for estimation of (1) viral, bacteria, and *Symbiodinium* abundance; (2) viral lytic production rate (3) cell physiological activity; and (4) bacterial community composition.

2.2. Enumeration of Symbiodinium, Bacteria, and Viruses

The fixed mucus was processed for viral and bacterial extraction and enumeration by using the potassium citrate method, as recommended by Leruste et al. [31], and adapted from Williamson et al. [42]. Briefly, 100 μL of mucus was eluted into 900 μL of 0.02-μm-pore-size-filtered, pH 7 solution of 1% citrate potassium (10 g potassium citrate, 1.44 g $Na_2HPO_4 \cdot 7H_2O$, and 0.24 g KH_2PO_4 per liter). All tubes were then vortexed at a moderate speed for 5 min before particles were stained and enumerated. The number of viruses and bacteria contained in duplicate subsamples (2 independent counts for each of the six corals) were determined after retention of the particles on 0.02-μm pore-size membranes (Anodisc, GE Healthcare, Little Chalfont, United Kingdom) and staining with the nucleic acid stain SYBR Gold [43]. On each slide, 300–500 bacteria and viruses were counted with an Olympus Provis-AX70 epifluorescence microscope (Olympus SAS, Rungis, France) in 20 fields under blue light excitation (488 nm). Symbiodinium cells, due to their photosynthetic pigments could be also enumerated on the same slides, under the blue light excitation.

2.3. Bacterial Physiological State

The proportion of respiring bacteria that have high rates of metabolism was determined using 5-cyano-2,3-ditolyl tetrazolium chloride (CTC), an indicator of the respiratory electron transport system activity [44]. A stock solution of 50 mmoL l-1 CTC (Tebu-bio SAS, Le Perray-en-Yvelines, France) was prepared at both sampling dates (day = 0 and day = 11), filtered through 0.01 mm filters and kept in the dark at 4 °C until use. CTC stock solution was then added to 0.45 mL of both duplicate fresh mucus and water samples (5 mmoL l-1 final CTC concentration) and incubated for 1.5 h at room temperature in the dark. Formaldehyde (3% final concentration) was used to stop the CTC reaction. Samples were flash frozen in liquid nitrogen and stored at −80 °C freezer until flow cytometer (FCM) analysis (FACSCALIBUR, BD Biosciences, Franklin Lakes, NJ, USA). The red fluorescence of CTC (FL3) and the light scatter SSC were used to discriminate the CTC+ cells from other cells or weak fluorescent particles. The percentage of CTC+ cells, based on triplicate analyses, was calculated relative to the total bacterial counts obtained by epifluorescence microscopy (see 2.2. Enumeration of Symbiodinium, Bacteria and Viruses).

2.4. Viral Lytic Production

The decay, i.e., the decrease in the viral concentration over time, was recorded after inhibition of new viral lytic production (VP) by the addition of potassium cyanide (KCN; final concentration of 2 mM) in both mucus and water samples [45,46]. All incubations for decay experiments were performed in duplicate at in situ temperature, for 12 h. Incubations were stopped after addition of formaldehyde (3% final concentration). Viral abundance was determined in KCN-treated and untreated water and coral mucus, by using SYBR Gold and epifluorescence microscopy. The difference between the abundance of viruses with and without KCN allows the estimation of VP.

2.5. DGGE Analysis of Bacteria Community Structure

The genetic diversity of both Eubacterial communities in water and mucus samples was estimated by using a fingerprinting technique: Denaturing gradient gel electrophoresis (DGGE) [47]. PowerSoil® DNA Isolation Kit (MO BIO, Carlsbad, CA, USA) was used to extract DNA from water and mucus samples. The DNA sequences were then subjected to touchdown PCR using following primers: 341F-GC [48] and 519R [49], which target bacterial 16S gene (178 bp). Thirty-five cycles of amplification were done starting at 93 °C initial denaturation of the dsDNAs followed by a second denaturation phase at 92 °C and the annealing step, which was done at the high temperature of 71 °C minimizing unspecific primer binding. After each cycle, the temperature was lowered by 1/2 of a degree until reaching the touchdown temperature of 61 °C, keeping that temperature for the last 15 cycles. The DGGE was performed with Ingeny U-Phor system (Ingeny, Waltham, MA, USA) in 0.5× TAE buffer (Euromedex, Souffelweyersheim, France) at 60 °C with a constant voltage of 80 V for 18 h. The DNA was then stained with SYBR-Gold. DNA bands were visualized on a UV trans-illumination table with the imaging system GelDoc® XR (Bio-Rad, Hercules, CA, USA) and analyzed using fingerprint and gel analysis Quantity One software (Bio-Rad, Hercules, CA, USA). Band matching was performed with 1.00% position tolerance and 1.00% optimization. After generating a band-matching table, we obtained the binary presence-absence matrix for all the detected bands. The matrix was used to calculate a distance matrix with Sorensen dissimilarity index, which subsequently was used for an ordination analysis—principal coordinate analysis (PCoA) (see data analysis).

2.6. Statistical Analysis

The differences among samples in all variables were tested by one-way ANOVA, followed by a *post-hoc* analysis Tukey-Kramer for pairwise comparisons of means between samples. All the parameters were normalized prior to test. A level of 0.05 was considered significant. The JMP 9.0 software (SAS institute, Cary, NC, USA) was used for these statistical analyses. In order to evaluate differences and variability in the bacterial community composition, a principal coordinate analysis (PCoA) was applied, using Sorensen dissimilarity matrices as inputs. The analysis and calculation were done using R software version 3.0.2 with the vegan, ade4 and betapart packages.

3. Results

The experiment was stopped after 11 days when clear signs of bleaching were observed in all the three replicates of *F. repanda* corals in the heat-treated aquarium. At that time, no visible trace of coral bleaching was detected in the control tank.

3.1. Abundances of Symbiodinium, Bacteria and Viruses

The heat treatment resulted in a large increase in the abundance of *Symbiodinium* in the mucus of *F. repanda* reaching up to 5.5×10^6 cells·mL^{-1} (one-way ANOVA, F = 12.7, $p = 0.002$), which was more than 250-fold higher than that observed in the control tank at the end of the experiment (Tukey-Kramer, $p < 0.05$). On the contrary, the micro-algal concentration remained relatively stable in untreated corals throughout the experiment (Tukey-Kramer, $p < 0.05$) (Figure 1A). The final abundance of bacteria was also greatly enhanced (by almost 130 times) in heat-stressed corals (Tukey-Kramer, $p < 0.05$) to reach an average value of 4.5×10^8 cells·mL^{-1} (Figure 1B). The pattern of viral abundance was highly comparable with that of bacteria, showing a final enhancement factor of almost nine between heat-treated and control tanks, respectively (Tukey-Kramer, $p < 0.05$) (Figure 1C). As for *Symbiodinium* cells, bacterial and viral abundance did not vary significantly in the control tank (Figure 1B,C).

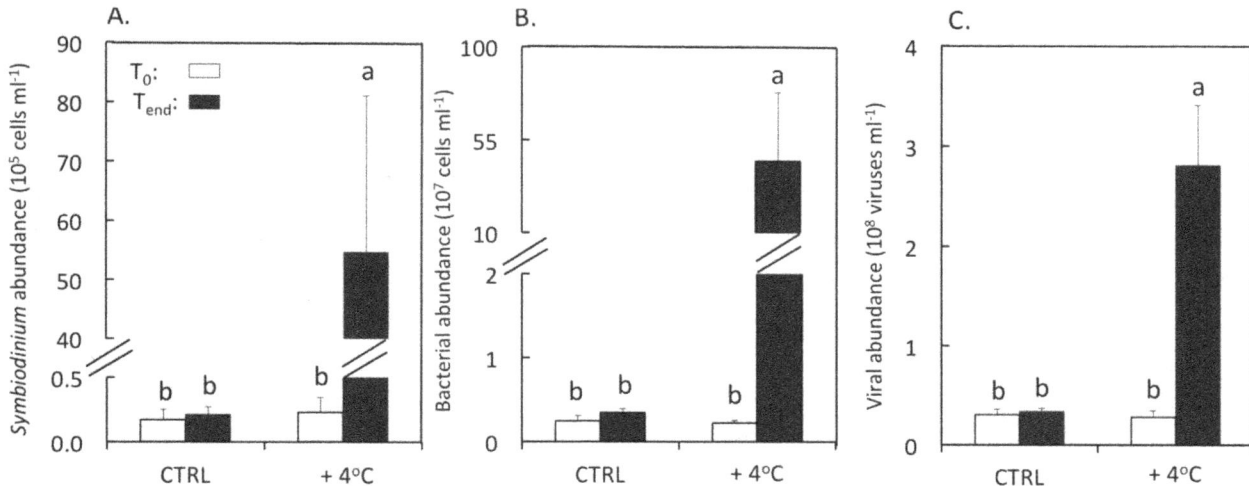

Figure 1. Abundances of *Symbiodinium* (**A**); bacteria (**B**), and viruses (**C**) in mucus samples of *F. repanda* in both control (CTRL) and heat-treatment (+4 °C) at the beginning (T_0) and at the end of the experiment (T_{end}). Error bars represent one standard deviation from the mean ($n = 3$).Histograms with the same letters are not significantly different at $p = 0.05$.

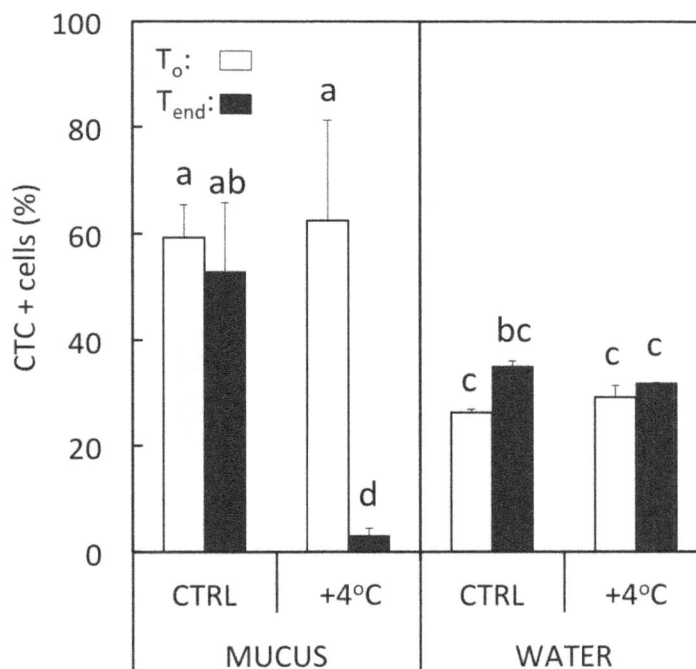

Figure 2. Percentages of CTC + cells in mucus samples of *F. repanda* in both control (CTRL) and heat-treatment (+4 °C). Error bars represent one standard deviation from the mean ($n = 3$). Histograms with the same letters are not significantly different at $p = 0.05$.

3.2. Physiological State of Bacteria

The percent of metabolically active respiring cells (CTC+) dramatically dropped from 62.4% to 3.1% in the mucus of heat stressed corals (one-way ANOVA, $F = 12.2$, $p = 0.0001$) (Figure 2). This value was 17-fold lower than that observed in the control aquarium (mean = 53.0%) at the end of the experiment (Tukey-Kramer, $p < 0.05$). In control treatment, coral mucus harbored a relatively constant proportion of active cells throughout the experiment (53.0%–59.3%).

3.3. Viral Lytic Production

The viral production rate in coral mucus was also strongly stimulated by the elevation of temperature, with values ultimately reaching up to 7.5×10^6 viruses $mL^{-1} \cdot h^{-1}$ (Figure 3), which was, on average, 4.9 times higher than that observed in the control tank (Tukey-Kramer, $p < 0.05$). In this aquarium, viral lytic production did not show any significant changes between the beginning and the end of the experiment (Tukey-Kramer, $p < 0.05$).

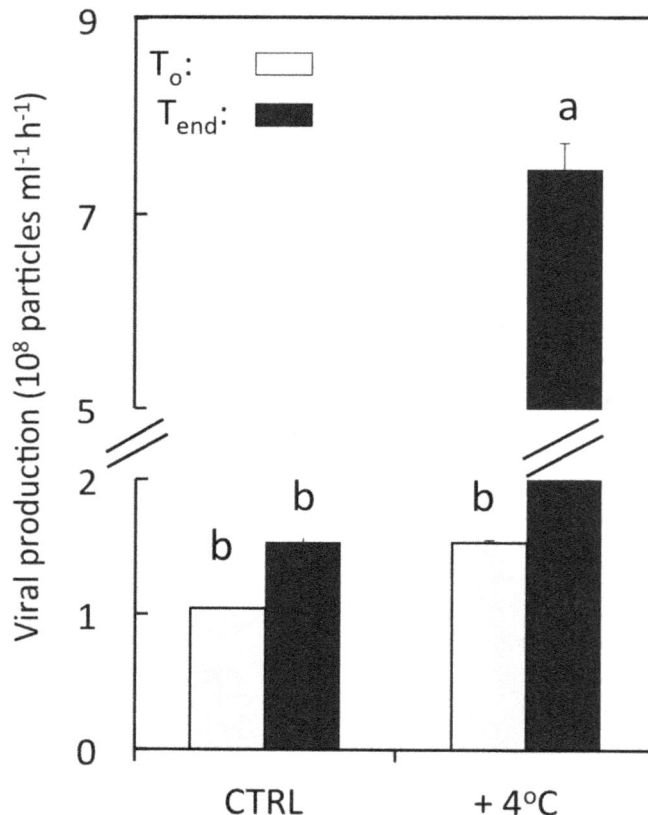

Figure 3. Viral lytic production in mucus samples of *F. repanda* in both control (CTRL) and heat-treatment (+4 °C). Error bars represent one standard deviation from the mean ($n = 3$). Histograms with the same letters are not significantly different at $p = 0.05$.

3.4. DGGE Analysis

The thermal stress also strongly impacted the community structure of epibiotic cells (Figure 4). The first principal coordinate (39.7% of total variations) showed clear discrimination between bleached and healthy corals at T_0. In the control tank, the structure of mucosal communities remained relatively stable over the experimental course (Figure 4). Conversely, a significant shift was observed in thermally stressed corals as shown by the length of the specific arrow.

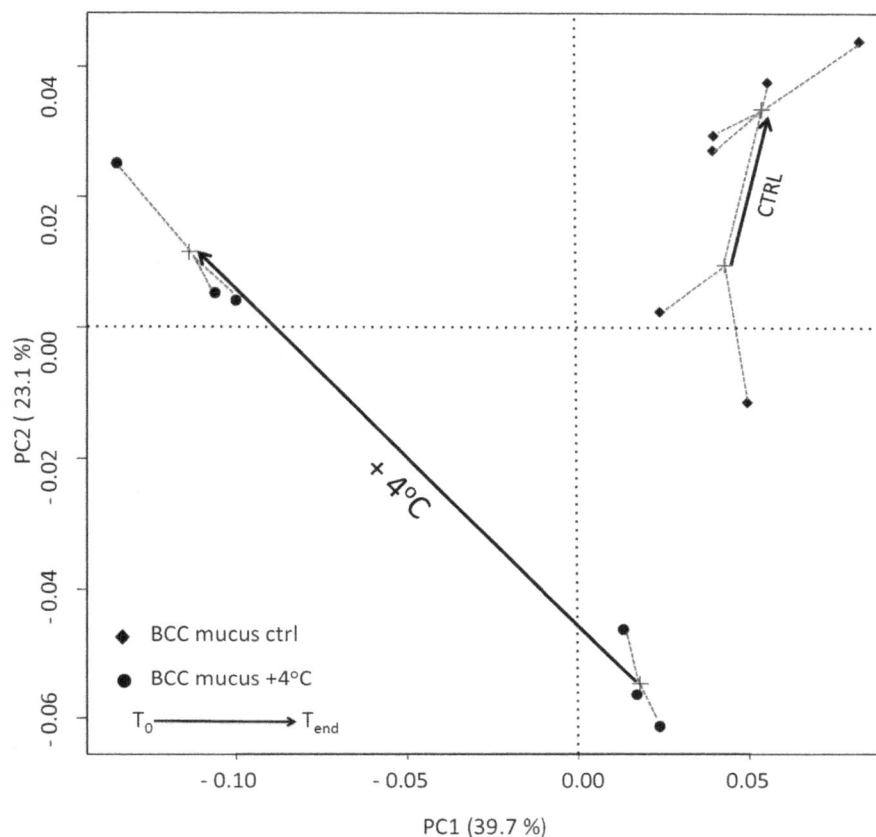

Figure 4. Principal component analysis (PCOA) obtained with the Sorensen dissimilarity index calculated from denaturing gradient gel electrophoresis (DGGE) presence/absence profile. Arrows, drawn from the calculated centroid of all replicates of each sample, show the evolution in the bacterial community structure between the beginning and the end of the experiment, in mucus samples of *F. repanda*, in both control (CTRL) and heat-treatment (+4 °C).

4. Discussion

Coral bleaching is typically characterized by the loss of intracellular endosymbionts (*Symbodinium*) from the coral tissue [2,50]. Here, the bleached *Fungia* observed at the end of the incubation, exhibited higher concentrations of *Symbodinium* cells in their mucus than healthy individuals, which clearly confirmed their expulsion from coral tissue to the mucus layer, before their ultimate transfer into the water column.

4.1. High Abundance but Low Metabolic Activity of Bleached-Coral-Associated Bacteria

The mucus of bleached corals contained a much higher abundance of bacteria than healthy ones. At first sight, such observations are not surprising as bacterial growth generally increases with temperature.

However, the thermal stress also resulted in a substantial drop of the physiological state of mucus associated bacterial cells, where the proportion of CTC+ respiring bacterial cells decreased from about 60% to 3%. As for all other physiological indicators, respiration is affected by temperature. After the optimum temperature, the Q_{10}-value (index of the temperature sensitivity of bacterial respiration) typically decreases and becomes negative, *i.e.*, respiration decreases with increasing temperature [51]. Our results indicate that the thermal tolerance of mucosal cells was probably low with a disruption point occurring between 28 °C and 32 °C. A significant reduction in the mineralization abilities of *Acropora millepora*'s associates was also reported after they faced a thermal increase of 1 °C to 3 °C [24,26]. Here, the proliferation of abundant but low active cells could be favored by the degradation of particular key active bacterial associates. For example, recent studies have demonstrated that the antibacterial activity of mucosal bacteria is altered at high temperature [13,15,52]. We then suspect that the decline in cell respiring activity in stressed corals may facilitate opportunist infections. In the absence of a strong line of defense, the alteration of the coral immune systems may then result in a higher susceptibility to colonization by the surrounding planktonic bacteria (pathogenic or not), which are typically less active than their epibiotic counterparts [32]. Alternatively, the drop in cell activity could be also simply explained by the physiological decline of the most thermo-sensitive cells due to their inability to cope with the increase in temperature.

In a changing environment, modifications to the abundance and physiological state of planktonic bacterial communities have been shown to induce dramatic changes in their structure [53]. Likewise, the observed changes in abundance and respiration of coral associated bacteria after they received the heat treatment were coupled with substantial changes in their community composition, as revealed by the principal coordinate analysis (Figure 4). Comparable shifts in bacterial community structure have been observed on several occasions when corals were subjected to elevated temperatures [24–26,54]. Vega-Thurber *et al.* [27], by using a pyrosequencing procedure, demonstrated that a controlled elevation of seawater temperature from a local ambient temperature of 25 °C to 30 °C can impair the microbiome of the coral *Porites compressa* to more pathogenic taxa. Such findings were corroborated by a recent study where Garren and her colleagues [55] showed that the mucus of heat-stressed corals could produce a high amount of dimethylsulfoniopropionate, which greatly attracts coral pathogens and causes coral bleaching [11]. In this study, the shift in the bacterial community composition could be also caused by changes in physicochemical properties of the mucus layer. In fact, coral mucus composition and production are quantitatively and qualitatively influenced by environmental factors such as temperature and/or irradiance [56,57]. As a result, particular substrates favoring the growth of the resident bacteria may disappear from the mucus and cause their decline or at least a metabolic depression. As suggested above, the surrounding bacteria may then find their chance to invade this new medium and outcompete with the native groups for the new nutrient sources [14].

4.2. High Viral Production and Abundance in the Mucus of Bleached F. repanda

One of the main findings of this research was the almost 5- and 9-fold increase in viral production rates (Figure 3) and abundance (Figure 1A), respectively, between healthy and bleached corals. These findings were expected as bacterial growth typically increases with temperature, which could naturally favor viral proliferation. Indeed, most of the viruses found in coral mucus were actually phages [31,32,58], and therefore their high occurrence in heat-treated corals is probably the result of a lytic activity on the numerous resident bacterial hosts. The high phage production measured in the bleached corals (compared to healthy ones) clearly confirmed this hypothesis. However, other explanations can be provided. For example, the members of the bacterial community found in bleached corals might be more susceptible to viral infections, in comparison with those in healthy ones where virus-bacteria interactions are presumably more stable and where lytic pressure is applied with parsimony to ultimately ensure coral persistence [35,36,40,59]. Also, we know that phage replication is dependent on host metabolism [60,61]. Therefore the high viral production rates measured in bleached corals might be the result of lytic infections of the most active mucosal cells, as demonstrated by Nguyen-Kim *et al.* [32].

Alternatively, the increase in the abundance and production of mucosal viruses could be due to the thermally-driven induction of the lysogens present in this organic layer. Nguyen-Kim *et al.* [62], reported that the proportion of lysogenized cells measured in the mucus of two different scleractinian corals (*Fungia repanda* and *Acropora formosa*) in a seasonal study was, on average, more than two-fold higher (mean = 8.5% of total bacterial counts) than that measured in the surrounding water (mean = 3.8%). It has been recently hypothesized that lysogeny, by conferring immunity to bacterial symbionts against other lytic surinfection, may therefore represent a vital strategy for corals for their subsistence [40].

The adhesive properties of the coral mucus layer may also represent another explanation for the high occurrence of viruses in bleached corals. Indeed, mucus typically acts as a trap for planktonic particles [63,64], and phages with their specific Ig-proteins located on their capsid are now known for their strong affinity with the mucin-protein of coral mucus [34].

Finally, a fraction of the viral pool might also come from the infected *Symbiodinium* after their expulsion from the coral tissues. Indeed, these endosymbiotic microalgae have been shown to harbor latent viruses, which can be activated by thermal or UV stresses [39,65] and this could be an additional explanation for the large concentrations of viruses in bleached corals.

Current research priorities include elucidating whether such highly abundant bacteria and viruses in bleached corals are the results of the proliferation of (1) the native mucosal associates after they were stimulated by the elevated temperature; (2) the planktonic communities after their adhesion to the mucus gel and/or (3) the opportunists surrounding cells (pathogens or not) taking advantage of the disruption of the bacterial assemblage.

The questions that also now need to be answered are whether coral bleaching occurs prior or after the reduction of cell activity and the shift in bacterial community structure in coral mucus, and whether the expulsion of *Symbiodinium* cells is ecologically linked with such mechanisms. The role of viruses in the structuring epibiotic bacteria also need to be clarified. Finally, further experiments should be conducted at a higher temporal resolution to better address these gaps.

5. Conclusions

Overall, our results seem to suggest that the thermal stress of +4 °C could be responsible for substantial changes in bacterial and viral traits, mostly resulting from the alteration of the physiological state of the native cells. The high levels of viral production measured in bleached corals may have strong implications for coral reef ecosystems including, for example, the enrichment of viruses in the water column through the continuous sloughing off of mucus, and which could in turn have a strong local influence on the bacterial stocks, diversity and functions in reef waters, but could also potentially interact with nearby hosts in other corals.

Acknowledgments

We thank the TOTAL foundation and the Hoa Sen Lotus French-Vietnamese Program for their financial support. Thanks to the French Institute of Research for Development (IRD) for PhD grant to HNK. We thank the Ecological Toxicity Laboratory of the Institute of Oceanography (Vietnam) and the MARBEC research unit laboratory (France) for lab space and material support to carry out the experiment. We are grateful to Lam Nguyen-Ngoc, Hai Doan-Nhu for their scientific and logistical support. We also thank Hoang Phan-Kim for diving assistant during the sampling, Callis Amid and Marie Olstedt for helping during the experiment.

Author Contributions

Hanh Nguyen-Kim, Yvan Bettarel, Thierry Bouvier and Huong Le-Lan designed and conducted the experiment with mesocosms; Hanh Nguyen-Kim, and Corinne Bouvier analyzed the samples and realized the statistical tests; Hanh Nguyen-Kim, Yvan Bettarel, Thierry Bouvier, and Ngoc Van-Bui co-wrote the paper.

Conflicts of Interest

The authors declare no conflict of interest.

References

1. Hoegh-Guldberg, O. Coral reef ecosystems and anthropogenic climate change. *Reg. Environ. Chang.* **2011**, *11*, S215–S227.
2. Brown, B.E. Coral bleaching: Causes and consequences. *Coral Reefs* **1997**, *16*, S129–S138.
3. Glynn, P.W.; Dcroz, L. Experimental evidence for high temperature stress as the cause of El Nino-coincident coral mortality. *Coral Reefs* **1990**, *8*, 181–191.
4. Jokiel, P.L.; Brown, E.K. Global warming, regional trends and inshore environmental conditions influence coral bleaching in hawaii. *Glob. Chang. Biol.* **2004**, *10*, 1627–1641.
5. Lesser, M.P.; Stochaj, W.R.; Tapley, D.W.; Shick, J.M. Bleaching in coral reef anthozoans: Effects of irradiance, ultraviolet radiation and temperature on the activities of protective enzymes against active oxygen. *Coral Reefs* **1990**, *8*, 225–232.
6. Coles, S.L.; Jokiel, P.L. Synergistic effects of temperature, salinity and light on hermatypic coral *Montipora verrucosa*. *Mar. Biol.* **1978**, *49*, 187–195.
7. Anthony, K.R.N.; Kline, D.I.; Diaz-Pulido, G.; Dove, S.; Hoegh-Guldberg, O. Ocean acidification causes bleaching and productivity loss in coral reef builders. *Proc. Natl. Acad. Sci. USA* **2008**, *105*, 17442–17446.
8. Guzman, H.M.; Jackson, J.B.C.; Weil, E. Short term ecological consequences of a major oil spill on panamanian subtidal reef corals. *Coral Reefs* **1991**, *10*, 1–12.

9. Vega Thurber, R.L.; Burkepile, D.E.; Fuchs, C.; Shantz, A.A.; McMinds, R.; Zaneveld, J.R. Chronic nutrient enrichment increases prevalence and severity of coral disease and bleaching. *Glob. Chang. Biol.* **2014**, *20*, 544–554.

10. Kushmaro, A.; Kramarsky-Winter, E. Bacteria as a Source of Coral Nutrition. In *Coral Health and Disease*; Rosenberg, E., Loya, Y., Eds.; Springer-Verlag: New York, NY, USA, 2004; pp. 231–241.

11. Rosenberg, E.; Kushmaro, A.; Kramarsky-Winter, E.; Banin, E.; Yossi, L. The role of microorganisms in coral bleaching. *ISME J.* **2009**, *3*, 139–146.

12. de Lima, L.A.; Migliolo, L.; Barreiro e Castro, C.; Pires, D.D.; Lopez-Abarrategui, C.; Goncalves, E.F.; Vasconcelos, I.M.; de Oliveira, J.T.A.; Otero-Gonzalez, A.D.J.; Franco, O.L.; *et al.* Identification of a novel antimicrobial peptide from brazilian coast coral *Phyllogorgia dilatata*. *Protein Pept. Lett.* **2013**, *20*, 1153–1158.

13. Kvennefors, E.C.E.; Sampayo, E.; Kerr, C.; Vieira, G.; Roff, G.; Barnes, A.C. Regulation of bacterial communities through antimicrobial activity by the coral holobiont. *Microb. Ecol.* **2012**, *63*, 605–618.

14. Rypien, K.L.; Ward, J.R.; Azam, F. Antagonistic interactions among coral-associated bacteria. *Environ. Microbiol.* **2010**, *12*, 28–39.

15. Shnit-Orland, M.; Sivan, A.; Kushmaro, A. Antibacterial activity of *Pseudoalteromonas* in the coral holobiont. *Microb. Ecol.* **2012**, *64*, 851–859.

16. Reshef, L.; Koren, O.; Loya, Y.; Zilber-Rosenberg, I.; Rosenberg, E. The coral probiotic hypothesis. *Environ. Microb.* **2006**, *8*, 2068–2073.

17. Rosenberg, E.; Koren, O.; Reshef, L.; Efrony, R.; Zilber-Rosenberg, I. The role of microorganisms in coral health, disease and evolution. *Nat. Rev. Microb.* **2007**, *5*, 355–362.

18. Ben-Haim, Y.; Rosenberg, E. A novel *Vibrio* sp. pathogen of the coral pocillopora damicornis. *Mar. Biol.* **2002**, *141*, 47–55.

19. Kushmaro, A.; Banin, E.; Loya, Y.; Stackebrandt, E.; Rosenberg, E. *Vibrio shiloi* sp. nov., the causative agent of bleaching of the coral oculina patagonica. *Int. J. Syst. Evolut. Microbiol.* **2001**, *51*, 1383–1388.

20. Kushmaro, A.; Rosenberg, E.; Fine, M.; Loya, Y. Bleaching of the coral oculina patagonica by vibrio ak-1. *Mar. Ecol. Prog. Ser.* **1997**, *147*, 159–165.

21. Toren, A.; Landau, L.; Kushmaro, A.; Loya, Y.; Rosenberg, E. Effect of temperature on adhesion of vibrio strain ak-1 to oculina patagonica and on coral bleaching. *Appl. Environ. Microbiol.* **1998**, *64*, 1379–1384.

22. Ainsworth, T.; Fine, M.; Roff, G.; Hoegh-Guldberg, O. Bacteria are not the primary cause of bleaching in the mediterranean coral *Oculina patagonica*. *ISME J.* **2008**, *2*, 67–73.

23. Leggat, W.; Hoegh-Guldberg, O.; Dove, S.; Yellowlees, D. Analysis of an est library from the dinoflagellate (*Symbiodinium* sp.) symbiont of reef-building corals. *J. Phycol.* **2007**, *43*, 1010–1021.

24. Bourne, D.; Iida, Y.; Uthicke, S.; Smith-Keune, C. Changes in coral-associated microbial communities during a bleaching event. *ISME J.* **2008**, *2*, 350–363.

25. Lins-de-Barros, M.M.; Cardoso, A.M.; Silveira, C.B.; Lima, J.L.; Clementino, M.M.; Martins, O.B.; Albano, R.M.; Vieira, R.P. Microbial community compositional shifts in bleached colonies of the brazilian reef-building coral siderastrea stellata. *Microb. Ecol.* **2013**, *65*, 205–213.

26. Littman, R.; Willis, B.L.; Bourne, D.G. Metagenomic analysis of the coral holobiont during a natural bleaching event on the great barrier reef. *Environ. Microbiol. Rep.* **2011**, *3*, 651–660.

27. Vega Thurber, R.; Willner-Hall, D.; Rodriguez-Mueller, B.; Desnues, C.; Edwards, R.A.; Angly, F.; Dinsdale, E.; Kelly, L.; Rohwer, F. Metagenomic analysis of stressed coral holobionts. *Environ. Microbiol.* **2009**, *11*, 2148–2163.

28. Fuhrman, J.A. Marine viruses and their biogeochemical and ecological effects. *Nature* **1999**, *399*, 541–548.

29. Suttle, C.A. Marine viruses—Major players in the global ecosystem. *Nat. Rev. Microbiol.* **2007**, *5*, 801–812.

30. Davy, J.E.; Patten, N.L. Morphological diversity of virus-like particles within the surface microlayer of scleractinian corals. *Aquat. Microb. Ecol.* **2007**, *47*, 37–44.

31. Leruste, A.; Bouvier, T.; Bettarel, Y. Enumerating viruses in coral mucus. *Appl. Environ. Microbiol.* **2012**, *78*, 6377–6379.

32. Nguyen-Kim, H.; Bouvier, T.; Bouvier, C.; Doan, N.H.; Nguyen, N.L.; Rochelle-Newall, E.; Desnues, C.; Reynaud, S.; Ferrier-Pages, C.; Bettarel, Y. High occurence of viruses in the mucus layer of scleractinian corals. *Environ. Microbiol. Rep.* **2014**, *6*, 675–682.

33. Weinbauer, M.G.; Ogier, J.; Maier, C. Microbial abundance in the coelenteron and mucus of the cold-water coral lophelia pertusa and in bottom water of the reef environment. *Aquat. Biol.* **2012**, *16*, 209–216.

34. Barr, J.J.; Auro, R.; Furlan, M.; Whiteson, K.L.; Erb, M.L.; Pogliano, J.; Stotland, A.; Wolkowicz, R.; Cutting, A.S.; Doran, K.S.; *et al.* Bacteriophage adhering to mucus provide a non-host-derived immunity. *Proc. Natl. Acad. Sci. USA* **2013**, *110*, 10771–10776.

35. Van Oppen, M.J.H.; Leong, J.A.; Gates, R.D. Coral-virus interactions: A double-edged sword? *Symbiosis* **2009**, *47*, 1–8.

36. Vega Thurber, R.L.; Correa, A.M.S. Viruses of reef-building scleractinian corals. *J. Exp. Mar. Biol. Ecol.* **2011**, *408*, 102–113.

37. Correa, A.M.; Welsh, R.M.; Thurber, R.L.V. Unique nucleocytoplasmic dsdna and +ssrna viruses are associated with the dinoglagellate endosymbionts of corals. *ISME J.* **2013**, *7*, 13–27.

38. Danovaro, R.; Bongiorni, L.; Corinaldesi, C.; Giovannelli, D.; Damiani, E.; Astolfi, P.; Greci, L.; Pusceddu, A. Sunscreens cause coral bleaching by promoting viral infections. *Environ. Health Perspect.* **2008**, *116*, 441–447.

39. Wilson, W.H.; Dale, A.L.; Davy, J.E.; Davy, S.K. An enemy within? Observations of virus-like particles in reef corals. *Coral Reefs* **2005**, *24*, 145–148.

40. Bettarel, Y.; Bouvier, T.; Nguyen, H.K.; Thu, P.T. The versatile nature of coral associated viruses. *Environ. Microbiol.* **2014**, *16*, doi:10.1111/1462-2920.12579.

41. Naumann, M.S.; Niggl, W.; Laforsch, C.; Glaser, C.; Wild, C. Coral surface area quantification-evaluation of established techniques by comparison with computer tomography. *Coral Reefs* **2009**, *28*, 109–117.

42. Williamson, K.E.; Wommack, K.E.; Radosevich, M. Sampling natural viral communities from soil for culture-independent analyses. *Appl. Environ. Microbiol.* **2003**, *69*, 6628–6633.

43. Patel, A.; Noble, R.T.; Steele, J.A.; Schwalbach, M.S.; Hewson, I.; Fuhrman, J.A. Virus and prokaryote enumeration from planktonic aquatic environments by epifluorescence microscopy with sybr green i. *Nat. Protoc.* **2007**, *2*, 269–276.

44. Sherr, B.F.; del Giorgio, P.; Sherr, E.B. Estimating abundance and single-cell characteristics of respiring bacteria via the redox dye ctc. *Aquat. Microb. Ecol.* **1999**, *18*, 117–131.

45. Bettarel, Y.; Desnues, A.; Rochelle-Newall, E. Lytic failure in cross-inoculation assays between phages and prokaryotes from three aquatic sites of contrasting salinity. *FEMS Microbiol. Lett.* **2010**, *311*, 113–118.

46. Fischer, U.R.; Velimirov, B. High control of bacterial production by viruses in a eutrophic oxbow lake. *Aquat. Microb. Ecol.* **2002**, *27*, 1–12.

47. Morrow, K.M.; Moss, A.G.; Chadwick, N.E.; Liles, M.R. Bacterial associates of two caribbean coral species reveal species-specific distribution and geographic variability. *Appl. Environ. Microbiol.* **2012**, *78*, 6438–6449.

48. Dar, S.A.; Kuenen, J.G.; Muyzer, G. Nested pcr-denaturing gradient gel electrophoresis approach to determine the diversity of sulfate-reducing bacteria in complex microbial communities. *Appl. Environ. Microbiol.* **2005**, *71*, 2325–2330.

49. Ovreas, L.; Forney, L.; Daae, F.L.; Torsvik, V. Distribution of bacterioplankton in meromictic lake saelenvannet, as determined by denaturing gradient gel electrophoresis of pcr-amplified gene fragments coding for 16s rrna. *Appl. Environ. Microbiol.* **1997**, *63*, 3367–3373.

50. Harvell, D.; Jordan-Dahlgren, E.; Merkel, S.; Rosenberg, E.; Raymundo, L.; Smith, G.; Weil, E.; Willis, B. Coral disease, environmental drivers, and the balance between coral and microbial associates. *Oceanography* **2007**, *20*, 172–195.

51. Pires, A.P.F.; Guariento, R.D.; Laque, T.; Esteves, F.A.; Farjalla, V.F. The negative effects of temperature increase on bacterial respiration are independent of changes in community composition. *Environ. Microbiol. Rep.* **2014**, *6*, 131–135.

52. Ritchie, K.B. Regulation of microbial populations by coral surface mucus and mucus-associated bacteria. *Mar. Ecol. Prog. Ser.* **2006**, *322*, 1–14.

53. Del Giorgio, P.A.; Bouvier, T.C. Linking the physiologic and phylogenetic successions in free-living bacterial communities along an estuarine salinity gradient. *Limnol. Oceanogr.* **2002**, *47*, 471–486.

54. Gilbert, J.A.; Hill, R.; Doblin, M.A.; Ralph, P.J. Microbial consortia increase thermal tolerance of corals. *Mar. Biol.* **2012**, *159*, 1763–1771.

55. Garren, M.; Son, K.; Raina, J.B.; Rusconi, R.; Menolascina, F.; Shapiro, O.H.; Tout, J.; Bourne, D.G.; Seymour, J.R.; Stocker, R. A bacterial pathogen uses dimethylsulfoniopropionate as a cue to target heat-stressed corals. *ISME J.* **2014**, *8*, 999–1007.

56. Lasker, H.R.; Peters, E.C.; Coffroth, M.A. Bleaching of reef coelenterates in san-blas islands, panama. *Coral Reefs* **1984**, *3*, 183–190.

57. Piggot, A.M.; Fouke, B.W.; Sivaguru, M.; Sanford, R.A.; Gaskins, H.R. Change in zooxanthellae and mucocyte tissue density as an adaptive response to environmental stress by the coral, montastraea annularis. *Mar. Biol.* **2009**, *156*, 2379–2389.

58. Marhaver, K.L.; Edwards, R.A.; Rohwer, F. Viral communities associated with healthy and bleaching corals. *Environ. Microbiol.* **2008**, *10*, 2277–2286.

59. Rohwer, F.; Seguritan, V.; Azam, F.; Knowlton, N. Diversity and distribution of coral-associated bacteria. *Mar. Ecol. Prog. Ser.* **2002**, *243*, 1–10.

60. Maurice, C.F.; Bouvier, C.; de Wit, R.; Bouvier, T. Linking the lytic and lysogenic bacteriophage cycles to environmental conditions, host physiology and their variability in coastal lagoons. *Environ. Microbiol.* **2013**, *15*, 2463–2475.

61. Weinbauer, M.G. Ecology of prokaryotic viruses. *FEMS Microbiol. Rev.* **2004**, *28*, 127–181.

62. Nguyen-Kim, H.; Bettarel, Y.; Bouvier, T.; Bouvier, C.; Doan-Nhu, H.; Nguyen-Ngoc, L.; Nguyen-Thanh, T.; Tran-Quang, H.; Brune, J. Coral mucus is a hot spot of viral infections. *Appl. Environ. Microbiol.* **2015**, *81*, doi:10.1128/AEM.00542–15.

63. Mayer, F.W.; Wild, C. Coral mucus release and following particle trapping contribute to rapid nutrient recycling in a northern red sea fringing reef. *Mar. Freshw. Res.* **2010**, *61*, 1006–1014.

64. Wild, C.; Huettel, M.; Klueter, A.; Kremb, S.G.; Rasheed, M.Y.M.; Jorgensen, B.B. Coral mucus functions as an energy carrier and particle trap in the reef ecosystem. *Nature* **2004**, *428*, 66–70.

65. Lohr, J.; Munn, C.B.; Wilson, W.H. Characterization of a latent virus-like infection of symbiotic zooxanthellae. *Appl. Environ. Microbiol.* **2007**, *73*, 2976–2981.

Domestication of Marine Fish Species: Update and Perspectives

Fabrice Teletchea

Research Unit Animal and Functionalities of Animal Products (URAFPA), University of Lorraine—INRA, 2 Avenue de la Forêt de Haye, BP 172, 54505 Vandoeuvre-lès-Nancy, France; E-Mail: fabrice.teletchea@univ-lorraine.fr

Academic Editor: Dean Jerry

Abstract: Domestication is a long and endless process during which animals become, generations after generations, more adapted to both captive conditions and humans. Compared to land animals, domestication of fish species has started recently. This implies that most farmed marine fish species have only changed slightly from their wild counterparts, and production is based partly or completely on wild inputs. In the past decades, global marine fish production has increased tremendously, particularly since the 1990s, to reach more than 2.2 million tons in 2013. Among the 100 marine fish species listed in the FAO's database in 2013, 35 are no longer produced, and only six have a production higher than 100,000 tons. The top ten farmed marine species accounted for nearly 90% of global production. The future growth and sustainability of mariculture will depend partly on our ability to domesticate (*i.e.*, control the life cycle in captivity) of both currently farmed and new species.

Keywords: domestication level; wild; domesticated; marine fish species; capture-based aquaculture; bottlenecks

1. Introduction

Domestication is a long and endless process during which animals become, generations after generations, more adapted to both captive conditions and humans [1–3]. Therefore, domestication should not be confused with taming, which is conditioned behavioral modification of wild-born animals [3,4].

During domestication, captive animals are progressively modified from their wild ancestors and at a certain moment are considered domesticated. Nevertheless, it is difficult to determine when captive animals have become domesticated, and such a decision is subjective and arbitrary [1]. According to most authors, a domesticated animal is bred in captivity and thereby modified from its wild ancestors in ways making it more useful to humans who control its reproduction and its food supply [4–6].

Domestication on land started about 12,000 years ago [4]. Over millennia, animal populations were modified by humans and changes in behavior, physiology and morphology occurred [6,7]. At the beginning of the twentieth century, modern breeding programs were initiated leading to dramatic changes in productivity, e.g., increase laying rate for laying hens or improved feed conservation ratio, meat yield and growth rate in broiler chickens [2,8,9]. As a result, thousands of genetically distinct livestock breeds have been created, and there is an apparent dichotomy between domesticated species and their wild congeners, which have sometimes gone extinct [6,10,11].

Compared to the domestication of land animals, the domestication of aquatic animals is a recent phenomenon [5]. Except for few species, such as the common carp (*Cyprinus carpio*) and the goldfish (*Carassius auratus*), the bulk of farming has started in the past century [5,12,13]. Most fish species farmed today are not much different from their wild conspecifics [9,10,13–15]. It is estimated that 90% of the global aquaculture industry is based on wild, undomesticated or non-selectively bred stocks [2,16]. Conversely, less than 10% of aquaculture production comes from selectively bred farm stocks [2]. The Atlantic salmon (*Salmo salar*) is an outlier, as almost 100% of the total production is based on selectively bred stocks [8,16]. Consequently, depending on the species considered, the control over aquaculture production can vary from managing only a portion of the life cycle to managing the complete life cycle in captivity [11,14,17–20]. In order to better describe the various fish production strategies, Teletchea and Fontaine [20] proposed a new classification based on the level of human control over the life cycle of farmed species and independence from wild inputs. This classification comprises five levels of domestication with one being the least domesticated to five being the most domesticated. Among the 250 species recorded in the Food and Agriculture Organization (FAO) database in 2009, 39 belong to level 1 (first trials of acclimatization to the culture environment), 75 to the level 2 (part of the life cycle closed in captivity, also known as capture-based aquaculture), 61 to the level 3 (entire life cycle closed in captivity with wild inputs), 45 to the level 4 (entire life cycle closed in captivity without wild inputs) and 30 to the level 5 (selective breeding programs are used focusing on specific goals) [20].

The main goal of the present study is to analyze the evolution of the aquaculture production of marine fish species since 1950, while updating the number of species per domestication level since 2009.

2. Materials and Methods

The central source of data about the world's fisheries and aquaculture operations is the United Nation's Food and Agriculture Organization (FAO). With fisheries catch and aquaculture production data going back to 1950, the FAO's database is an invaluable source of temporal information about the quantity, value, and geographic location of global seafood production [19]. Nevertheless, concerns have been raised in the past two decades about the quality of the data, mainly due a lack of clarity and transparency in terms of what is and is not being reported as "aquaculture", if reported at all [19,21].

However, a full discussion of these concerns lay outside the scope of this paper (instead, see [19,22]). In the present study, I choose to focus on marine fish species, thus excluding diadromous species, such as Atlantic salmon [23]. The domestication level for all marine fish species listed in the FAO in 2013 (n = 100 species) was determined based on [20] for species already in the database in 2009 (n = 87), and on the literature for "new" species (n = 17). Group of species were excluded.

3. Results

3.1. Evolution of Global Marine Fish Aquaculture Production

Global marine fish production increased slightly from 1950 up to the beginning of the 1970s. Then, the production increased steadily up to the 1990s; thereafter, it rose tremendously (Figure 1). However, more than half of the production is not identified at the species level in the FAO database (Figure 1). One group called "Marine fish nei (not elsewhere included)" totaled 621,275 tons, which is more than one-quarter of the global production in 2013.

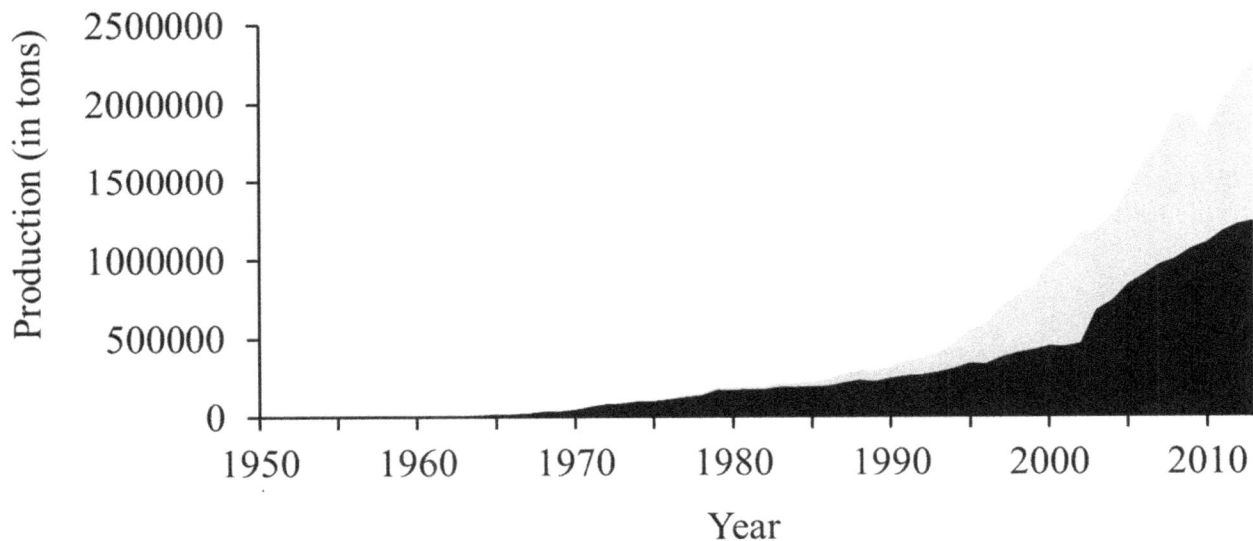

Figure 1. Evolution of global aquaculture production of marine fish species. Group of species, not identified at the species level (**upper** part, in grey), identified at the species level (**lower** part, in black).

In 2013, global marine fish production reached more than 2.2 million tons (Figure 1), which represents about half the production of diadromous fish and about 6% of freshwater fish production (Figure 2). This implies that despite its strong increase, marine fish aquaculture remains small compared to non-marine fish production.

Marine fish are mainly produced in Asia (83.1%), followed by Europe (9.2%) and Africa (7.1%) (Figure 3a). Asia is also the main producer of diadromous fish (39.0%); yet the production is more evenly distributed in the world, with Europe (37.8%) and the Americas (21.9%) (Figure 3b). For freshwater fish, almost the entire production (93.8%) is realized in Asia (Figure 3c).

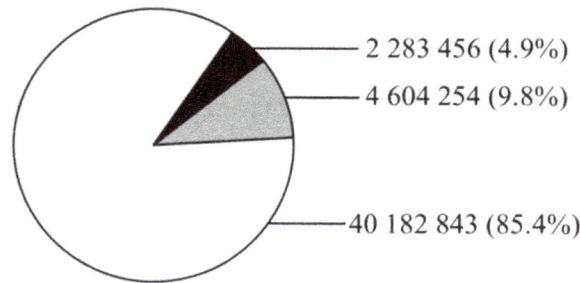

Figure 2. Comparison of the aquaculture production of marine fish species (**black**) with those of diadromous (**grey**) and freshwater (**white**) species in 2013. The first number is the total production followed by the percentage in parentheses.

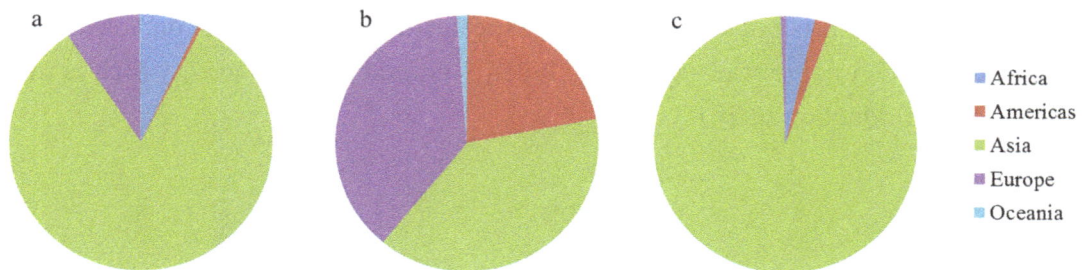

Figure 3. Main aquaculture regions of fish species in 2013: (**a**) marine fish species; (**b**) diadromous fish species; and (**c**) freshwater fish species.

3.2. Evolution of the Number of Farmed and Domesticated Fish Species

The number of farmed species has strongly increased in the past decades to reach 65 in 2013 (Figure 4). Since the mid-1990s, the number has doubled, despite slight decreases in 1997, 1998, 2006, 2007, 2009, 2011 and 2012.

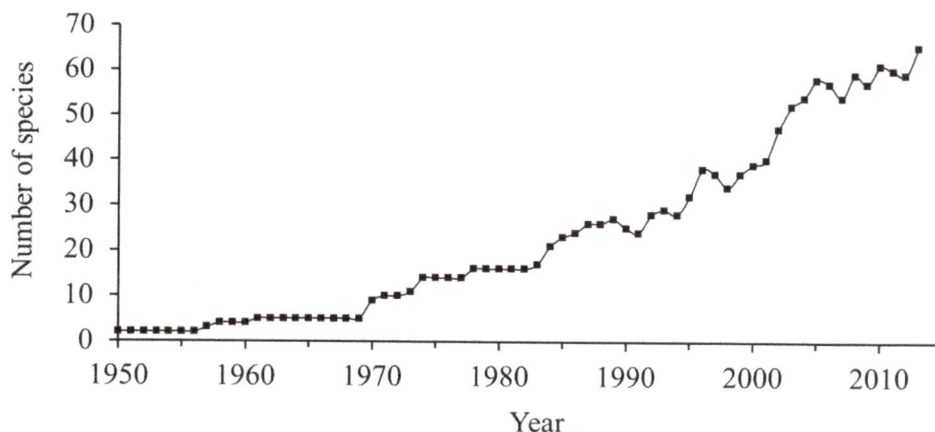

Figure 4. Evolution of the number of marine fish species farmed per year.

The comparison between 2009 and 2013 showed that 87 species were listed in 2009 and 100 in 2013. Four species listed in 2009 are no longer present in 2013, the striped weakfish (*Cynoscion striatus*), the eastern pomfred (*Schuettea scalaripinnis*), the streamerish (*Agrostichthys parkeri*) and the rice-paddy eel (*Pisodonophis boro*). Seventeen "new" species were present (in bold in Appendix). Among the 100 species listed in the FAO database, 35 were no longer produced in 2013, 24 had a production less than 100 tons, 13 had a production between 100 and 1000 tons, 13 species between 1001 and 10,000, nine species between 10,001 and 100,000, and only six have a production higher than 100,000 tons. The domestication level of marine species ranged from one to five (Figure 5). The domestication level for each species is provided in the Appendix. There are only slight differences between 2009 and 2013: only the numbers of species at levels 1, 2 and 3 have increased.

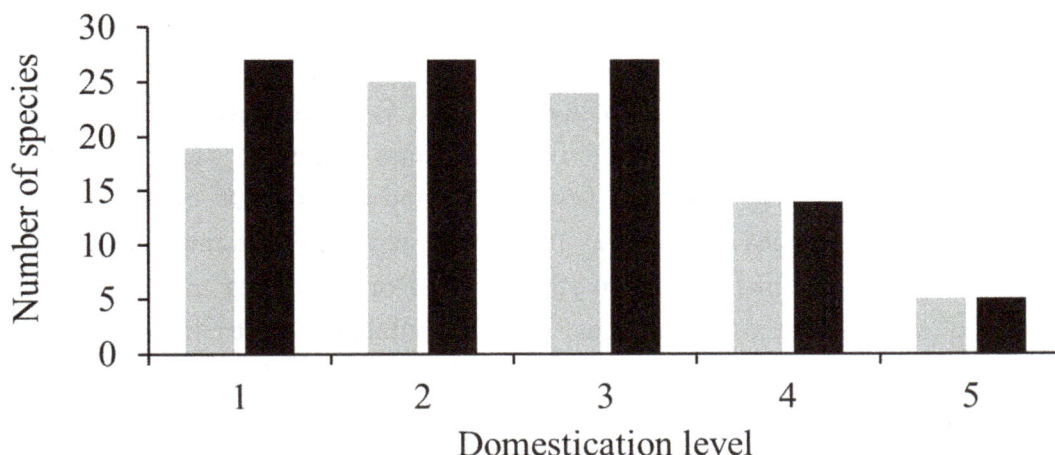

Figure 5. Domestication level of marine fish species in 2009 (**grey**) and 2013 (**black**).

The top ten marine farmed species in 2013 totaled 86.5% of the global production, which was 1,241,149 tons (when excluding groups not identified at the species level). These species are almost exclusively produced in Asia, except for the two leading species, gilthead seabream (*Sparus aurata*) and European sea bass (*Dicentrarchus labrax*), for which about half of the production is in Europe (Figure 6).

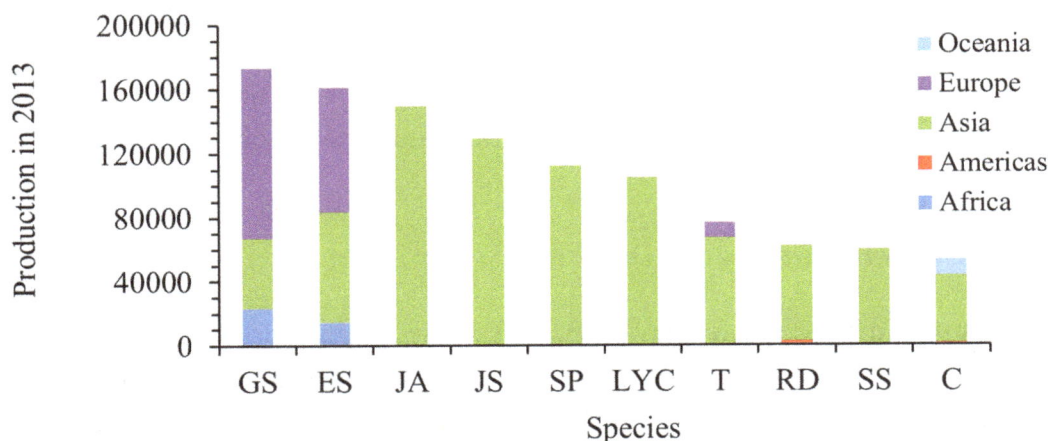

Figure 6. Aquaculture production of the top ten farmed marine species. Gilthead seabream (GS); European seabass (ES); Japanese amberjack (JA); Japanese seabass (JS), subnose pompano (SP); large yellow croaker (LYC); turbot (T); red drum (RD); silver seabream (SS) and cobia (C) (see Appendix for scientific names).

4. Discussion

4.1. Evolution of Marine Fish Aquaculture Production

Aquaculture, the farming of aquatic organisms, including fish, shellfish and mollusks (excluding plants), is the fastest growing food production system globally, with an increase in production of about 9.3% per year since 1985 [14]. In 2012, aquaculture production has reached 66.7 million tons [2]. Despite a strong increase in the past two decades, the production of marine fish species is still small compared to global aquaculture, and particularly mariculture, which is dominated by mollusks and crustaceans [7]. In comparison, fish is totally dominant in freshwater aquaculture (>99%) [7].

When compared with the two other groups of fish listed in the FAO database, it appears that marine fish species represent a very small amount of global fish production (Figure 2). This is partly due to the fact that farming of most marine fish species is very recent, and thus the life cycle is controlled in captivity for only a handful of species (see below). In addition, several constraints have restricted the expansion of aquaculture of marine fish species, particularly in North America and Europe (Figure 3), among which limited areas sheltered from ocean swells, regulatory restrictions on sites, other competitive factors, such as tourism and port development, the relatively high costs (e.g., investments in infrastructure, maintenance, cost and transport of feed), and the high developmental costs and risks associated with off-shore aquaculture technologies [7,9,24–28].

Today, the marine environment contributes less than two percent of the human food supply [27,29]. It is largely because the development of controlled food production in the ocean lags several millennia behind that on land [12,27,29]. The space used for mariculture production is estimated at about 0.01 million km², or about 0.04% of the global shelf area [29]. Mariculture production is concentrated in a selected number of countries (e.g., China, Spain, Greece, Norway, Chile, and Scotland), particularly in sheltered bays and lagoons [29]. In future decades, it is anticipated that mariculture would increase significantly [9,27,29]. The FAO forecasts that mariculture will reach 54 million metric tons to 70 million metric tons by year 2020 [29]. However, further development of mariculture will run into major bottlenecks concerning the availability, suitability, and cost of feed; space availability; and adverse environmental impacts, which must be overcome if mariculture is to become a major component of global food production [28,29].

4.2. Evolution of the Number of Farmed and Domesticated Marine Species

The number of farmed marine species has strongly increased in the past decades, as observed for other fish groups [5,13,28]. However, only 10 percent of the marine species listed in the FAO in 2013 accounted for nearly 90% of global production (Figure 6). Eight of the top ten farmed species have reached the domestication level 4 ($n = 5$) or 5 ($n = 3$). This implies that the entire life cycle of these species has been closed in captivity without the need to use wild inputs, and for three of them, breeding programs have been developed: five for the gilthead seabream (*Sparus aurata*), three for the European sea bass (*Dicentrarchus labrax*) and two for the turbot (*Psetta maxima*) [25]. However, reaching level 4 and 5 does not necessarily imply that the entire aquaculture production of the 19 marine fish species classified at these two domestication levels (Figure 5) is based on domesticated or genetically improved stocks only [11,20]. For instance, even though sea bass has reached the level 5, most farms still rely today on wild broodstock for reproduction or, to lesser extent, from first-generation (F1) individuals and rarely from selected F2 or F3 fish [11].

Achieving full life cycle in captivity over several generations, which could then be called domesticated fish, thus appears an important progress in mastering the sustainability and the increase of production [17]. However, domestication (and notably selection) of a new species is a long and difficult process [30] that requires, among others, broodstock management (production of high quality broodstock, gonad and gamete development, ovulation/oviposition in females and ejaculation of milt in males), incubation of eggs, and rearing of larvae and juveniles [20,25,31]. Therefore, domestication needs access to specific skills, knowledge, and technology, and both long-term public and private funding [2,8,9,13]. This explains why it is has been primarily carried out in developed countries, notably in Europe [25,32].

For all the top ten farmed species, except the large yellow croaker (*Larimichthys croceus*), the aquaculture production now strongly exceeds capture fisheries (Figure 7). This may be caused by competition between these two sources, because a wild caught fish can commonly be sold at a relatively low price, but cannot be cultured at this low price for a profit. As capture fisheries decline because of overharvesting, the prices of target species often increase dramatically. Under these conditions, aquaculture can thrive, thereby further reducing the value of that capture fishery [14].

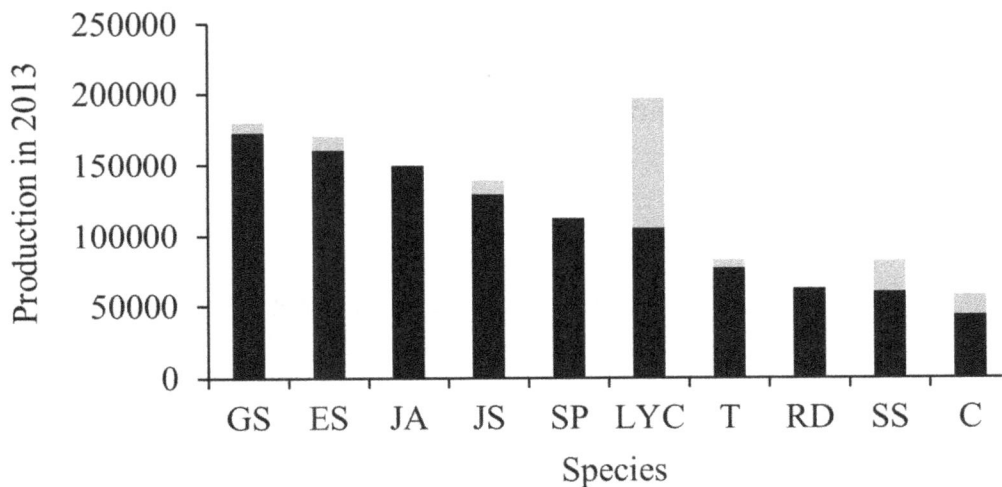

Figure 7. Global production of the top ten farmed marine species, aquaculture (**black**) and fisheries (**grey**). Gilthead seabream (GS); European seabass (ES); Japanese amberjack (JA); Japanese seabass (JS); subnose pompano (SP); large yellow croaker (LYC); turbot (T); red drum (RD); silver seabream (SS) and cobia (C). The fisheries production of JA, SP and RD are 0, 79, and 204 tons respectively (see Appendix for scientific names).

Two among the top ten farmed species, the Japanese amberjack (*Seriola quinqueradiata*) and the snubnose pompano (*Trachinotus blochii*), have only reached level 2, which implies that the entire aquaculture production is based on wild input. This method of production, known as capture-based aquaculture, consists of growing and fattening individuals removed from wild populations [17]. Tuna fattening and much of the marine cage culture in Asia, relies directly on wild-caught small pelagic fish with relatively low market price [19,28,31]. The aquaculture process transforms fish protein from low to high value for human consumption [28]. However, such systems can only function as long as survival and sustainable utilization of the affected wild stocks are warranted [17,19,31]. Therefore, capture-based aquaculture can only be seen as a transitory form of fish production, viable only as long as the wild resource is still available for seed withdrawal [17,31]. Efforts have to be made to domesticate species (*i.e.*, the closing of the life cycle in captivity) to allow a reliable production, independent of wild inputs, and then improve desirable traits through selective breeding [2,25,31].

One-third of the marine species listed in FAO in 2013 are no longer produced, and 50 percent more have a production less than 1000 tons. Nearly all species with a production inferior to 1000 tons have a domestication level between 1 and 3 (Appendix). This highlights that for most species, farming corresponds to one or a few years of aquaculture trial before being abandoned [15,20]. The main reason why numerous attempts with new species fail is that premature attempts to develop industrial enterprises were based on overly optimistic speculation about market demand, rather than on biological and technical knowledge and adequate information about economic feasibility [20,27].

5. Conclusions

Compared to the domestication of land animals, the domestication of aquatic animals, and particularly marine fish species, is a recent phenomenon. Mariculture of fish has only started a few decades ago, and today only a handful of species can be considered domesticated. In contrast, for numerous species, farming was only performed for a few years before being stopped. The future growth and sustainability of mariculture will depend partly on our ability to domesticate (*i.e.*, control the life cycle in captivity) of both currently farmed and new species.

Acknowledgments

The author would like to thank two anonymous reviewers for their comments on a previous draft of the manuscript.

Conflicts of Interest

The author declares no conflict of interest.

Appendix

Table A1. Domestication level for marine fish species listed in the 2013 FAO report (*n* = 100 species). Group of species were excluded. Species on the list for the first time are in bold (*n* = 17). Aquaculture production is in tons. When no reliable scientific information was found, species were arbitrarily considered to belong to level 1 when their production was less than five continuous years (*n* = 15) to the level 2 when their production was between five and up to 10 continuous years (*n* = 3), and to the level 3 when their production was greater than 10 years (*n* = 3).

Scientific Name	Common Name	Production in 2013	Domestication Level	Main Reference
Anarhichas lupus	Atlantic wolffish	0	1	Gunnarsson *et al.*, 2009, [33]

Table A1. *Cont.*

Atherina boyeri	**Big-scale sand smelt**	**0**	**1**	**Dulcic *et al.*, 2008, [34]**
Bolbometopon muricatum	**Green humphead parrotfish**	**1**	**1**	
Carangoides malabaricus	**Malabar trevally**	**387**	**1**	
Caranx hippos	Crevalle jack	0	1	
Caranx sexfasciatus	**Bigeye trevally**	**1**	**1**	
Centropristis striata	Black seabass	0	1	Rezek *et al.*, 2010, [35]
Chaetodipterus faber	**Atlantic spadefish**	**0**	**1**	**Gaspar 2005, [36]**
Dentex tumifrons	Yellowback seabream	0	1	
Dicentrarchus punctatus	**Spotted seabass**	**2**	**1**	**Ly *et al.*, 2012, [37]**
Labrus bergylta	Ballan wrasse	25	1	Muncaster *et al.*, 2010, [38]
Lethrinus miniatus	Trumpet emperor	45	1	
Lutjanus bohar	Two-spotted red snapper	0	1	
Megalops atlanticus	**Tarpon**	**0**	**1**	
Micropogonias furnieri	Whitemouth croaker	5	1	Velloso and Pereira Jr. 2010, [39]
Mugil liza	Lebranche mullet	7	1	
Muraenesox cinereus	Daggertooth pike conger	0	1	
Mycteroperca bonaci	**Black grouper**	**2**	**1**	
Pagrus major	**Japanese seabream**	0	**1**	
Platichthys flesus	European flounder	0	1	Engel-Sørensen *et al.*, 2004, [40]
Pleurogrammus azonus	Okhotsk atka mackerel	0	1	
Pomatomus saltatrix	Bluefish	0	1	
Scophthalmus rhombus	Brill	0	1	Cruzado *et al.*, 2004, [41]
Siganus canaliculatus	White-spotted spinefoot	1	1	Xu *et al.*, 2011, [42]
Siganus javus	Streaked spinefoot	0	1	

Table A1. *Cont.*

Siganus rivulatus	Marbled spinefoot	0	1	El Dakar *et al.*, 2011, [43]
Valamugil seheli	Bluespot mullet	0	1	Belal 2004, [44]
Acanthopagrus berda	Goldsilk seabream	0	2	Liao *et al.*, 2001, [45]
Argyrosomus japonicus	**Japanese meagre**	**130**	**2**	**Mirimin and Roodt-Wilding, 2015, [46]**
Boleophthalmus pectinirostris	**Great blue spotted mudskipper**	**0**	**2**	**Zhang *et al.*, 1989, [47]**
Centropomus undecimalis	Common snook	0	2	Carter *et al.*, 2010a, [48]
Eleutheronema tetradactylum	Fourfinger threadfin	4173	2	Liao *et al.*, 2001, [45]
Epinephelus areolatus	Areolate grouper	47	2	Ottolenghi *et al.*, 2004, [49]
Epinephelus coioides	Orange-spotted grouper	492	2	Ottolenghi *et al.* 2004, [49]
Epinephelus fuscoguttatus	Brown-marbled grouper	86	2	Ottolenghi *et al.*, 2004, [49]
Epinephelus lanceolatus	**Giant grouper**	**36**	**2**	**Peng *et al.*, 2015, [50]**
Epinephelus malabaricus	Malabar grouper	68	2	Ottolenghi *et al.*, 2004, [49]
Gnathanodon speciosus	Golden trevally	58	2	Liao *et al.*, 2001, [45]
Liza ramada	Thinlip grey mullet	0	2	Marino *et al.*, 1999, [51]
Lutjanus goldiei	Papuan black snapper	0	2	
Lutjanus guttatus	Spotted rose snapper	2	2	García-Ortega 2009, [52]
Lutjanus johnii	John's snapper	278	2	Liao *et al.*, 2001, [45]
Miichthys miiuy	**Mi-iuy (brown) croaker**	**0**	**2**	**An *et al.*, 2012, [53]**
Mugil soiuy	So-iuy mullet	905	2	
Plectropomus maculatus	Spotted coralgrouper	7	2	
Polydactylus sexfilis	Sixfinger threadfin	0	2	Deng *et al.*, 2011, [54]
Psammoperca waigiensis	Waigieu seaperch	5,704	2	Pham *et al.*, 2010, [55]
Seriola quinqueradiata	Japanese amberjack	149,766	2	Bilio 2007b, [56]
Seriola rivoliana	**Longfin yellowtail**	**400**	**2**	**Roo *et al.*, 2014, [57]**

Table A1. *Cont.*

Thunnus albacares	Yellowfin tuna	171	2	Wexler *et al.*, 2011, [58]
Thunnus maccoyii	Southern bluefin tuna	3,482	2	Carter *et al.*, 2010b, [59]
Thunnus thynnus	Atlantic bluefin tuna	3,445	2	Carter *et al.*, 2010b, [59]
Trachinotus blochii	Snubnose pompano	112,499	2	Liao *et al.*, 2001, [45]
Trachinotus carolinus	Florida pompano	350	2	Pfeiffer and Riche 2001, [60]
Cromileptes altivelis	Humpback grouper	2	3	Hong and Zhang 2003, [61]
Dentex dentex	Common dentex	54	3	Suquet *et al.*, 2009, [62]
Diplodus sargus	White seabream	24	3	Suquet *et al.*, 2009, [62]
Diplodus vulgaris	Common two-banded seabream	0	3	Suquet *et al.*, 2009, [62]
Epinephelus akaara	Hong Kong grouper	0	3	Hong and Zhang 2003, [61]
Epinephelus tauvina	Greasy grouper	5,354	3	Hong and Zhang 2003, [61]
Evynnis japonica	Crimson seabream	0	3	
Liza vaigiensis	Squaretail mullet	0	3	
Lutjanus argentimaculatus	Mangrove red snapper	5,357	3	Hong and Zhang 2003, [61]
Lutjanus russelli	Russell's snapper	13	3	Hong and Zhang 2003, [61]
Melanogrammus aeglefinus	Haddock	0	3	Roselund and Skretting 2006, [63]
Pagellus bogaraveo	Blackspot seabream	2	3	Suquet *et al.*, 2009, [62]
Pagellus erythrinus	Common pandora	0	3	Suquet *et al.*, 2009, [62]
Platax orbicularis	**Orbicular batfish**	**8**	**3**	**Coeurdacier and Gasset 2013, [64]**
Pollachius pollachius	Pollack	0	3	Roselund and Skretting 2006, [63]
Pseudocaranx dentex	White trevally	3,155	3	
Rhabdosargus sarba	Goldlined seabream	3	3	Hong and Zhang 2003, [61]
Sciaena umbra	Brown meagre	0	3	Bilio 2007a, [65]
Seriola dumerili	Greater amberjack	0	3	Hong and Zhang 2003, [61]

Table A1. *Cont.*

Solea senegalensis	Senegalense sole	640	3	Imsland 2010, [66]
Solea solea	Common sole	45	3	Imsland 2010, [66]
Sparidentex hasta	Sobaity seabream	551	3	Teng *et al.*, 1999, [67]
Takifugu obscurus	**Obscure pufferfish**	**4,860**	**3**	**Kim *et al.*, 2010, [68]**
Takifugu rubripes	**Tiger pufferfish**	**19,359**	**3**	**Wu *et al.*, 2015, [69]**
Thunnus orientalis	Pacific bluefin tuna	16,624	3	Carter *et al.*, 2010b, [59]
Trachurus japonicas	Japanese jack mackerel	958	3	Masuda 2006, [70]
Umbrina cirrosa	Shi drum	1,070	3	Suquet *et al.* 2009, [62]
Acanthopagrus latus	Yellowfin seabream	0	4	Hong and Zhang 2003, [61]
Acanthopagrus schlegeli	Blackhead seabream	1,161	4	Hong and Zhang 2003, [61]
Anarhichas minor	Spotted wolfish	0	4	Le François *et al.* 2010, [71]
Argyrosomus regius	Meagre	6,659	4	Lazo *et al.*, 2010, [72]
Diplodus puntazzo	Sharpsnout seabream	250	4	Suquet *et al.*, 2009, [62]
Hippoglossus hippoglossus	Atlantic halibut	1,485	4	Imsland 2010, [66]
Larimichthys croceus	Large yellow croaker	105,230	4	Bilio 2007b, [56]
Lateolabrax japonicus	Japanese seabass	129,334	4	Hong and Zhang 2003, [61]
Mugil cephalus	Flathead grey mullet	12,245	4	Hong and Zhang 2003, [61]
Pagrus auratus	Silver seabream	59,616	4	Suquet *et al.*, 2009, [62]
Pagrus pagrus	Red porgy	350	4	Suquet *et al.*, 2009, [62]
Rachycentron canadum	Cobia	43,395	4	Bilio 2007a,b, [56,65]
Sciaenops ocellatus	Red drum	62,197	4	Lazo *et al.*, 2010, [72]
Sebastes schlegeli	Korean rockfish	23,757	4	Bilio 2007b, [56]
Dicentrarchus labrax	European seabass	161,059	5	Jobling *et al.*, 2010, [73]
Gadus morhua	Atlantic cod	4,252	5	Björnsson *et al.*, 2010, [74]

Table A1. *Cont.*

Paralichthys olivaceus	Bastard halibut	39,445	5	Bilio 2007b, [56]
Psetta maxima	Turbot	76,998	5	Hulata 2001, [75]
Sparus aurata	Gilthead seabream	173,062	5	Jobling and Perruzi 2010, [76]

References

1. Price, E.O. Behavioral development in animals undergoing domestication. *Appl. Anim. Behav. Sci.* **1999**, *65*, 245–271.

2. Olesen, I.; Bentsen, H.B.; Phillips M.; Ponzoni R.W. Can the global adoption of genetically improved farmed fish increase beyond 10%, and how? *J. Mar. Sci. Eng.* **2015**, *3*, 240–266.

3. Zeder, M.A. Core questions in domestication research. *Proc. Natl. Acad. Sci. USA* **2015**, *112*, 3191–3198.

4. Diamond, J. Evolution, consequences and future of plant and animal domestication. *Nature* **2002**, *418*, 700–707.

5. Balon, E.K. About the oldest domesticates among fishes. *J. Fish Biol.* **2004**, *65*, 1–27.

6. Mirkena, T.; Duguma, G.; Haile, A.; Tibbo, M.; Okeyo, A.M.; Wurzinger, M.; Sölkner, J. Genetics of adaptation in domestic farm animals: A review. *Livest. Sci.* **2010**, *132*, 1–12.

7. Olsen, Y. How can mariculture better help feed humanity? *Front. Mar. Sci.* **2015**, *2*, doi:10.3389/fmars.2015.00046.

8. Gjedrem, T. The first family-based breeding program in aquaculture. *Rev. Aquac.* **2010**, *2*, 2–15.

9. Gjedrem, T.; Robinson, N.; Morten, R. The importance of selective breeding in aquaculture to meet future demands for animal proteins: A review. *Aquaculture* **2012**, *350–353*, 117–129.

10. Lind, C.E; Ponzoni R.W.; Nguyen, N.H.; Khaw, H.L. Selective breeding in fish and conservation of genetic resources for aquaculture. *Reprod. Domest. Anim.* **2012**, *47*, 255–263.

11. Teletchea, F. Domestication and genetics: What a comparison between land and aquatic species can bring. In *Evolutionary Biology. Biodiversification from Genotype to Phenotype*; Pontarotti, P., Ed.; Springer: Cham, Switzerland, 2015; pp. 389–402.

12. Duarte, C.M.; Marbá, N.; Holmer M. Rapid domestication of marine species. *Science* **2007**, *316*, 382–383.

13. Bartley, D.M.; Nguyen, T.T.T.; Halwart, M.; de Silva, S.S. Use and exchange of aquatic genetic resources in aquaculture: Information relevant to access and benefit sharing. *Rev. Aquac.* **2009**, *1*, 157–162.

14. Diana, J.S. Aquaculture production and biodiversity conservation. *BioScience* **2009**, *59*, 27–37.

15. Hedgecock, D. Aquaculture, the next wave of domestication. In *Biodiversity in Agriculture, Domestication, Evolution, and Sustainability*; Gepts, P., Famula, T.R., Bettinger, R.L., Brush, S.B., Damania, A.B., McGuire, P.E., Qualset, C.O., Eds.; Cambridge University Press: Cambridge, UK, 2012; pp. 538–548.

16. Gjedrem, T. Genetic improvement for the development of efficient global aquaculture: A personal opinion review. *Aquaculture* **2012**, *344–349*, 12–22.

17. Bilio, M. Plenary lecture: the future of capture and culture fisheries. In Proceedings of the 7th International Symposium, Keeping and Creating Diversity in Fish Production—Innovative Marine Life Science for Three Es, Edibles Environment and Education, in 21st century, Hokkaido, Japan, 17–19 November 2008; pp. 5–24.

18. Teletchea, F. Qu'est-ce qu'un poisson domestique? Implications pour le développement futur de l'aquaculture. *Ethnozootechnie* **2012**, *90*, 7–12.

19. Klinger, D.H.; Turnipseed, M.; Anderson, J.L.; Asche, F.; Crowder, L.B.; Guttormsen A.G.; Halpern, B.S.; O'Connor, M.I.; Sagarin, R.; Selkoe, K.A.; *et al.* Moving beyond the fished or farmed dichotomy. *Mar. Policy* **2013**, *68*, 369–374.

20. Teletchea, F.; Fontaine, P. Levels of domestication in fish: Implications for the sustainable future of aquaculture. *Fish Fish.* **2014**, *15*, 181–195.

21. Watson, R.; Pauly, D. Systematic distortions in world fisheries catch trends. *Nature* **2001**, *414*, 534–536.

22. Campbell, B.; Pauly D. Mariculture: A global analysis of production trends since 1950. *Mar. Pol.* **2013**, *39*, 94–100.

23. Available online: http://www.fao.org/fishery/statistics/global-aquaculture-production/query/en (accessed on 10 July 2015).

24. De Silva, S.S. Aquaculture: A newly emergent food production sector and perspectives of its impacts on biodiversity and conservation. *Biodivers. Conserv.* **2012**, *21*, 3187–3220.

25. Migaud, H.; Bell G; Cabrita E.; McAndrew B.; Davie, A.; Bobe, J.; Herráez, M.P.; Carillo, M. Gamete quality and broodstock management in temperate fish. *Rev. Aquac.* **2013**, *5*, S194–S223.

26. Natale, F.; Hofherr, J.; Fiore, G.; Virtanen, J. Interactions between aquaculture and fisheries. *Mar. Policy* **2013**, *38*, 205–113.

27. Sorgeloos, P. Aquaculture: The blue biotechnology of the future. *World Aquac.* **2013**, *44*, 16–25.

28. Bostock, J.; McAndrew, B.; Richards, R.; Jauncey, K.; Telfer, T.; Lorenzen, K.; Little, D.; Ross, L.; Handisyde, N.; Gatward, I.; *et al.* Aquaculture: Global status and trends. *Philos. Trans. R. Soc. B* **2010**, *365*, 2897–2912.

29. Duarte, C.M.; Holmer, M.; Olsen, Y.; Soto, D.; Marbà, N.; Guiu, J.; Black, K.; Karakassis, I. Will the oceans help feed humanity? *BioScience* **2009**, *59*, 967–976.

30. Larson, G.; Fuller, D.Q. The evolution of animal domestication. *Annu. Rev. Ecol. Evol. Syst.* **2014**, *66*, 115–136.

31. Mylonas, C.C.; de la Gándara, F.; Corriero, A.; Belmonde Ríos, A. Atlantic Bluefin tuna (*Thunnus thynnus*) farming and fattening in the Mediterranean sea. *Rev. Fish. Sci.* **2010**, *18*, 266–280.

32. Trujillo, P.; Piroddi, C.; Jacquet J. Fish farms at sea: The ground truth from google Earth. *PLoS ONE* **2012**, *7*, e30546.

33. Gunnarsson, S.; Sigurdsson, S.; Thorarensen, H.; Imsland, A.K. Cryopreservation of sperm from spotted wolffish. *Aquac. Int.* **2009**, *17*, 385–389.

34. Dulcic, J.; Grubisic, L.; Pallaoro, A.; Glamuzina, B. Embryonic and larval development of big-scale sand smelt *Atherina boyeri* (Atherinidae). *Cybium* **2008**, *32*, 27–32.

35. Rezek, T.C.; Watanabe, W.O.; Harel, M.; Seaton, P.J. Effects of dietary docosahexaenoic acid (22:6n-3) and arachidonic acid (20:4n-6) on the growth, survival, stress resistance and fatty acid composition in black sea bass *Centropristis striata* (Linnaeus 1758) larvae. *Aquac. Res.* **2010**, *41*, 1302–1314.

36. Gaspar, A.G. Induced spawning and rear larvae of spadefish *Chaetodipterus faber* in Margarita Island, Venezuala. In *Proceedings of the 47th Gulf and Caribbean Fisheries Institute*; Gulf and Caribbean Fisheries Institute: Margarita Island, Venezuala, 1995; Volume 48, pp. 15–24.

37. Ly, C.-L.; Vergnet, A.; Molinari, N.; Fauvel, C.; Bonhomme, F. Fitness of early life stages in F1 interspecific hybrids between *Dicentrarchus labrax* and *D. punctatus*. *Aquat. Living Resour.* **2012**, *25*, 67–75.

38. Muncaster, S.; Andersson, E.; Kjesbu, O.S.; Taranger, G.L.; Skiftesvik, A.B.; Norberg, B. The reproductive cycle of female Ballan wrasse *Labrus bergylta* in high latitude, temperate waters. *J. Fish Biol.* **2010**, *77*, 494–511.

39. Velloso, A.L.; Pereira, J., Jr. Influence of ectoparasitism on the welfare of *Micropogonias furnieri*. *Aquaculture* **2010**, *310*, 43–46.

40. Engell-Sørensen, K.; Støttrup, J.G.; Holmstrup, M. Rearing of flounder (*Platichthys flesus*) juveniles in semiextensive systems. *Aquaculture* **2004**, *230*, 475–491.

41. Cruzado, I.H.; Herrera, M.; Quintana, D.; Rodiles, A.; Navas, J.I.; Lorenzo, A.; Almansa, E. Total lipid and fatty acid composition of brill eggs *Scophthalmus rhombus* L. relationship between lipid composition and egg quality. *Aquac. Res.* **2004**, *42*, 1011–1025.

42. Xu, S.; Zhang, L.; Wu, Q.; Liu, X.; Wang, S.; You, C.; Li, Y. Evaluation of dried seaweed *Gracilaria lemaneiformis* as an ingredient in diets for teleost fish *Siganus canaliculatus*. *Aquacult. Int.* **2011**, *19*, 1007–1018.

43. El-Dakar, A.Y.; Shalaby, S.M.; Saoud, I.P. Dietary protein requirement of juvenile marbled spinefoot rabbitfish *Siganus rivulatus*. *Aquac. Res.* **2011**, *42*, 1050–1055.

44. Belal, I.E.H. Replacement of corn with mangrove seeds in bluespot mullet *Valamugil seheli* diets. *Aquac. Nutr.* **2004**, *10*, 25–30.

45. Liao, I.C.; Su, H.M.; Chang, E.Y. Techniques in finfish larviculture in Taiwan. *Aquaculture* **2001**, *200*, 1–31.

46. Mirimin, L.; Roodt-Wilding, R. Testing and validating a modified CTAB DNA extraction method to enable molecular parentage analysis of fertilized eggs and larvae of an emerging South African aquaculture species, the dusky kob *Argyrosomus japonicas*. *J. Fish Biol.* **2015**, *86*, 1218–1223.

47. Zhang, Q.-Z.; Hong W.S.; Dai Q.N.; Zhang J.; Cai Y.-Y.; Huang J.L. Studies on induced ovulation, embryonic development and larval rearing of the mudskipper (*Boleophthalmus pectinirostris*). *Aquaculture* **1989**, *83*, 375–385.

48. Carter, C.; Glencross, B.; Katersky, R.S.; Bermudes, M. Chapter 14: The Snooks (Family: Centropomidae). In *Finfish Aquaculture Diversification*; Le François, N., Jobling, M., Carter, C., Blier, P., Eds.; CABI: Oxfordshire, UK, 2010; pp. 323–336.

49. Ottolenghi, F.; Silvestri, C.; Giodano, P.; Lovatelli, A.; New, M.B. *Capture-Based Aquaculture, the Fattening of Eels, Groupers, Tuna and Yellowtails*; FAO: Rome, Italy, 2004.

50. Peng, C.; Ma, H.; Su, Y.; Wen, W.; Feng, J.; Guo, Z.; Qiu, L. Susceptibility of farmed juvenile giant grouper *Epinephelus lanceolatus* to a newly isolated grouper iridovirus (genus *Ranavirus*). *Vet. Microbiol.* **2015**, *177*, 270–279.

51. Marino, G.; Ingle, E.; Cataudella, S. Status of aquaculture in Italy. *Cah. Opt. Méditerr.* **1999**, *43*, 117–126.

52. García-Ortega, A. Nutrition and feeding research in the spotted rose snapper (*Lutjanus guttatus*) and bullseye puffer (*Sphoeroides annulatus*), new species for marine aquaculture. *Fish Physiol. Biochem.* **2009**, *35*, 69–80.

53. An, H.S.; Kim, E.M.; Lee, J.W.; Kim, D.J.; Kim, Y.C. New polymorphic microsatellite markers in the Korean mi-iuy croaker, *Miichthys miiuy*, and their application to the genetic characterization of wild and farmed populations. *Anim. Cells Syst.* **2012**, *16*, 41–49.

54. Deng, D.-F.; Ju, Z.Y.; Dominy, W.; Murashige, R.; Wilson, R.P. Optimal dietary protein levels for juvenile Pacific threadfin (*Polydactylus sexfilis*) fed diets with two levels of lipid. *Aquaculture* **2011**, *316*, 25–30.

55. Pham, H.Q.; Nguyen, A.T.; Nguyen, M.D.; Arukwe, A. Sex steroid levels, oocyte maturation and spawning performance in Waigieu seaperch (*Psammoperca waigiensis*) exposed to thyroxin, human chorionic gonadotropin, luteinizing hormone releasing hormone and carp pituitary extract. *Comp. Biochem. Phys. A* **2010**, *155*, 223–230.

56. Bilio, M. Controlled reproduction and domestication in aquaculture—The current state of the art, Part II. *Aquac. Eur.* **2007**, *32*, 5–23.

57. Roo, J.; Fernandez-Palacios, H.; Hernandez-Cruz, C.M.; Mesa-Rodriguez, A.; Schuchardt, D.; Izquierdo, M. First results of spawning and larval rearing of longfin yellowtail *Seriola rivoliana* as a fast-growing candidate for European marine finfish aquaculture diversification. *Aquac. Res.* **2010**, *45*, 689–700.

58. Wexler, J.B.; Margulies, D.; Scholey, V.P. Temperature and dissolved oxygen requirements for survival of yellowfin tuna, *Thunnus albacares*, larvae. *J. Exp. Mar. Biol. Ecol.* **2011**, *404*, 63–72.

59. Carter, C.; Nowak, B.; Clarke, S. Chapter 20: The Tunas (Family: Scombridae). In *Finfish Aquaculture Diversification*; Le François, N., Jobling, M., Carter, C., Blier, P., Eds.; CABI: Oxfordshire, UK, 2010; pp. 432–449.

60. Pfeiffer, T.J.; Riche, M.A. Evaluation of a low-head recirculating aquaculture system used for rearing Florida pompano to market size. *J. World Aquac. Soc.* **2011**, *42*, 198–208.

61. Hong, W.; Zhang, Q. Review of captive bred species and fry production of marine fish in China. *Aquaculture* **2003**, *227*, 305–318.

62. Suquet, M.; Divanach, P.; Hussenot, J.; Coves, D.; Fauvel, C. Pisciculture marine de «nouvelles espèces» d'élevage pour l'Europe. *Cah. Agric.* **2009**, *2–3*, 148–156.

63. Roselund, G.; Skretting, M. Worldwide status and perspective on gadoid culture. *ICES J. Mar. Sci.* **2006**, *63*, 194–197.

64. Coeurdacier, J.L.; Gasset, E. Dossier récapitulatif du développement de l'élevage du *Platax orbicularis* en Polynésie Française. Avaliable online: http://archimer.ifremer.fr/doc/00134/24490/ (accessed on 15 July 2013).

65. Bilio, M. Controlled reproduction and domestication in aquaculture—The current state of the art, Part I. *Aquac. Eur.* **2007**, *32*, 5–14.

66. Imsland, A.K. Chapter 21: The Flatfishes (Order: Pleuronectiformes). In *Finfish Aquaculture Diversification*; Le François, N., Jobling, M., Carter, C., Blier, P., Eds.; CABI: Oxfordshire, UK, 2010; pp. 450–496.

67. Teng, S.-K.; El-Zahr, C.; Al-Abdul-Elah, K.; Almatar, S. Pilot-scale spawning and fry production of blue-fin porgy, *Sparidentex hasta* (Valenciennes), in Kuwait. *Aquaculture* **1999**, *178*, 27–41.

68. Kim, J.H.; Rhee, J.-S.; Lee, J.S.; Dahms, H.U.; Lee, J.; Han, K.N.; Lee, J.S. Effect of cadmium exposure on expression of antioxidant gene transcripts in the river pufferfish, *Takifugu obscurus* (Tetraodontiformes). *Comp. Biochem. Phys. C* **2010**, *152*, 473–479.

69. Wu, F.; Tang, K.; Yuan, M.; Shi, X.; Shakeela, Q.; Zhang, X.H. Studies on bacterial pathogens isolated from diseased torafugu (*Takifugu rubripes*) cultured in marine industrial recirculation aquaculture system in Shandong Province, China. *Aquac. Res.* **2015**, *46*, 736–744.

70. Masuda, R. Ontogeny of anti-predator behavior in hatchery-reared jack mackerel *Trachurus japonicus* larvae and juveniles: Patchiness formation, swimming capability, and interaction with jellyfish. *Fish. Sci.* **2006**, *72*, 1225–1235.

71. Le François, N.; Tveiten, H.; Halfyard, L.C.; Foss, A. Chapter 19. The Wolffishes (Family: Anarhichadidae). In *Finfish Aquaculture Diversification*; Le François, N., Jobling, M., Carter, C., Blier, P., Eds.; CABI: Oxfordshire, UK, 2010; pp. 417–431.

72. Lazo, J.P.; Holt, J.G.; Fauvel, C.; Suquet, M.; Quéméner, L. Chapter 18. Drum-Fish or Croakers (Family: Sciaenidae). In *Finfish Aquaculture Diversification*; Le François, N., Jobling, M., Carter, C., Blier, P., Eds., CABI: Oxfordshire, UK, 2010; pp. 397–416.

73. Jobling, M.; Peruzzi, S.; Woods, C. Chapter 15. The Temperate Basses (Family: Moronidae). In *Finfish Aquaculture Diversification*; Le François, N., Jobling, M., Carter, C., Blier, P., Eds.; CABI: Oxfordshire, UK, 2010; pp. 337–360.

74. Björnsson, B.; Litvak, M.; Trippel, E.A.; Suquet, M. Chapter 13. The Codfishes (Family: Gadidae). In *Finfish Aquaculture Diversification*; Le François, N., Jobling, M., Carter, C., Blier, P., Eds.; CABI: Oxfordshire, UK, 2010; pp. 290–322.

75. Hulata, G. Genetic manipulations in aquaculture: A review of stock improvement by classical and modern technologies. *Genetica* **2001**, *111*, 155–173.

76. Jobling, M.; Peruzzi, S. Chapter 16. Seabreams and Porgies (Family: Sparidae). In *Finfish Aquaculture Diversification*; Le François, N., Jobling, M., Carter, C., Blier, P., Eds.; CABI: Oxfordshire, UK, 2010; pp. 361–373.

5

Design Optimization for a Truncated Catenary Mooring System for Scale Model Test

Climent Molins [1,*], Pau Trubat [1], Xavi Gironella [2] and Alexis Campos [1]

[1] Department of Civil and Environmental Engineering, Universitat Politècnica de Catalunya—BarcelonaTech, Jordi Girona 1-3, Campus Nord C1-206, 08018 Barcelona, Spain; E-Mails: pau.trubat.casal@upc.edu (P.T.); alexis.campos@pupc.edu (A.C.)

[2] Department of Civil and Environmental Engineering, Universitat Politècnica de Catalunya—BarcelonaTech, Jordi Girona 1-3, Campus Nord D1- 111A, 08018 Barcelona, Spain; E-Mail: xavi.gironella@upc.edu

* Author to whom correspondence should be addressed; E-Mail: climent.molins@upc.edu

Academic Editor: Bjoern Elsaesser

Abstract: One of the main aspects when testing floating offshore platforms is the scaled mooring system, particularly with the increased depths where such platforms are intended. The paper proposes the use of truncated mooring systems to emulate the real mooring system by solving an optimization problem. This approach could be an interesting option when the existing testing facilities do not have enough available space. As part of the development of a new spar platform made of concrete for Floating Offshore Wind Turbines (FOWTs), called Windcrete, a station keeping system with catenary shaped lines was selected. The test facility available for the planned experiments had an important width constraint. Then, an algorithm to optimize the design of the scaled truncated mooring system using different weights of lines was developed. The optimization process adjusts the quasi-static behavior of the scaled mooring system as much as possible to the real mooring system within its expected maximum displacement range, where the catenary line provides the restoring forces by its suspended line length.

Keywords: scale moorings; scale tests; wind turbines; floating offshore platform; optimization

---

1. Introduction

Floating offshore wind energy research is focused on developing new platform concepts that fit the necessary requirements of the stability for a wind turbine design and also present competitive construction and operational costs.

As in the Oil and Gas (O&G) Industry, into the design stage of the new platform concepts, the motion and loads of the platform have to be assessed and well predicted in several load combinations to ensure the reliability of the structure and the mooring system. The main approaches to predict the whole platform behavior are the numerical simulation and the physical scale models. However, it is still widely accepted in the offshore industry that model testing is the most reliable procedure to validate the results and to be the final benchmark for the design of a platform.

The physical model testing is mainly performed in the ocean engineering basins, where the environmental conditions such as waves, currents and wind can be reproduced [1,2]. Furthermore, there are others facilities that can reproduce the ocean situations like wave flumes that are not commonly used due to their highly restrictive dimensions. On the other hand, the usage of these installations would help the development of the offshore wind technology allowing the performance of the model tests in more places and reducing costs in that research field.

Some wind offshore platforms model tests have been performed in wave flumes despite their limited width dimension. One example is a Tension Leg Platform (TLP) prototype tested in the CEHINAV (Canal de Ensayos Hidrodinámicos) [3]. In this particular case, a flume is a suitable place for testing because of the inner configuration of a TLP, a buoyant platform moored with vertical tethers. Even a spread mooring system does not seem to fit well in a wave flume; Krawkosky et al [4] tested a spar scale model with a four line mooring system in a flume. The azimuthal angles between two adjacent mooring lines were of 90 degrees. The lines were scaled in two different ways, the lines placed in the longitudinal direction of the tank are well-scaled using proper tethers, while in the transverse direction the mooring lines are modeled as two constant forces. The forces were applied by two ropes hanging on both sides of the flume with weights on their extremes. However, this simplification does not allow changing the waves relative direction to the platform position because the scaled mooring system only works in the longitudinal direction. Furthermore, a three line mooring system could not be scaled in the same way because the different symmetry between the mooring system and the wave flume. Then, in order to perform tests with different wave directions using the same scaled mooring system, an equivalent system with shorter radius to anchor should be designed. In such a mooring system, the line length should be truncated to allow placing all mooring lines in the wave flume. Other scale models have been tested in bigger basins, [5] uses truncated mooring lines attached to springs because the depth of the basin does not match the model in the selected scale. Another solution to overcome the width basin constraint was adopted by [6], where the mooring line segment, constantly resting on the seabed during the tests, was removed. This lead to a shorter radius to anchor using the well scaled prototype mooring system.

Truncated mooring systems are a common scale method used in the O&G industry [7–10]. Since the tank basins sizes do not allow performing tests in the common scales for the upcoming ultra-deep waters, new systems have to be conceived to manage this challenge. The truncated passive system is the most widely used and feasible method of the hybrid model testing methods, which uses a combination of physical model tests and numerical modeling. In the passive method, all the model characteristics like platform properties, wave height, current velocity, *etc.*, are well scaled except the working depth and the mooring shape. This method uses an equivalent truncated mooring system for the scale tests and the results are used to interpret and adjust the model in order to perform a full depth numerical model.

Stansberg [9] states the challenges for the development of the truncated passive methods. The new truncated mooring system set-up has to guarantee the following aspects: (1) the motion response should have the same behavior as the results of the full-depth mooring system and (2) the truncated mooring system should present the most similar physical properties as the full-depth system. To achieve the correct design of the truncated mooring system, Stansberg presents the following rules ordered by priority.

- Model the total horizontal restoring force
- Model the quasi-static coupling between vessel responses
- Model a "representative" level of mooring and riser system damping, and current force
- Model a "representative" single line tension characteristics (at least quasi-static)

In the field of the truncated mooring system design, optimization models to better adjust the truncated system approach to the real one commonly solve the problem. Zhang [10] proposed an annealing simulation algorithm for hybrid discrete variables (ASFHDV) to optimize the static response of a single catenary and the whole catenary system static response in one direction. Further investigations propose an optimization model that accounts for the mooring-induced damping generated by the transverse motion of the mooring line due to the low-frequency surge oscillation using a genetic algorithm [7]. In order to improve the behavior of the truncated mooring line, Qiao [8] proposes the connection of viscous dampers joined to the mooring line to simulate the whole damping of the real mooring line. These models have been validated and widely used, but the quasi-static approach could underestimate the tension in the mooring lines due to dynamics when those are important [11]. On the other hand, new methods have recently been developed to take into account the line dynamics and obtain a more realistic system behavior using the real scaled mooring line in the upper sections, where the line dynamics are more important, and using external actuators that replicate the truncated line segments behavior [12].

The main contribution of this paper is the design of a truncated mooring system to replace the prototype mooring system, which cannot be installed due to basin constraints. The truncated mooring system is designed as a simple mooring line composed of two materials without any other external systems as springs. This new mooring system allows the study of several wave approaching directions to the whole structure, float and moorings. Tests results and the comparison with numerical simulations are also presented.

First, the real model and the scale model due to the basin constrains are presented. Then, the calculation of the static mooring forces and the optimization problem are described. Finally, the optimization and the experimental results, with the comparison with numerical simulations, are discussed.

2. Real Model

The monolithic concrete spar, the so-called Windcrete [13], is a prototype floating platform for wind turbines developed in AFOSP (Alternative Floating Platform Designs for Offshore Wind Turbines using Low Cost Materials) within a KIC-InnoEnergy innovation project [14–16].

The spar prototype is designed as a monolithic concrete structure from the top of the tower to the bottom of the buoy, thus joints are avoided to ensure water-tightness and a good fatigue behavior. The structure, for a 5 MW wind turbine, is composed of three parts: first, the buoy, composed of a cylinder with a diameter of 13 m and a height of 120 m; second, the transition segment, which is a cone of 10 m high, these two parts are the submerged ones, therefore the total draft of the structure is 130 m. The third part is the emerged tower that reaches 87.6 m above the SWL. A sketch of the concept and its hydrodynamic characteristics are shown in Figure 1 and Table 1. The moorings system is connected to the platform at the fairleads located 60 m above the bottom with a draft of 70 m, near the Centre of Gravity (COG) to reduce the coupling motions between the surge and pitch.

In this study, the Windcrete is considered to be placed in a 265 m depth sea location. The mooring system is configured to provide enough restoring force to maintain the platform motion in a relative offset and to prevail over the wave and wind loads. In order to achieve simplicity in the model, the prototype mooring system is composed of three equispaced chain mooring lines with the same cross section. The main characteristics of the line are defined in the Table 2.

3. Scale Model

Model tests were performed in the ICTS-CIEM (Investigation flume and offshore experimentation) inside LIM (Maritime Engineering Laboratory) at the UPC (Universitat Politècnica de Catalunya—BarcelonaTech). The flume is equipped with a wave generator that can generate waves from 10 cm up to 160 cm height. The flume is 100 meters long with a cross section 3.5 m wide and 5 m high. The flume and the wave paddle are shown in Figures 2 and 3, respectively.

According to the prototype's sizes and the flume height, the selected scale is 1:100. Then, the scaled depth would be 2.65 m. The flume width does not allow the direct scaling of the catenary lines because the common anchor radius should be between two and four times the total depth, requiring a 10 m wide channel. Furthermore, if several wave directions are studied, the mooring system should be able to rotate in z direction allowing the wave to impact on the platform from different relative direction with the mooring system. Then, the mooring system should not be connected through the flume wall. For these reasons, a truncated mooring system is used in the model scale test, reducing the radius to anchor distance, and therefore the total length of the lines. Figure 4 shows the cross section of the wave flume with both the scale prototype mooring system and the truncated one.

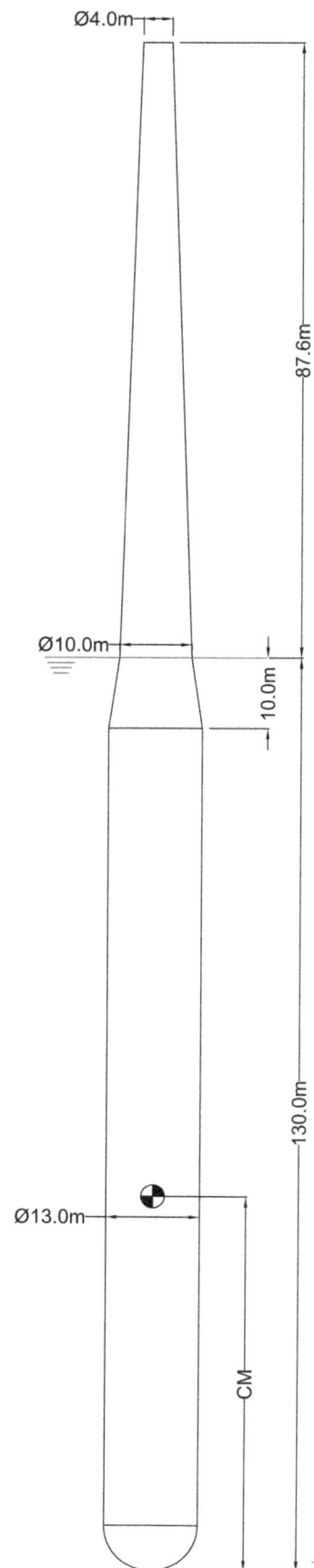

Figure 1. Sketch of Windcrete concept.

Table 1. Hydrostatic characteristics of Windcrete.

Property	Value
Displaced Volume [m³]	1.69×10^4
Draft [m]	130.0
Concrete mass [kg]	8.71×10^6
Ballast mass [kg]	8.34×10^6
Wind turbine mass [kg]	3.50×10^5
CM [m]	53.34
CB [m]	63.97
Metacentric height [m]	10.57

Table 2. Prototype mooring line characteristics.

Depth [m]	265
Draft to fairlead [m]	70
Mooring depth	195
Radius to anchor [m]	660
Line length [m]	732.93
Line mass per unit length [kg/m]	150.3

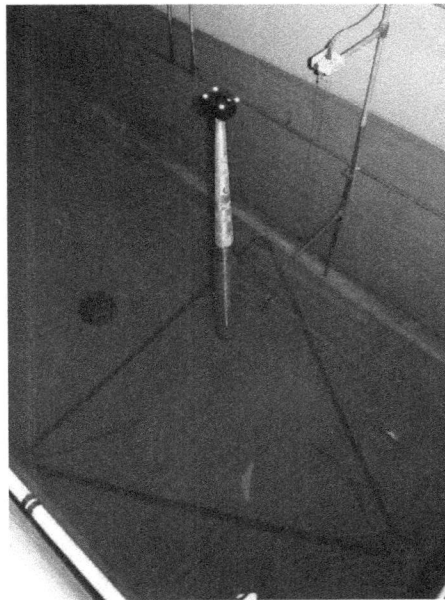

Figure 2. Wave flume.

The truncated mooring system is defined by the radius to anchor, the line length and the materials that compose the different segments of the mooring line. The radius to anchor is previously defined as the maximum radius allowed by the channel width taking account the margins for a proper installation operation of the mooring system. Furthermore, if the truncated mooring line were composed of a unique cross section, the necessary weight to achieve the restoring forces of the prototype system would cause huge vertical forces on the floating platform. For this reason, two different chain sections are chosen to design the mooring line. The heaviest line section is positioned at the bottom, connected to the anchor, providing the restoring horizontal force. The upper section, a light segment connected to the platform, reduces the total line payload due to its light weight. The exact properties of the lines are obtained through an optimization problem to fit the responses between the prototype mooring system and the truncated one, which is presented in next sections.

Figure 3. CIEM wave paddle.

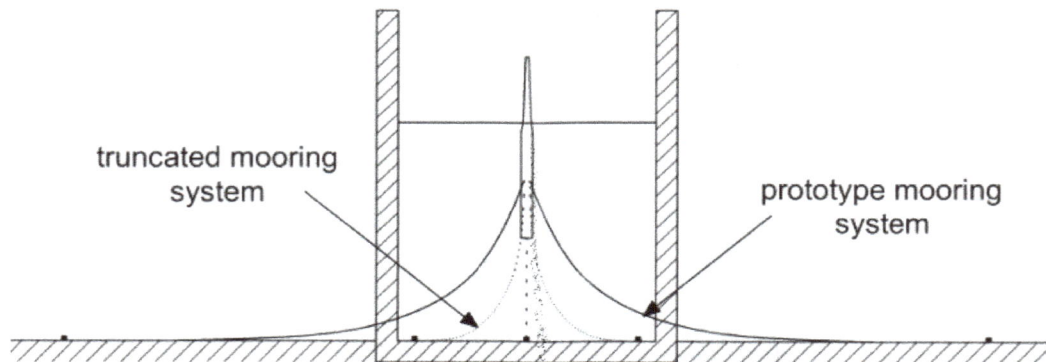

Figure 4. Prototype *vs.* truncated mooring system.

4. Static Mooring Lines Forces

The static catenary line can be described by the equations deduced from applying equilibrium on the whole forces acting on a line segment. As is shown in Figure 5, a catenary segment is subjected to the inner line tensions (T), the gravity forces accounted by the weight per unit length of the line (ω) and the hydrodynamic forces, which are the transversal drag forces per unit length (F) and the normal drag forces per unit length (D).

The static equilibrium of the line segment leads to the following equations:

$$dT = \left[\omega \sin(\phi) - F\left(1 + \frac{T}{EA}\right)\right] ds \qquad (1)$$

$$T \cdot d\phi = \left[\omega \cos(\phi) + D\left(1 + \frac{T}{EA}\right)\right] ds \qquad (2)$$

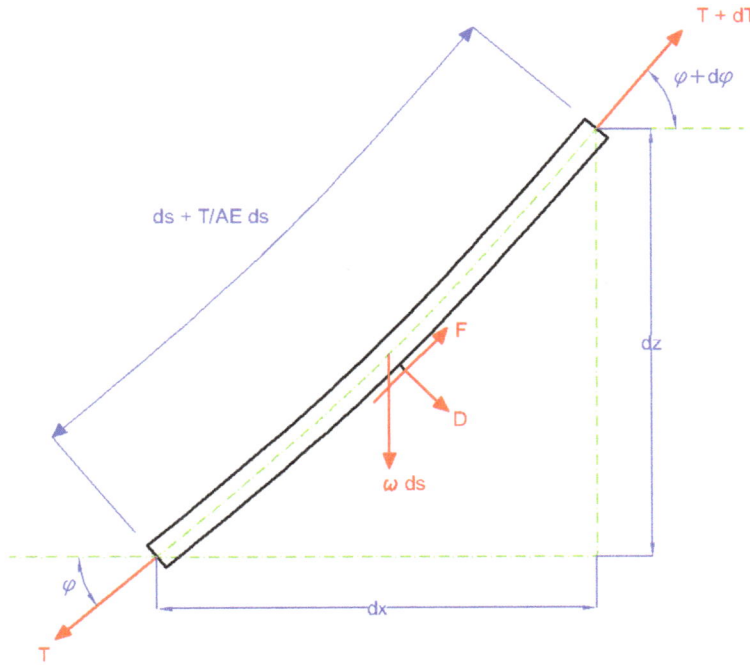

Figure 5. Segment line forces scheme.

To solve Equations (1) and (2), the mooring line is discretized in n + 1 nodes forming n line elements. The nodes are described as $N = [N_1, N_2, ..., N_i, ..., N_{n+1}]$ and the segments as $S = [S_1, S_2, ..., S_i, ..., S_n]$. The properties that define each element are the weight per unit length ω_i, its length l_i and the longitudinal stiffness EA_i. Figure 6 shows a sketch of the mooring line composed by n segments.

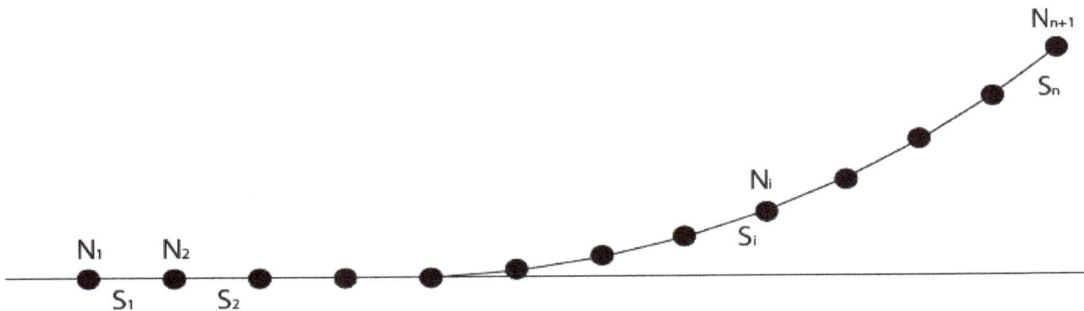

Figure 6. Mooring line piecewise scheme.

Without taking into account the hydrodynamic forces, the tensions in each node are evaluated from the tensions of the previous node applying the external forces placed on the centre of each element. Equations (3)–(5) express the tensions of the $(i+1)^{th}$ node $(\vec{T}_{i+1})$ as a function of the tensions of ith node $(\vec{T}_i)$ in the 3DOF (Degrees of freedom) (Figure 7).

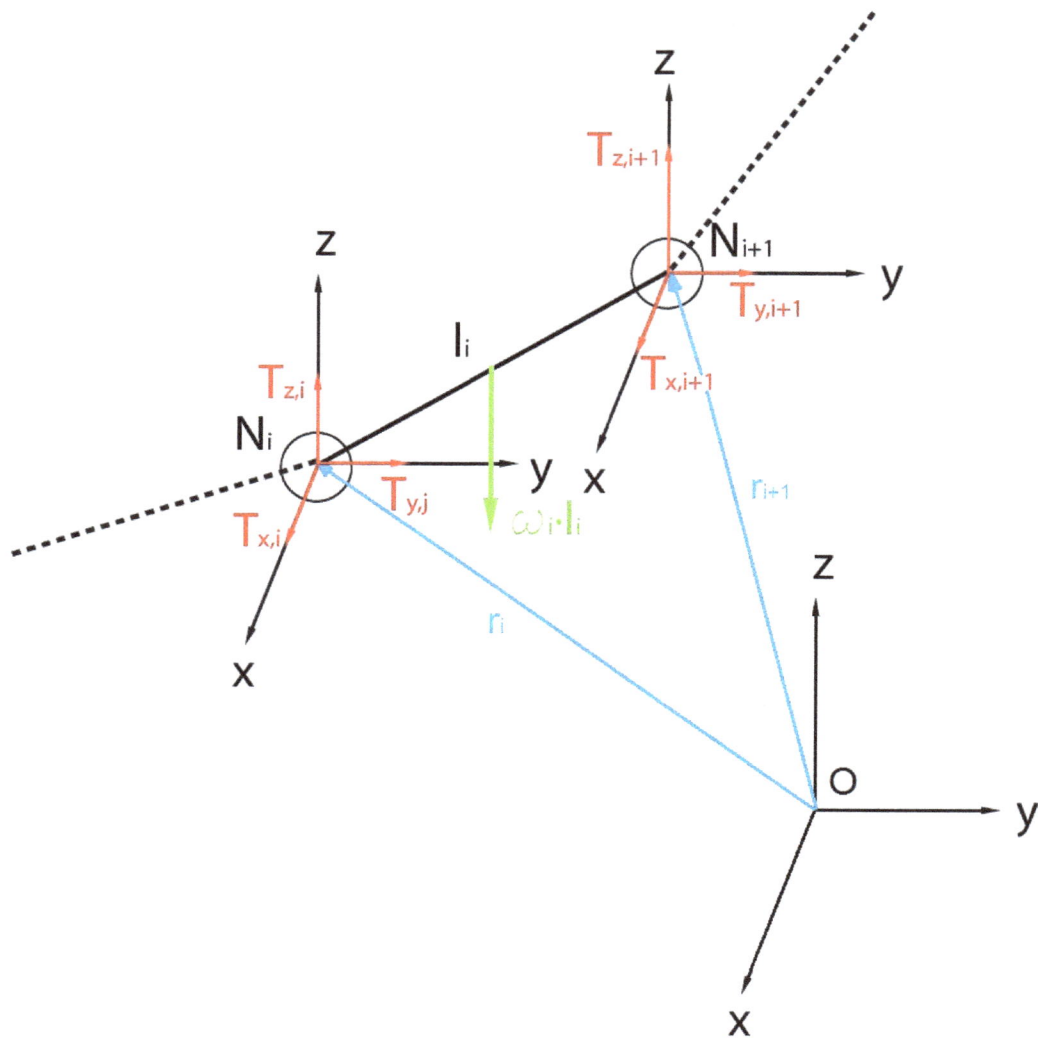

Figure 7. Force scheme of two consecutive nodes.

$$T_{X_{i+1}} = T_{X_i} \qquad (3)$$

$$T_{Y_{i+1}} = T_{Y_i} \qquad (4)$$

$$T_{Z_{i+1}} = T_{Z_i} + \omega_i l_i \qquad (5)$$

The position of the $(i+1)^{th}$ node $(\vec{r}_{i+1})$, as shown in Figure 7, can be expressed from the position of the ith node $(\vec{r}_i)$ (Equation (7)). Where $\vec{v}_{d,i}$ (Equation (6)) is the directional vector, and ε_i (Equation (8)) is the strain of the element.

$$\vec{v}_{d_i} = \frac{\vec{T}_i}{|T_i|} \qquad (6)$$

$$\vec{r}_{i+1} = \vec{r}_i + [(1 + \varepsilon_i)l_i]\vec{v}_{d,i} \qquad (7)$$

$$\varepsilon_i = \frac{|T_i|}{AE_i} \qquad (8)$$

The computing procedure follows the scheme proposed by [7]. The position of the nodes are evaluated from the initial approach of the tension force in the node n + 1 ($\vec{T}_{n+1}$) and the position of the anchor ($\vec{r}_1$). If the assessed/estimated position of the fairlead ($\vec{r}_{n+1}$) is not close enough to the known value, a new approximation of ($\vec{T}_{n+1}$) is applied using a non-linear solving method.

By using this method, the mooring line response is evaluated at any fairlead position and the forces on the top of the line are determined. This method can be modified to take into account the hydrodynamic forces in the static analysis. In addition, more complex geometries, such as delta connections, and external elements, such as clumps weights or buoys, can be included.

The mooring system response is obtained by combining the forces on the fairleads of all mooring lines.

5. Optimization Problem

The objective of the optimization problem is to determine a new catenary system presenting a similar static behavior of the prototype when the radius to anchor is reduced. This kind of problem can be expressed as a minimization problem, where the static response of the truncated mooring system has to fit with the prototype ones in a non-scale scenario. Then the properties of the truncated mooring system will be well scaled for the tests.

The optimization problem can be expressed mathematically as:

$$\min\left[f(X)\right] \tag{9}$$

Subjected to the constraints:

$$g_i(X) \leq 0$$
$$h_j(X) = 0 \tag{10}$$

where:

$f(X): \mathbb{R}^n \longrightarrow \mathbb{R}$ is the objective function to be minimized over the variables X

$g_i(X) < 0$; $i = 1,...,m$ are the inequality constraints

$h_j(X) = 0$; $j = 1,...,p$ j = 1,...,p are the equality constraints

The design variables are the parameters that define the mooring line. In order to reduce the complexity of the optimization problem and the final design, the mooring truncated line would have two different segments. Each segment is defined by its lengths and chain diameter. Then, the design variables are the components of the vector X defined as $X = [d_1, d_2, l_1, l_2]$, where d is the diameter, l the segment length and the subscript defines the segment. The weight per unit length (ω) and the longitudinal stiffness (EA) of the segment line can be calculated from the chain diameter (d[mm]), using Equations (11) and (12).

$$\omega = 0.1875d^2 [\text{N/m}] \tag{11}$$

$$EA = 90000d^2 [\text{N}] \tag{12}$$

The objective function—to be minimized—is formulated from the prototype mooring line force responses (F_i) that have to be emulated. The objective function is evaluated in several points of the surge work range (x_j) since it is the main movement direction. The responses emulated are the horizontal and vertical restoring forces of the mooring system and the line tension. As shown in Figure 8, the objective function is expressed as the difference between the response of the prototype mooring system and the truncated one. This distance has to be minimized as a function of the mooring variables (X).

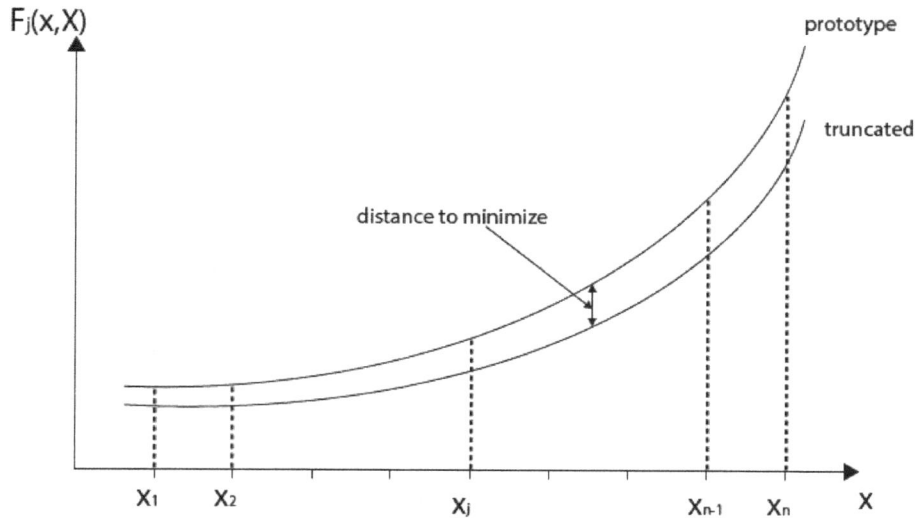

Figure 8. Prototype *vs.* truncated mooring system response.

The objective function can be expressed as the sum of the single objective function of each property multiplied by a weight factor Equation (13). The sum of all weight factors is 1: $\sum_{i=1}^{n} \omega_i = 1$.

$$f(X) = \omega_i \sum_{i=1}^{n} f_i(X) \tag{13}$$

Each single objective function (Equation (14)) is expressed as the root mean square value (rms) of the dimensionless difference between the prototype response and the truncated one, which are evaluated in the n selected surge points:

$$f_i(X) = \sqrt{\frac{1}{n} \sum_{j=1}^{n} \left(\frac{F_{i,prototype}(x_j) - F_{i,truncated}(x_j, X)}{F_{i,prototype}(x_j)} \right)^2} \tag{14}$$

The optimization problem is solved using the GlobalSearch Algorithm in *Matlab* [17], which uses a scatter-search mechanism for generating start points. From the starting points, GlobalSearch examines the trial points and choose the ones that can generate a better solution. Then, the chosen points are evaluated by a local minimization solver. The process ends when all the trial-points have been evaluated [18].

6. Truncated Mooring System Design for Windcrete

The truncated mooring system was designed for a radius to anchor of 140 m. The objective function was evaluated for a surge excursion ranging from −40 to 25 m. Surge excursion is the main platform motion that depends on the mooring system response. The surge interval ensures a horizontal response for a 600 kN mean wind force. The asymmetry of the mooring system (Figure 9) produces an asymmetry response in surge.

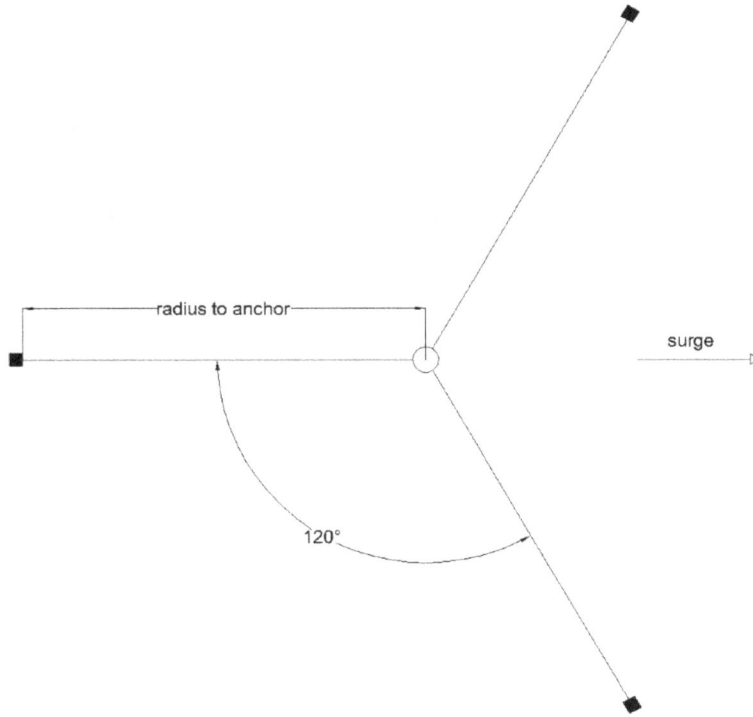

Figure 9. Mooring system sketch.

The chosen characteristics of the prototype to be minimized in the multi-objective function are: the tension of one line, the horizontal mooring system response and the vertical mooring system response. Therefore the multi-objective function can be described as Equation (15).

$$f(X) = \omega_T f_T(X) + \omega_{x,s} f_{x,s}(X) + \omega_{z,s} f_{z,s}(X))$$ (15)

where f_T is the function to optimize the tension of a line; and $f_{x,s}$ and $f_{z,s}$ are the functions which describe the difference between the response prototype and truncated mooring system in x and z direction, respectively. Since the surge response is mainly depends on the mooring system, the largest weight factor $\omega_{x,s}$ is given to it. Selected weights are consistent with the references [7,10]. The weighting factors used for the multi-objective problem are 0.2, 0.6 and 0.2 for ω_T $\omega_{x,s}$ and $\omega_{z,s}$ respectively. A sensitivity analysis of the variation of the weight factors was performed. It confirmed that low variations of the weight factors lead to a similar solution of the optimization problem.

The constraints applied to the problem are the length of the line, which is defined between its taut (Equation (16)) and completely slack (Equation (17)) shape. These constraints are expressed as a function of the radius to anchor and the mooring depth. Other constraints are the minimum and maximum diameter of the lines, defined by Equations (18) and (19), respectively, which are fixed to get a feasible scalable chain for the tests.

$$l_1 + l_2 \geq \sqrt{140^2 + 195^2} = 240.1m$$ (16)

$$l_1 + l_2 \leq 140 + 195 = 335m$$ (17)

$$d_{min} > 60mm$$ (18)

$$d_{max} \leq 250mm \tag{19}$$

The solution of the optimization problem yields to a truncated catenary system composed by lines of two different segments with different weight per unit length. The segment a is the lower one and is linked to the anchor, while the segment b is the upper one and is connected to the platform. The segment a is heavier and shorter than segment b. With this configuration, the stiffness of the restoring force of the catenary system is mainly provided by the segment a, while the segment b contributes to reduce the suspended weight of the mooring line. The properties of each segment of the mooring line as a result of the optimization problem are shown in the Table 3.

Table 3. Truncated mooring line characteristics.

Segment a	Diameter [mm]	200.1
	Length [m]	80
	Line mass per unit length [kg/m]	878.6
Segment b	Diameter [mm]	58.4
	Length [m]	177
	Line mass per unit length [kg/m]	74.7

The response of the optimized catenary system is presented in Figures 10 and 11. Figure 10 shows the comparison between the horizontal and the vertical response on both mooring systems. The horizontal response of the truncated system fits well with the prototype response. However, for large offsets, the responses start to diverge. The mooring system vertical force component is larger for the truncated one. The reduction of the radius to anchor implies an increment of the suspended weight on the platform for a similar horizontal force. Then, a deeper draft would be expected in the platform during the tests: of about 0.5 m. The line tension is well fitted along the whole surge excursion studied in the optimization problem (Figure 11).

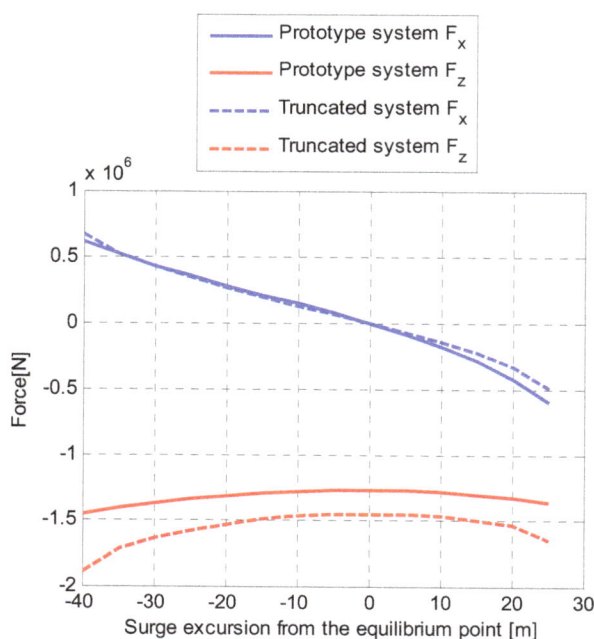

Figure 10. Mooring system response curves.

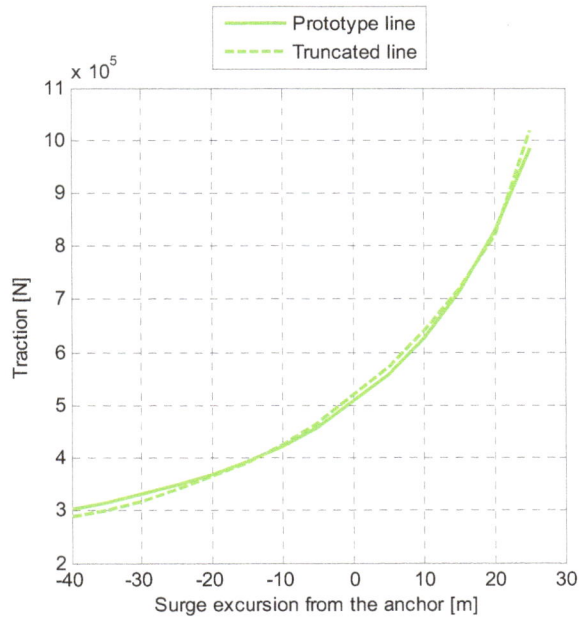

Figure 11. Traction mooring line response.

7. Experimental Results

The model is scaled at 1:100 factor using the Froude similitude. The platform model is made of aluminum to adjust the density of the material to be close to concrete, simplifying the fit of the rest of the platform parameters. The mooring lines are composed of two chain segments that adjusted to fit the weight per meter length computed in the optimization problem. The scale model placed inside the flume attached to the mooring system is shown in Figure 12.

Figure 12. Scale model inside the flume.

The scaled prototype mooring line response were validated by checking the horizontal line response for a depth of 1.95 m and an excursion between 90 and 180 cm from the anchor. The results obtained in the static verification and the numerical results are plotted in Figure 13. The figure shows good agreement between both responses.

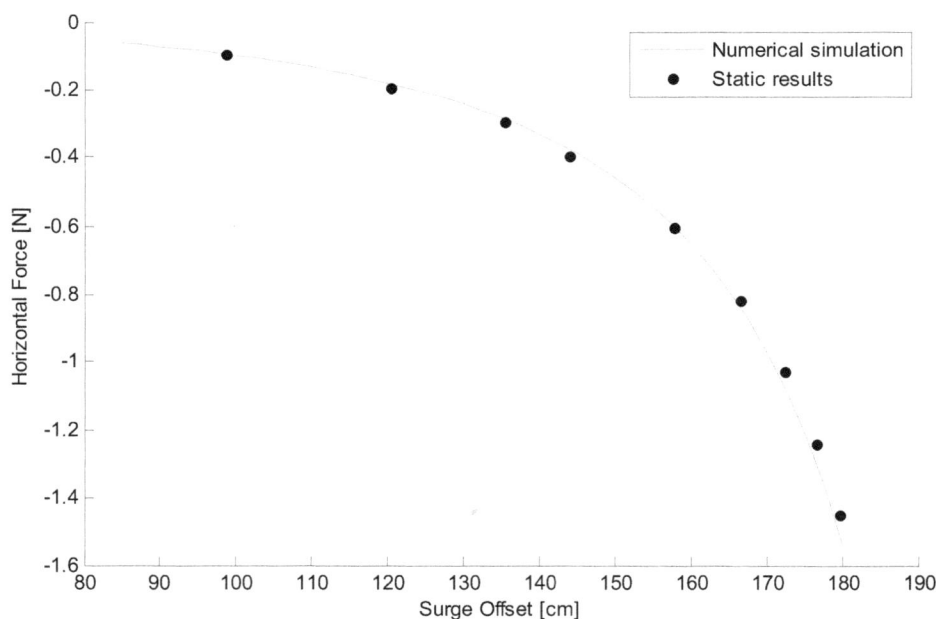

Figure 13. Comparative horizontal force response.

Dynamic tests were carried out in several sea states with regular and irregular waves, and an almost constant wind force on the top of the nacelle. The experimental results were measured from the nacelle motion by an optical system which can track the 6 DOF [19].

Figure 14 shows the surge and heave response comparison between the experiment, the simulation with the truncated lines and the simulation of the platform equipped with the prototype mooring system. These responses correspond to a regular wave of 14 cm height and a period of 1.5 s, and a constant force on the top of the platform of 0.6 N to simulate the wind. The experimental results show good agreement between the test and the numeric simulation with the truncated lines in terms of mean offset, mean draft and also with the wave amplitude movement. However, some differences can be seen due to a low frequency movement that occurred during the test. This disturbance was produced by a long wave reflection in the longitudinal direction of the flume. In the transverse direction, no reflections were noticed. The simulation of the prototype mooring system shows a shorter total surge excursion than the truncated one. This difference is explained by two effects. First, the stiffness of the prototype mooring system for positive excursions is higher than the truncated ones, as is shown in Figure 10. Second, there is a loss of stiffness due to a draft increase of about 0.5 cm. In this situation, the lower depth of the fairlead position requires an increased excursion to achieve the same horizontal force. The draft increase can be noticed in the heave response (Figure 14b) as a decrease of the mean heave position of the truncated mooring compared to the prototype one.

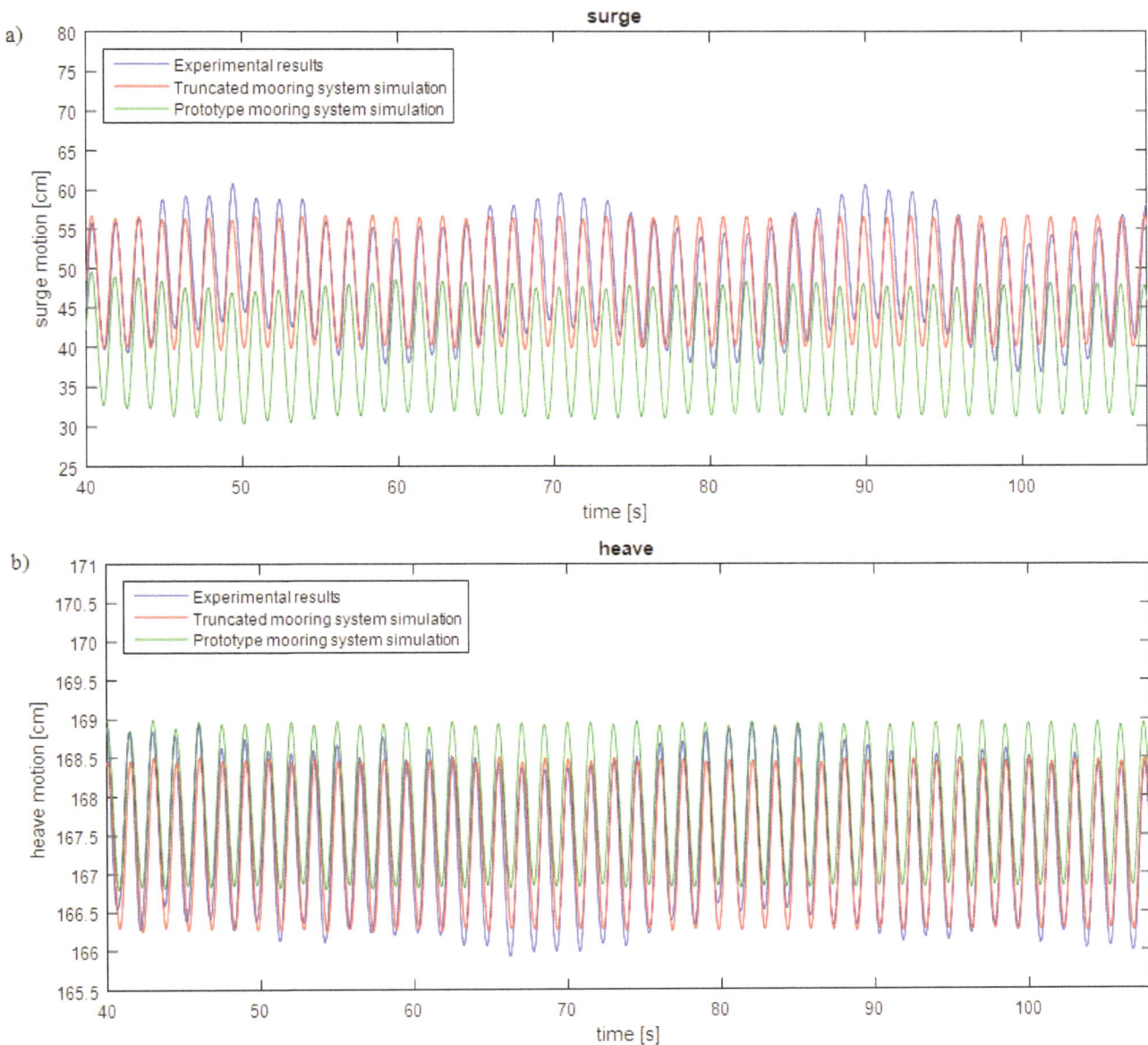

Figure 14. Comparison surge (**a**) and heave (**b**) nacelle motion response.

The results of a Fast Fourier Transform (FFT) of the surge and heave platform responses are shown in Figure 15. Both diagrams show clearly the peak motion due to the wave excitation at a frequency of 0.66 Hz (period of 1.5 s) and the amplitudes match very well.

The main differences between the simulations and the experimental results are the excitation of the low frequency surge motion of the platform during the experiments, as already discussed. This affects the heave response, which FFT (Figure 15b) presents as two small peaks: at 0.05 Hz, the natural surge frequency, and at 0.337 Hz, the heave natural frequency.

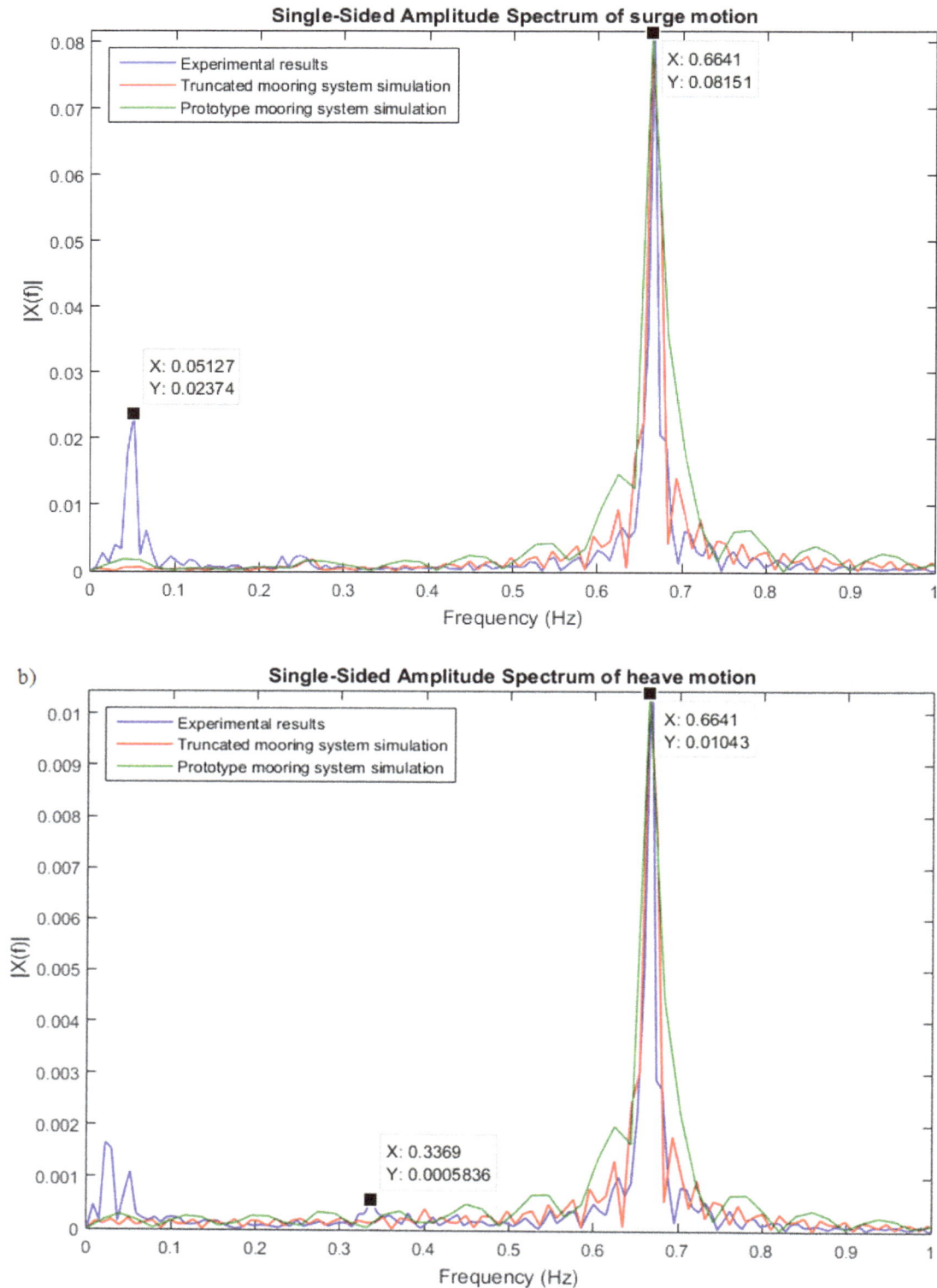

Figure 15. Surge (**a**) and heave (**b**) motion Fast Fourier Transform functions.

8. Conclusions

This paper describes and assesses the use of a truncated mooring system to emulate the real one in scaled experiments when there is a limitation of the available width in the flume. In this case, where the radius to anchor has to be reduced, an optimization problem based on the static mooring system response helped to fit the horizontal and vertical mooring responses and the traction line. The optimization problem is evaluated in the surge work range because is the main platform motion that depends on the mooring system response.

The truncated mooring line stiffness is obtained by using two different line sections: the bottom one is the heaviest and provides the horizontal mooring line stiffness. The upper section is lighter than the prototype mooring and reduces the vertical force over the platform.

The truncated catenary presents almost the same traction response as the completed prototype mooring system (differences less than 5%). On the other hand, the horizontal stiffness of the truncated system differs from the prototype, particularly for large excursions. In addition, the truncated mooring system presents a higher vertical force on the platform that lead to an increment of the draft.

Experiments are compared to numerical simulations with the real and prototype mooring system. The experimental results show good agreement with the numerical simulations. Some differences are noticed in the mean surge excursion of the truncated mooring system, which is larger. This is a consequence of the lower surge stiffness and an extra loss of stiffness due to the increase of the draft. Despite this, surge and heave responses due to wave loads are well predicted.

Acknowledgments

The work presented in this paper has been developed during the KIC Innoenergy (EIT) project AFOSP (Alternative Floating Offshore Support Platform). Its financial support is greatly appreciated.

Author Contributions

C.M. and P.T. conceived and designed the optimization problem, and wrote the article. A.C. and X.G. contributed in the tests and in the text of the article.

Conflicts of Interest

The authors declare no conflict of interests.

References

1. Tomasicchio, G.R.; Armenio, E.; Alessandro, F.D.; Fonseca, N.; Mavrakos, S.A.; Penchev, V.; Schüttrumpf, H.; Voutsinas, S.; Kirkegaard, J.; Jensen, P.M. Design of a 3D physical and numerical experiment on floating off-shore wind turbines. In Proceedings of the 32th International Conference on Coastal Engineering, Santander, Spain, 1–6 July 2012.
2. Krivtsov, V.; Linfoot, B. Basin Testing of Wave Energy Converters in Trondheim: Investigation of Mooring Loads and Implications for Wider Research. *J. Mar. Sci. Eng.* **2014**, *2*, 326–335.
3. Amate, J.; Víctor, L.; Martín, D.D.; García, L.; Pablo, M.; Alonso, G. Iberdrola Ingeniería TLPWIND A smart way to drive costs down. In Proceedings of the EWEA 2014 Scientific, Fira de Barcelona Gran Via, Spain, 10–13 March 2014.

4.   Kraskowski, M.; Zawadzki, K.; Rylke, A. A Method for Computational and Experimental Analysis of the Moored Wind Turbine Seakeeping. In Proceedings of the 18th Australasian Fluid Mechanics Conference, Launceston, Australia, 3–7 December 2012.

5.   Damiani, L.; Musci, E.; Tomascchio, G.R.; D'Alessandro, F. Spar buoy numerical model calibration and verification. In Proceedings of the VI International Conference on Computational Methods in Marine Engineerint Marine, Rome, Italy, 15–17 June 2015; pp. 814–825.

6.   Harnois, V.; Weller, S.D.; Johanning, L.; Thies, P.R.; le Boulluec, M.; le Roux, D.; Soulé, V.; Ohana, J. Numerical model validation for mooring systems: Method and application for wave energy converters. *Renew. Energy* **2015**, *75*, 869–887.

7.   Fan, T.; Qiao, D.; Ou, J. Optimized Design of Equivalent Truncated Mooring System Based on Similarity of Static and Damping Characteristics Governing Equation of Mooring Line. In Proceedings of the Twenty-Second (2012) International Offshore and Polar Engineering Conference, Rodos, Greece, 17–23 June 2012; Volume 4, pp. 959–966.

8.   Qiao, D.; Ou, J. Truncated model tests for mooring lines of a semi-submersible platform and its equivalent compensated method. *J. Mar. Sci. Technol.* **2014**, *22*, 125–136.

9.   Stansberg, C.T.; Oritsland, O.; Ormberg, H. Challenges in Deep Water Experiments: Hybrid Approach. In Proceedings of the 20th International Conference on Offshore Mechanics and Arctic Engineering, OMAE2001/OFT-1352, Rio de Janeiro, Brazil, 3–8 June 2001; pp. 1–9.

10. Zhang, H.; Huang, S.; Guan, W. Optimal design of equivalent water depth truncated mooring system based on baton pattern simulated annealing algorithm. *China Ocean Eng.* **2014**, *28*, 67–80.

11. Stansberg, C.T.; Karlsen, S.I.; Ward, E.G. OTC 16587 Model Testing for Ultradeep Waters. In Proceedings of the Offshore Technology Conference, Houston, TX, USA, 3–6 May 2004; pp. 1–9.

12. Argyros, A.; Langley, R.S.; Ahilan, R.V. Simplifying Mooring Analysis for Deepwater Systems using Truncation. In Proceedings of the Twenty-first (2011) International Offshore and Polar Engineering Conference, Maui, HI, USA, 19–24 June 2011; Volume 8, pp. 195–202.

13. Molins, C.; Rebollo, J.; Campos, A. Estructura Flotante de Hormigón Prefabricado Para Soporte de Aerogeneradores. WO 2013/093160 A1, 27 June 2013.

14. Molins, C.; Campos, A.; Sandner, F.; Matha, D. Monolithic concrete off-shore floating structure for wind turbines. In Proceedings of the EWEA 2014 Scientific, Fira de Barcelona Gran Via, Spain, 10–13 March 2014; pp. 107–111.

15. Matha, D.; Sandner, F.; Molins, C.; Campos, A.; Cheng, P.W. Efficient preliminary floating offshore wind turbine design and testing methodologies and application to a concrete spar design. Philosophical Transactions of the Royal Society of London A: Mathematical, Physical and Engineering Sciences **2015**, *373*, doi:10.1098/rsta.2014.0350.

16. Campos, A.; Molins, C.; Gironella, X.; Trubat, P.; Alarcón, D. Experimental rao's analysis of a monolithic concrete spar structure for offshore floating wind turbines. In Proceedings of the 34th International Conference on Ocean, Offshore and Arctic Engineering, OMAE 2015, St. John's, NL, Canada, 31 May–5 June 2015; pp. 1–9.

17. Mathworks. *Global Optimization Toolbox User ' s Guide R 2015 b*; Mathworks: Natick, MA, USA, 2015.

18. Saxén, A.; Bernander, K.B. *Parallel Global Optimization ABB's Metal Process Models Using Matlab*; Project report; Uppsala University: Uppsala, Sweden, January 2014.

19. Campos, A.; Molins, C.; Gironella, X.; Alarcón, D.; Trubat, P. Experiments on a scale model of a monolithic concrete spar for floating wind turbines. In Proceedings of the EWEA Ofsshore 2015 Copenhagen, Copenhagen, Denmark, 10–12 March 2015; p. 10.

6

Human Genotoxic Study Carried Out Two Years after Oil Exposure during the Clean-up Activities Using Two Different Biomarkers

Gloria Biern [1], Jesús Giraldo [2], Jan-Paul Zock [3,4,5], Gemma Monyarch [1], Ana Espinosa [3,4,5], Gema Rodríguez-Trigo [6,7,8], Federico Gómez [8,9], Francisco Pozo-Rodríguez [8,10], Joan-Albert Barberà [8,9] and Carme Fuster [1,*]

[1] Unitat de Biologia Cel·lular i Genètica Mèdica, Facultat de Medicina, Universitat Autònoma de Barcelona (UAB), 08193-Bellaterra, Barcelona, Spain; E-Mails: gloria.biern@gmail.com (G.B.); taguca@gmail.com (G.M.)

[2] Unitat de Bioestadística i Institut de Neurociències, Facultat de Medicina, Universitat Autònoma de Barcelona (UAB), 08193-Bellaterra, Barcelona, Spain; E-Mail: jesus.giraldo@uab.es

[3] Centre de Recerca en Epidemiologia Ambiental (CREAL), 08003-Barcelona, Spain; E-Mails: jpzock@creal.cat (J.-P.Z.); aespinosa@creal.cat (A.E.)

[4] Universitat Pompeu Fabra, 08002-Barcelona, Spain

[5] CIBER Epidemiología y Salud Pública (CIBERESP), 28029-Madrid, Spain

[6] Servicio de Neumología, Hospital Clínico San Carlos, 28040-Madrid, Spain; E-Mail: grodriguezt@salud.madrid.org

[7] Facultad de Medicina, Universidad Complutense, 28040-Madrid, Spain

[8] CIBER Enfermedades Respiratorias (CIBERES), Bunyola, 07004-Mallorca, Spain; E-Mails: fpgomez@clinic.ub.es (F.G.); fpozo@h12o.es (F.P.-R.); jbarbera@clinic.ub.es (J.-A.B.)

[9] Departamento de Medicina Respiratòria, Hospital Clínic-Institut d'Investigacions Biomèdiques August Pi I Sunyer (IDIBAPS), 08036-Barcelona, Spain

[10] Departamento de Medicina Respiratoria, Unidad Epidemiologia Clínica, Hospital 12 de Octubre, 28047-Madrid, Spain

* Author to whom correspondence should be addressed; E-Mail: carme.fuster@uab.es

Academic Editor: Merv Fingas

Abstract: Micronuclei, comet and chromosome alterations assays are the most widely used biomarkers for determining the genotoxic damage in a population exposed to genotoxic chemicals. While chromosome alterations are an excellent biomarker to detect short- and long-term genotoxic effects, the comet assay only measures early biological effects, and furthermore it is unknown whether nuclear abnormalies, such as those measured in the micronucleus test, remain detectable long-term after an acute exposure. In our previous study, an increase in structural chromosome alterations in fishermen involved in the clean-up of the *Prestige* oil spill, two years after acute exposure, was detected. The aim of this study is to investigate whether, in lymphocytes from peripheral blood, the nuclear abnormalies (micronucleus, nucleoplasmic bridges and nuclear buds) have a similar sensitivity to the chromosome damage analysis for genotoxic detection two years after oil exposure in the same non-smoker individuals and in the same peripheral blood extraction. No significant differences in nuclear abnormalies frequencies between exposed and non-exposed individuals were found ($p > 0.05$). However, chromosome damage, in the same individuals, was higher in exposed *vs.* non-exposed individuals, especially for chromosome lesions ($p < 0.05$). These findings, despite the small sample size, suggest that nuclear abnormalities are probably less-successful biomarkers than are chromosome alterations to evaluate genotoxic effects two or more years after an exposure to oil. Due to the great advantage of micronucleus automatic determination, which allows for a rapid study of hundreds of individuals exposed to genotoxic chemical exposure, further studies are needed to confirm whether this assay is or is not useful in long-term genotoxic studies after the toxic agent is no longer present.

Keywords: micronucleus test; chromosome damage; nuclear abnormities; chromosome alterations; oil exposure; genotoxicity; *Prestige* catastrophe

---

1. Introduction

Significant marine oil spills, approximately namely 14 accidents involving large oil tankers, have occurred in regions with a high population density in the last five decades [1,2]. When a sizeable spill occurs, there are usually a large number of individuals, in general local inhabitants, who collaborate in clean-up tasks to minimize the negative ecological and economic impact. So, for example, more than 300,000 people were involved in the clean-up activities after the wreck of the oil tanker *Prestige*, in November 2002. Although there are relatively few studies which have focused on the repercussions of acute oil exposure for human health, there is growing concern about the chemical exposure that clean-up activities involve and their potential health effects. Direct contact with oil or its vapors can cause skin rash and eye redness, and prolonged exposure can cause nausea, dizziness, headache, respiratory problems and psychiatric disorders [1,2]. Moreover, due to certain volatile organic oil compounds, in particular benzene, being carcinogenic in humans [3], it is very important to determine if exposure to oil during clean-up tasks is associated with genotoxic effects in the short- (less than 12 months) and long-term (more than one year). So far, only a few human genotoxic studies in oil exposed populations have been published, most after the wreck of the *Prestige* [4–14]. In general, these studies revealed increased genomic damage in exposed individuals during the clean-up tasks [4–11]. Nevertheless, only two research groups have carried out long-term genotoxic studies after oil exposure to the *Prestige* [12–15], with conflicting results. In one group, the authors described an increase of structural chromosome alterations in highly exposed *vs.* non-exposed individuals two and six years after exposure [12–14], this follow-up study reveal that chromosome damage persisted at least for the six years. Yet in another group, the study detected no genotoxic effects to be present seven years after exposure using other biomarkers (comet, micronucleus and T-cell receptor mutation assays) [15]. With the exception of T-cell receptor mutation assays, the sensibility of the two other biomarkers to detect long-term genotoxic effects has not been tested.

The micronucleus test, comet assay and chromosome alterations have been the most common biomarkers to determine genetic damage in any population exposed occupationally or environmentally to genotoxic chemicals, e.g., oil exposure during clean-up tasks [4–18]. A micronucleus is the result of chromosome breakage (acentric fragment) and/or loss (whole chromosome) caused by errors in DNA repair or in chromosome segregations not included in the main nucleus that are surrounded individually by the nuclear membrane [19]. The micronucleus test is performed by cytokinesis-block assays using cytochalasin B, which allows to be analyzed other nuclear abnormalies, such as nucleoplasmic bridges and nuclear buds in binuclear cells. The nucleoplasmic bridges indicate the occurrence of reorganizations in which chromatids or chromosomes are pulled to opposite poles during anaphase, resulting in dicentric or ring chromosomes. The nuclear buds are characterized by having the same morphology as a micronucleus, but they remain connected to the main nucleus and represent the process of elimination of amplified DNA or of the DNA repair complex and possibly excess chromosomes from aneuploid cells. Recently, it has been described that nucleoplasmic bridges and nuclear buds are also useful biomarkers for monitoring genetic damage by detecting and quantifying DNA damage and chromosome instability [20–23]. The comet biomarker is based on how a genotoxic agent will produce DNA-strand breaks and measures the extent of DNA migration in electrophoresis [24] and has been frequently used because it is a fast and easy method to assess DNA breaks with excellent sensitivity. Finally, chromosome alterations are any change in the normal structure or number of chromosomes. In general, their analyses for genotoxic studies include lesions (gaps and breaks of one or both chromatids) and structural alterations (such as acentric fragments, deletions, translocations, dicentrics, rings, marker chromosomes) resulting in direct DNA breakage, errors in synthesis or repair of DNA, and have been widely used biomarkers since the 1970s [16,23,24]. Although comet and nuclear anomaly assays are probably less resolute and less informative than is metaphasic chromosome analysis, in the last several years both tests have been used extensively in studies to evaluate genotoxic damage in large populations of exposed individuals because they are much easier and faster [19].

The evaluation of chromosome alterations requires cell cultures, while the evaluation of nuclear abnormalities requires cells in division but comet assay can be performed without the use of proliferative cells. For a long time, peripheral lymphocyte cultures have been the most widely employed in human genotoxic studies, however the introduction of nuclear abnormalities and comet assays as biomarkers allows for the use of alternative cell types, such as epithelial cells [25,26]. Epithelial cells can be used as early-effect biomarkers; nasal epithelial cells are replaced approximately once every 30 days and buccal epithelial cells one every 10–14 days [26] *vs.* peripheral blood lymphocytes, which serve as long-effect biomarkers and are renewed around every 4 to 6 years [27]. The great advantage of using, for example, buccal epithelial cells *vs.* lymphocytes is the easy and minimally invasive collection of samples, but the most important disadvantages are the discrepancies which come from using blood cells.

In contrast to the body of research regarding the use of different biomarkers to determine the genotoxic effect when the agent is present, there is scarce information to determine long-term effects after an acute exposure, with chromosome damage being the biomarker most frequently used. Given that the comet test indicates early biological effects [28,29]; it is probably not an ideal biomarker for long-term studies after acute genotoxic exposure. Although the usefulness of nuclear abnormalies, especially the micronucleus test, for short-term genotoxic studies is unquestionable, its sensibility for long-term studies, when the toxic agent is missing, has not been demonstrated yet. The main objective of this study is to determine whether nuclear abnormalies remain useful biomarkers for detecting genotoxic effects two years after Prestige oil exposure, comparing their results with those detected by chromosome alterations analyses.

2. Material and Methods

2.1. Study Population

The present study was performed on randomly selected subsamples of individuals included in a previous genotoxic study [12,13]. It was conducted using peripheral blood lymphocytes from individuals who had participated in clean-up activities of the *Prestige* oil spill. Only fishermen were included in our study in order to minimize other occupational sources of exposure that could act as confounders. A questionnaire survey including information about participation in clean-up tasks, health problems, lifestyle, history of cancer, medication, smoking status, fertility, age, and gender among 6780 fishermen one year after exposure was performed [30]. The selection criteria of individuals highly exposed and non-exposed to the oil was established from this information, described previously [30]. In brief, exposed local fishermen who participated for at least 15 days in clean-up activities of the *Prestige* oil spill, for four or more hours per day, during the first two months (when exposure presumably was greatest) were included as highly-exposed subjects for the study. Non-exposed individuals were selected from fishermen who had not participated in clean-up tasks for reasons other than those related to health. All exposed and non-exposed individuals were non-smokers (current smokers and ex-smokers were excluded), fertile and without a history of cancer, A new questionnaire and face-to-face interview, in order to verify the answers, was performed in a mobile unit that traveled to the participants' villages on the same day in which the samples were obtained two years after the spill. In the present work, a total of 20 exposed and 20 non-exposed individuals were studied, randomly selected from 91 exposed and 46 non-exposed individuals in which chromosome damage was analyzed [12,13]. Figure 1 shows the flow diagram of the study. The exposed group consisted of 9 men and 11 women with an average age of 48.2 years (ranging from 32.2 to 62.2; SEM = 1.9). The non-exposed group consisted of 3 men and 17 women with an average age of 53.1 years (ranging from 36.6 to 58.8; SEM = 1.3). No significant relationship between sex and group was found according to Fisher's exact test ($p = 0.0824$). The difference in age was found to be statistically significant both by Student's t-test and Wilcoxon sum of ranks. In principle, as a higher age could be associated with a higher propensity to present genotoxic abnormalities, the distribution of age in the sample could make it more difficult to statistically prove the association between abnormalities and oil exposure. Thus, because there are more older individuals included in the non-exposed group, age distribution should not favor the hypothesis of finding abnormalities in the exposed group. The collection, transport and processing of the samples were performed between 22 and 27 months after the *Prestige* disaster.

The project was approved by the Ethics Committee on Clinical Research of Galicia, and all participants provided written, informed consent.

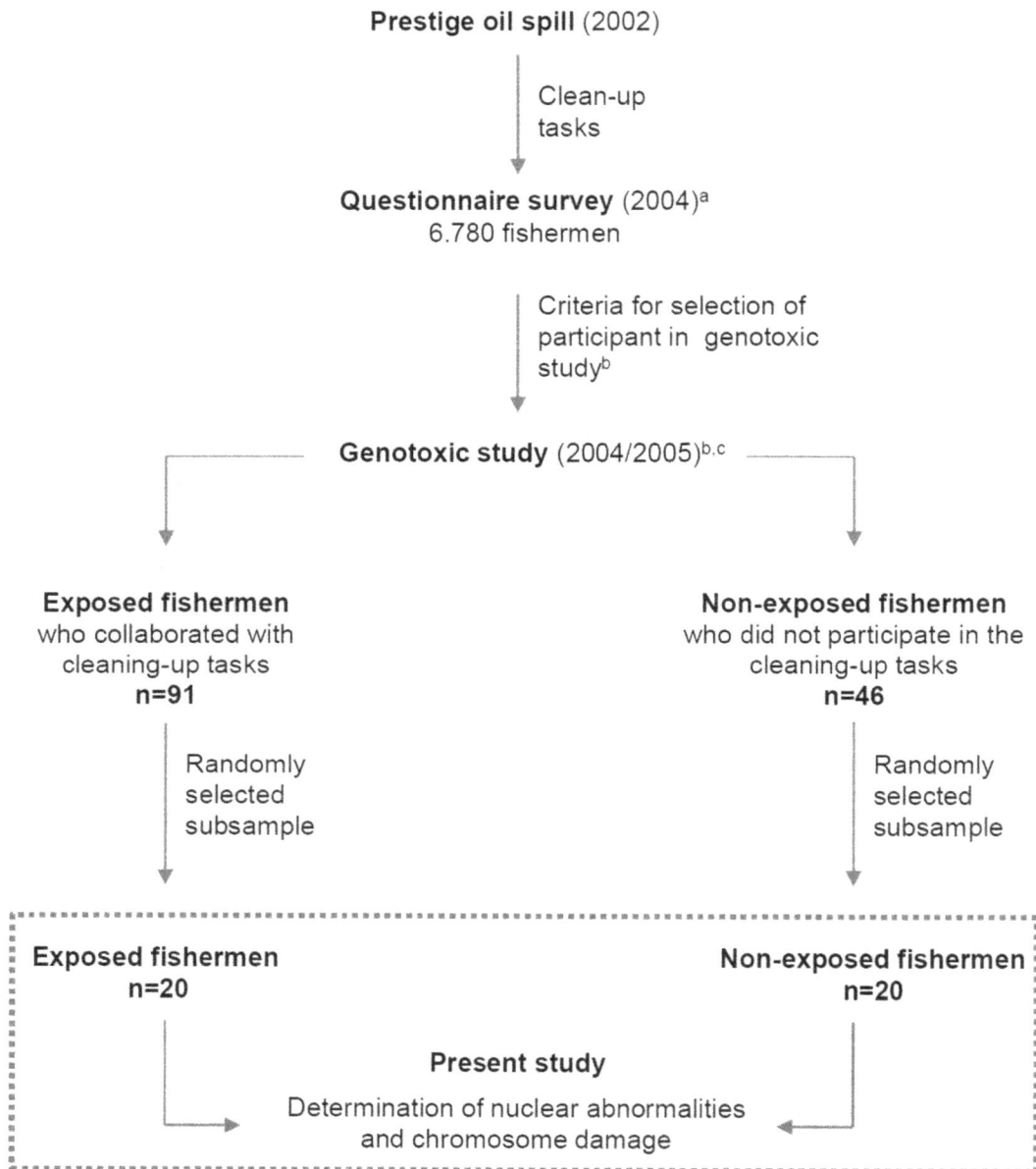

Figure 1. Flow diagram of the study. [a] Detailed description in Zock *et al.* [30]; [b] Detailed description in Rodriguez-Trigo *et al.* [12]; and [c] Detailed description in Monyarch *et al.* [13].

2.2. Cytogenetic Analysis

Peripheral blood was obtained in same extraction and later cultured at 37 °C in supplemented RPMI-1640 medium (GIBCO Invitrogen Cell Culture, Invitrogen; Carlsbad, CA, USA) according to standard procedures.

For the cytokinesis-block nuclear abnormalies test, peripheral blood was cultured, in duplicate, for 44 h and then cytochalasin-B was added to a final concentration of 6 µg/mL. Cells were harvested by centrifugation after 72 h of culture and submitted to middle-hypotonic treatment with 0.075 mM KCl at 4 °C. Cells were fixed in Carnoy (methanol-acetic acid 3:1 v/v), placed on dry slides, and stained with Leishman according to standard procedures. The micronucleus, nucleoplasmic bridges and nuclear buds in binucleated cells were identified according to the criteria of the HUMN project [31] and were evaluated by scoring 1000 binucleated cells (500 from each culture) using an Olympus Bx60 microscope. The cytogenetic-block proliferation index was calculated by the relation between total of cells with 1, 2, 3 and 4 micronuclei vs. total of cells analyzed.

For chromosome breakage analyses involved in lesions and structural chromosome alterations, analyzed in published studies [12,13], peripheral blood was cultured, in duplicate, for 72 h and then harvested according to standard procedures. For chromosome lesions, a minimum at 100 metaphases were analyzed in each individual (50 from each culture). For structural chromosome alterations, at least 25 banded metaphases were karyotyped in each participant. Criteria for cytogenetic evaluations were determined according to the International System for Human Cytogenetic Nomenclature [32].

2.3. Statistical Analysis

A generalized estimating equation, GEE [12,13,33,34], was used for assessing the differences between the exposed and non-exposed groups for the micronucleus, nucleoplasmic bridges and nuclear buds, chromosome lesions and structural chromosome alterations. The GEE approach is an extension of generalized linear models designed to account for repeated, within-individual measurements. This method is particularly indicated for when the normality assumption is not reasonable, as happens, for instance, with discrete data. The GEE model was used instead of the classic Fisher exact test because the former takes into account the possible within-individual correlation, whereas the latter assumes that all observations are independent. Since several metaphases were analyzed per individual, the GEE model is more appropriate. Statistical significance was set at $p < 0.05$. Statistical analyses were carried out with SAS/STAT release 9.02 (SAS Institute Inc.; Cary, NC, USA). The GEE model was fitted using the REPEATED statement in the GENMOD procedure. The conservative Type 3 statistics score was used for the analysis of the effects in the model. Sex and age were found not to be statistically significant when included in the GEE model and therefore were removed from the analysis.

3. Results

A satisfactory cell growth in all cultures was observed. A total of 40,000 binucleate cells, 4260 metaphases and 1100 karyotypes were analyzed in lymphocytes from exposed and non-exposed individuals respectively. All individuals had normal karyotypes (46,XX or 46,XY), except two cases, one with a polymorphic inversion of chromosome 9, inv(9)(p11q12), in an exposed individual (E14) and another case with an increased length of the heterochromatin on the long arm of the Y chromosome, Yqh+, in a non-exposed individual (NE5).

Cell growth in cytochalasin-B cultures (from exposed and non-exposed individuals) showed a cytogenetic-block proliferation index ranging between 30% and 60%. No significant statistical differences were found in the micronucleus or nuclear buds between exposed and non-exposed individuals ($p = 0.4774$ and $p = 0.2356$, respectively), and nucleoplasmic bridges were marginally influential ($p = 0.08$).

Chromosome lesions were higher in exposed rather than in non-exposed individuals ($p = 0.0231$), but structural chromosome alterations were only marginally ($p = 0.0972$). Marker chromosomes, unbalanced translocations and deletions were the structural chromosome alterations most frequently observed in both groups of individuals, and ring chromosomes and acentric fragments were only detected in exposed individuals. Numerical chromosome alterations (such as trisomies and monosomies) were excluded in these analyses because they can be attributed to the failure of the chromosome spread due to non-specific techniques having been applied to remaining cell membranes.

Table 1 and Figure 2 show the nuclear abnormalies (micronucleus, nucleoplasmic bridges and nuclear buds) and chromosome damage (lesions and structural alterations) observed in the same individuals. Cytogenetic results for each individual are found in Table 2, showing high inter-individual genotoxic variability for all biomarkers analyzed (micronucleus, nucleoplasmic bridges, nuclear buds, chromosome lesions and structural chromosome alterations) in exposed and in non-exposed individuals. With the exception of the degree of oil exposure no other associations were found between genotoxic damage and the different factors analyzed in the present study.

Table 1. Nuclear abnormalities and chromosome damage detected in same individuals exposed and non-exposed to oil.

	Exposed	Non-Exposed	p-Value
Total Individuals, No.	20	20	
Total Binucleate Cells, No.	20.000	20.000	
Binucleated cells with micronucleus, No. (%)	457 (2.3)	514 (2.6)	0.4774
1 micronucleus/cell, No.	399	450	
2 micronucleus/cell, No.	49	53	
3 micronucleus/cell, No.	9	11	
Nucleplasmic bridges, No. (%)	131 (0.65)	98 (0.49)	0.08
Nuclear buds, No. (%)	106 (0.53)	68 (0.34)	0.2356
Total Metaphases Analyzed (Uniform Stain), No.	2112	2.148	
Chromosome lesion, No. (%)	28 (1.3)	7 (0.3)	0.0231
Total Metaphases Karyotyped (G-Banded), No.	537	563	
Structural chromosome alterations, No. (%)	36 (6.7)	16 (2.8)	0.0972
Balanced, No.	1	3	
Unbalanced, No.	35	13	

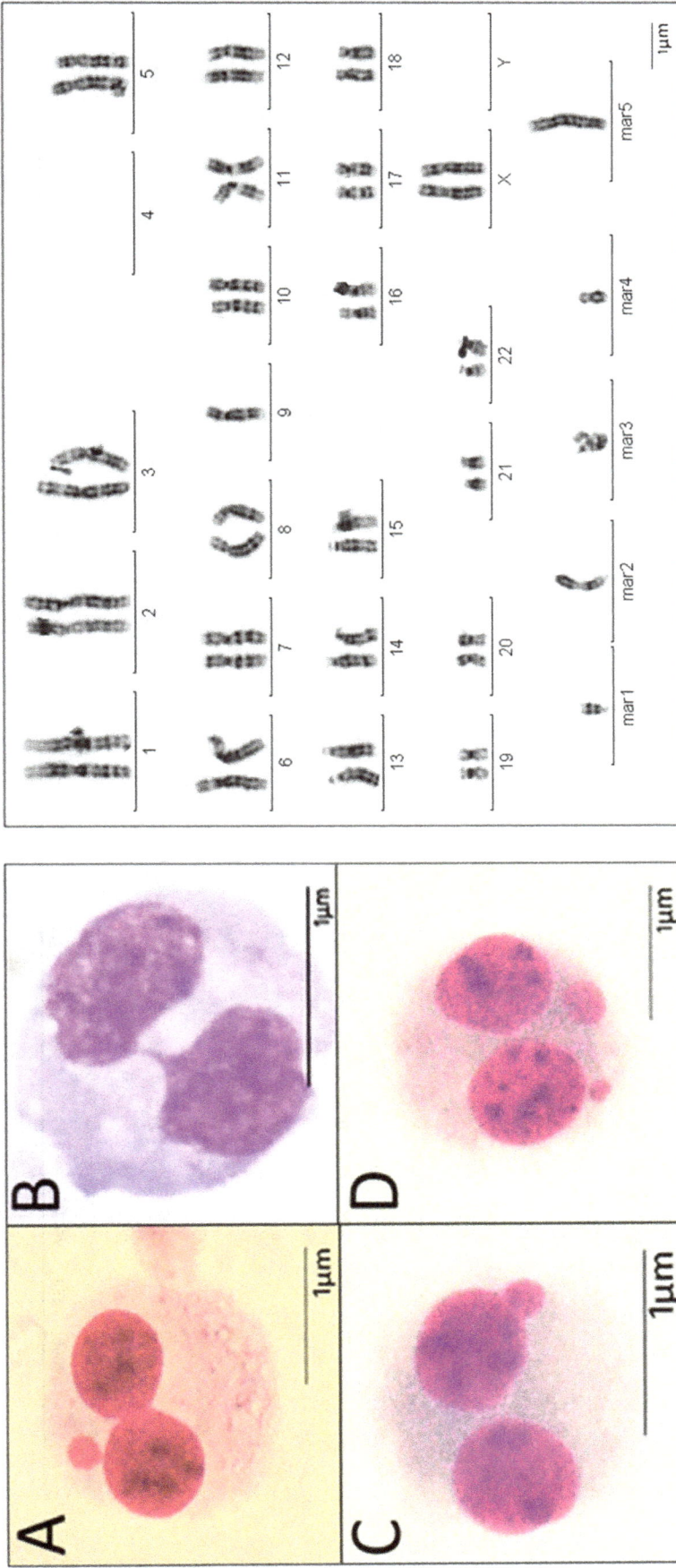

Figure 2. Genotoxic study using nuclear abnormalities and chromosome damage as biomarkers. Binuclear cells (left) showing: (**A**) one micronucleus, (**B**) nucleoplasmatic bridges, (**C**) nuclear buds, (**D**) two micronuclei; and Karyotype (right) with five marker chromosomes.

Table 2. Cytogenetic results in individuals exposed and non-exposed to oil.

Type of individuals	Sex	Age	Binucleate cells								Chromosomal Lesions			Structural Chromosomal Alteration			
			Cells	MN	MN 0	MN 1	MN 2	MN ≥3	NBUP	NPB	Cells	Lesions	Karyotypes	Structural Alterations	Balanced Structural Alterations	Unbalanced Structural Alterations	Type of Structural Alteration
Exposed																	
E1	woman	49,02	1000	29	971	27	2	0	4	7	100	1	27	5	0	5	t(13;20)(q32;q12); t(9;?), t(10;18),mar; del(2)(q12)
E2	man	54,01	1000	23	977	20	3	0	2	4	116	3	28	0	0	0	
E3	man	44,02	1000	21	979	19	2	0	5	3	105	6	29	1	0	1	ace
E4	man	31,26	1000	17	983	16	1	0	7	1	105	2	25	2	0	2	ace,ace
E5	woman	51,90	1000	25	975	22	3	0	11	8	103	0	26	0	0	0	
E6	man	35,92	1000	17	983	16	1	0	3	8	108	1	30	2	0	2	ace; add(12)(qter)
E7	woman	49,62	1000	43	957	37	4	2	9	5	100	2	25	6	0	6	ace; t(4;16)(q13,p13.3); mar,mar,mar,mar
E8	man	56,45	1000	43	957	37	4	2	19	6	102	1	29	0	0	0	
E9	woman	50,51	1000	11	989	9	2	0	16	14	100	6	26	1	0	1	ace
E10	man	52,50	1000	9	991	8	1	0	2	17	102	1	26	0	0	0	
E11	woman	54,86	1000	32	968	24	8	0	4	5	110	0	25	0	0	0	
E12	woman	44,89	1000	37	963	30	5	2	12	2	112	0	26	1	0	1	mar
E13	man	57,02	1000	18	982	18	0	0	9	3	100	1	31	1	1	0	t(13;11)(p25;q23)
E14	woman	46,13	1000	18	982	16	1	1	7	3	104	1	26	1	0	1	mar
E15	woman	62,17	1000	14	986	11	3	0	4	3	103	0	25	3	0	3	t(7;10); r, r
E16	woman	48,99	1000	10	990	8	2	0	1	1	112	0	25	1	0	1	t(X;4)(q21;p16)
E17	woman	58,37	1000	8	992	8	0	0	1	3	105	0	27	1	0	1	del(7)(q33)
E18	woman	36,90	1000	31	969	26	3	2	4	4	112	0	30	1	0	1	mar
E19	man	52,98	1000	38	962	34	4	0	5	6	106	2	26	1	0	1	del(7)(p15)
E20	man	54,94	1000	13	987	13	0	0	6	3	107	1	25	9	0	9	mar,mar; ace, ace; mar, mar,mar,mar,mar
			20000	457	19543	399	49	9	131	106	2112	28	537	36	1	35	

Table 2. *Cont.*

Non-Exposed																	
NE1	woman	54,33	1000	22	978	19	3	0	3	0	103	0	25		0	0	0
NE2	woman	50,60	1000	22	978	17	4	1	0	0	100	3	34	mar,mar,mar,mar	0	0	0
NE3	woman	57,52	1000	13	987	10	2	1	4	2	105	0	25	del(1)(q21)	4	0	4
NE4	woman	57,19	1000	14	986	13	1	0	16	0	101	1	25		1	0	1
NE5	man	58,78	1000	22	978	19	2	1	1	8	108	0	26	del(2)(q21); del(1)(q23), mar	3	0	3
NE6	woman	57,61	1000	59	941	52	6	1	2	6	100	0	25	t(13;14)(q14,q32)	1	1	0
NE7	woman	36,56	1000	5	995	3	2	0	5	7	107	0	25	del(1)(q32)	1	0	1
NE8	woman	53,22	1000	30	970	26	3	1	1	7	110	0	33	mar	1	1	1
NE9	woman	58,58	1000	50	950	43	6	1	5	1	113	0	43		0	0	0
NE10	woman	56,00	1000	26	974	22	1	3	9	1	107	0	26		0	0	0
NE11	woman	56,85	1000	10	990	10	0	0	3	4	120	1	25	mar	1	1	1
NE12	woman	46,36	1000	38	962	32	5	1	9	5	130	0	26		0	0	0
NE13	woman	55,82	1000	12	988	10	2	0	2	6	101	0	25		0	0	0
NE14	woman	45,84	1000	26	974	24	2	0	8	4	108	2	25		0	0	0
NE15	man	48,56	1000	41	959	33	7	1	8	3	107	0	25		0	0	0
NE16	woman	55,54	1000	23	977	23	0	0	5	2	102	0	26		0	0	0
NE17	woman	58,62	1000	36	964	32	4	0	8	4	112	0	26	del(9)(q21)	1	0	1
NE18	man	57,15	1000	38	962	37	1	0	6	3	105	0	43		0	0	0
NE19	woman	49,98	1000	18	982	16	2	0	3	1	108	0	26	t(8;13)((q24.1;q31); t(2;5)	2	0	1
NE20	woman	46,31	1000	9	991	9	0	0	0	4	101	0	29	t(3;8)(q27;q13)	1	0	0
			20000	514	19486	450	53	11	98	68	2148	7	563		16	3	13

Abbreviations: ace: acentric fragment; add: additional material of unknown origin; del: deletion; mar: marker chromosome; p: short arm; q: long arm; qter: terminal long arm; t: translocation; MN: micronuclei; NBUP: nucleoplasmic bridges; NPB: nuclear buds. Commas indicate the beginning of new metaphase.

4. Discussion

Toxic agents can induce complex changes in the genome, and to-date there is no single biomarker to detect all types of these alterations, probably due to different molecular mechanisms being involved [29,35,36]. It is therefore probable that not all genotoxic biomarkers are equally useful for long-term evaluation after exposure.

To date, very few long-term genotoxic studies after an accidental oil exposure have been previously published [12–15]. In all of these studies only individuals highly exposed to oil were included, yet the findings obtained were not coincident. While an increase of structural chromosome alterations in exposed individuals two and six years after exposure was observed [12–14], no genotoxic effects using other biomarkers (comet, micronucleus and T-cell receptor mutation assays) after seven years were detected [15]. The T-cell receptor mutation assay, used in Laffon's study [15], is an excellent biomarker for long-term studies because it provides information about the genotoxic effect which have occurred several months to several years after exposure [37]. However, the comet assay, also a successful biomarker employed by the authors, indicates early biological effects [28,29], so is probably not the most suitable test to evaluate long-term genotoxic effects. In relation to nuclear abnormalies, with the exception of Laffon *et al.* [15] and the present study, no other long-term genotoxic analyses have been performed, and in both studies no genotoxicity was detected two and seven years after oil exposure. Thus, we are not sure that this biomarker is still valid for long-term analysis.

The main differences in the above referred studies [12–15] were the individuals included in the study, the time following oil exposure (two, six and seven years), and the biomarkers used (mainly chromosome damage *vs.* micronuclei and nuclear abnormalies). In order to minimize the dispersion of these factors, we have analyzed these same biomarkers two years after oil exposure in the same individuals in which chromosome alterations were observed. Our results showed no differences in micronuclei and nuclear abnormalies between those exposed and non-exposed to oil. Thus, if we had only used these biomarkers, our findings would have suggested that the genotoxic effect has disappeared two years after oil exposure, long before the seven years as described Laffon *et al.* [15] However, the present study shows an increase of chromosome damage in the same exposed individuals, in which no differences for nuclear abnormalies were found, indicating that genotoxic damage does persist two years after acute oil exposure. Additionally, an increase of chromosome damage in a high number of exposed individuals was previously reported two and six years after oil exposure [12–14]. Despite the very strict selection criteria for exposed individuals and the small sample size analyzed, the present findings, supported by those reported previously [12–15], suggest that micronuclei and nuclear abnormalies are probably less-successful biomarkers than are chromosome damage to evaluate genotoxic effect more than two years after acute oil exposure when the toxic agent is no longer present. Moreover, our results indicate that chromosome damage is more informative than micronuclei and nuclear abnormalies because acentric fragments (which corresponds to the origin of the micronucleus) and ring chromosomes (corresponding to nucleoplasmic bridges) were detected in exposed individuals two years after oil exposure. It is relevant to note that smokers were excluded in present study because an association between smoking and chromosome damage has been described [33,38]. Finally, due to the fact that the micronucleus test and other nuclear abnormalies can be detected automatically *versus* chromosome analysis, and moreover this test is much easier, faster and allows for analysis of a large number of cells and does not require as much extensive personnel training, further studies are needed to confirm these preliminary results in larger samples.

5. Conclusions

To date, no information has been published regarding whether the micronucleus test remains suitable several years after the toxic agent is no longer present. For this reason, in the present study we evaluated the utility of nuclear abnormalies, including micronucleus test, to assess the genotoxic oil effect two years after the wreck of the *Prestige* comparing these results with those obtained from chromosome alterations analyses in the same non-smoker individuals and in the same peripheral blood extraction. Our results showed no differences in nuclear abnormalies between those exposed and non-exposed, however the chromosome damage was higher in exposed individuals. These features were compared with previous a report derived from long-term genotoxic studies after an accidental oil exposure. The main findings are:

- nuclear abnormalies (micronucleus, nucleoplasmic bridges and nuclear buds) in binucleated cells may not detect genotoxic effects more than two years after acute oil exposure when the toxic agent is no longer present;
- chromosome damage (chromosome lesion and structural chromosome alterations) in metaphases cells is a useful biomarker for assessing genotoxic effect two years after acute oil exposure using the same peripheral blood extraction in which nuclear abnormalies were analyzed; and
- comparative study using nuclear abnormalies and chromosome damage analyses emphasizes the need to use appropriate biomarker for detection of genotoxic effect in individuals involved in toxic accidents.

Despite the reduced number of individual analyzed, the present study suggests that micronuclei and nuclear abnormalies are probably less-successful biomarkers for the evaluation of long-term genotoxic oil effects when the toxic agent is no longer present. However, due to the fact that with the micronucleus test these and other nuclear abnormalies can be detected automatically, further studies are needed to confirm these preliminary results.

Acknowledgments

The kind participation of the fishermen's cooperatives and the efforts made by Antonio Devesa are gratefully acknowledged. The authors wish to thank Ana Souto Alonso, Marisa Rodríguez Valcárcel, Luisa Vázquez Rey, Emma Rodríguez (Complexo Hospitalario Universitario A Coruña), and Ana Utrabo, Angels Niubó (Universitat Autònoma de Barcelona). The investigators are greatly indebted to J. Ancochea and J.L. Alvarez-Sala, past presidents of SEPAR, for their initiative and support. For this study, grants were provided from the Health Institute Carlos III FEDER (PI03/1685), Sociedad Española de Neumología y Cirugía Torácica (SEPAR), Cowmissionat per a Universitats I Recerca from Generalitat de Catalunya (14SGR903), Centro de Investigación en Red de Enfermedades Respiratorias (CIBERES). The different sponsors were not involved in the design of the study, sample collection, cytogenetic analysis, or interpretation of the data or preparation/revision of the manuscript.

Author Contributions

Conception and design of the experiments: C.F. Analysis and interpretation of the data: G.B., G.M., C.F. Statistical analysis: J.G., A.E., J.P.Z. Composition and revision of the manuscript: J.G., J.P.Z., G.R.T., F.P.G., F.P.R., J.A.B. and C.F.

Conflicts of Interest

The authors declare no conflict of interest.

References

1. Aguilera, F.; Méndez, J.; Pásaro, E.; Laffon, B. Review of the effects of exposure to spilled oils on human health. *J. Appl. Toxicol.* **2010**, *30*, 291–301.
2. Goldstein, B.D.; Osofsky, H.J.; Lichtveld, M.Y. The Gulf oil spill. *N. Engl. J. Med.* **2011**, *364*, 1334–1348.
3. IARC. *Occupational Exposures in Petroleum Refining: Crude Oil and Major Petroleum Fuels IARC. Monographs on the Evaluations of Carcinogenic Risk to Humans*; International Agency for Research on Cancer: Lyon, France, 1989; Volume 45.
4. Clare, M.G.; Yardley-Jones, A.; Maclean, A.C.; Dean, B.J. Chromosome analysis from peripheral blood lymphocytes of workers after an acute exposure to benzene. *Br. J. Ind. Med.* **1984**, *41*, 249–253.
5. Cole, J.; Beare, D.M.; Waugh, A.P.; Capulas, E.; Aldridge, K.E.; Arlett, C.F.; Green, M.H.; Crum, J.E.; Cox, D.; Garner, R.C.; *et al.* Biomonitoring of possible human exposure to environmental genotoxic chemicals: Lessons from a study following the wreck of the oil tanker Braer. *Environ. Mol. Mutagen.* **1997**, *30*, 97–111.
6. Laffon, B.; Fraga-Iriso, R.; Perez-Cadahia, B.; Méndez, J. Genotoxicity associated to exposure to Prestige oil during autopsies and cleaning of oil-contaminated birds. *Food Chem. Toxicol.* **2006**, *44*, 1714–1723.
7. Perez-Cadahia, B.; Laffon, B.; Pasaro, E.; Méndez, J. Genetic damage induced by accidental environmental pollutants. *Sci. World J.* **2006**, *6*, 1221–1237.
8. Pérez-Cadahía, B.; Lafuente, A.; Cabaleiro, T.; Pasaro, E.; Méndez, J.; Laffon, B. Initial study on the effects of Prestige oil on human health. *Environ. Int.* **2007**, *33*, 176–185.
9. Pérez-Cadahía, B.; Laffon, B.; Porta, M. Relationship between blood concentrations of heavy metals and cytogenetic and endocrine parameters among subjects involved in cleaning coastal areas affected by the Prestige tanker oil spill. *Chemosphere* **2008**, *7*, 447–455.
10. Pérez-Cadahía, B.; Laffon, B.; Valdiglesias, V.; Pasaro, E.; Mendez, J. Cytogenetic effects induced by Prestige oil on human populations: The role of polymorphisms in genes involved in metabolism and DNA repair. *Mutat. Res.* **2008**, *653*, 117–123.
11. Perez-Cadahia, B.; Mendez, J.; Pasaro, E.; Lafuente, A.; Cabaleiro, T.; Laffon, B. Biomonitoring of human exposure to Prestige oil: Effects on DNA and endocrine parameters. *Environ. Health Insights* **2008**, *2*, 83–92.

12. Rodríguez-Trigo, G.; Zock, J.P.; Pozo-Rodríguez, F.; Gómez, F.P.; Monyarch, G.; Bouso, L.; Coll, M.D.; Verea, H.; Antó, J.M.; Fuster, C.; *et al.* SEPAR-Prestige Study Group. Health changes in fishermen 2 years after clean-up of the Prestige oil spill. *Ann. Intern. Med.* **2010**, *153*, 489–498.

13. Monyarch, G.; de Castro-Reis, F.; Zock, J.P.; Giraldo, J.; Pozo-Rodríguez, F.; Espinosa, A.; Rodríguez-Trigo, G.; Verea, H.; Castaño-Vinyals, G.; Gómez, F.P.; *et al.* Chromosomal bands affected by acute oil exposure and DNA repair errors. *PLoS ONE* **2013**, *8*, e81276.

14. Hildur, K.; Templado, C.; ZocK, J.P.;, Giraldo, J.; Pozo-Rodríguez, F.; Frances, A.; Monyarch, G.; Rodríguez-Trigo, G.; Rodriguez-Rodriguez, E.; Souto, A.; *et al.* Follow-up genotoxic study: Chromosome damage two and six years after exposure to the *Prestige* oil spill. *PLoS ONE* **2015**, *10*, doi:10.1371/journal.pone.0132413.

15. Laffon, B.; Aguilera, F.; Ríos-Vázquez, J.; Valdiglesias, V.; Pásaro, E. Follow-up study of genotoxic effects in individuals exposed to oil from the tanker Prestige, seven years after the accident. *Mutat. Res.* **2014**, *760*, 10–16.

16. Mateuca, R.; Lombaert, N.; Aka, P.V.; Decordier, I.; Kirsch-Volders, M. Chromosomal changes: Induction, detection methods and applicability in human biomonitoring. *Biochimie* **2006**, *88*, 1515–1531.

17. Valverde, M.; Rojas, E. Environmental and occupational biomonitoring using the Comet assay. *Mutat. Res.* **2009**, *681*, 93–109.

18. DeMarini, D.M. Genotoxicity biomarkers associated with exposure to traffic and near-road atmospheres: A review. *Mutagenesis* **2013**, *28*, 485–505.

19. Fenech, M.; Kirsch-Volders, M.; Rossnerova, A.; Sram, R.; Romm, H.; Bolognesi, C.; Ramakumar, A.; Soussaline, F.; Schunck, C.; Elhajouji, A.; *et al.* HUMN project initiative and review of validation, quality control and prospects for further development of automated micronucleus assays using image cytometry systems. *Int. J. Hyg. Environ. Health* **2013**, *216*, 541–552.

20. Norppa, H.; Bonassi, S.; Hansteen, I.L.; Hagmar, L.; Strömberg, U.; Rössner, P.; Boffetta, P.; Lindholm, C.; Gundy, S.; Lazutka, J.; *et al.* Chromosomal aberrations and SCEs as biomarkers of cancer risk. *Mutat. Res.* **2006**, *600*, 37–45.

21. Bonassi, S.; Norppa, H.; Ceppi, M.; Strömberg, U.; Vermeulen, R.; Znaor, A.; Cebulska-Wasilewska, A.; Fabianova, E.; Fucic, A.; Gundy, S.; *et al.* Chromosomal aberration frequency in lymphocytes predicts the risk of cancer: Results from a pooled cohort study of 22,358 subjects in 11 countries. *Carcinogenesis* **2008**, *29*, 1178–1183.

22. Fenech, M.; Kirsch-Volders, M.; Natarajan, A.T.; Surralles, J.; Crott, J.W.; Parry, J.; Norppa, H.; Eastmond, D.A.; Tucker, J.D.; Thomas, P. Molecular mechanisms of micronucleus, nucleoplasmic bridge and nuclear bud formation in mammalian and human cells. *Mutagenesis* 2011, *26*, 125–132.

23. Kirsch-Volders, M.; Bonassi, S.; Knasmueller, S.; Holland, N.; Bolognesi, C.; Fenech; M.F. Commentary: Critical questions, misconceptions and a road map for improving the use of the lymphocyte cytokinesis-block micronucleus assay for *in vivo* biomonitoring of human exposure to genotoxic chemicals-A HUMN project perspective. *Mutat. Res.* 2014, *759*, 49–58.

24. Azqueta, A.; Collins, A.R. The essential comet assay: A comprehensive guide to measuring DNA damage and repair. *Arch. Toxicol.* **2013**, *87*, 949–968.

25. Torres-Bugarín, O.; Zavala-Cerna, M.G.; Nava, A.; Flores-García, A.; Ramos-Ibarra, M.L. Potential Uses, Limitations, and Basic Procedures of Micronuclei and Nuclear Abnormalities in Buccal Cells. *Dis. Markers* **2014**, *2014*, doi:10.1155/2014/956835.

26. Rojas, E.; Lorenzo,Y.; Haug, K.; Nicolaissen, B.; Valverde, M. Epithelial cells as alternative human biomatrices for comet assay. *Front. Genet.* **2014**, *5*, doi:10.3389/fgene.2014.00386.

27. Sprent, J.; Tough, D.G. Turnover of native and memry phenotype T cells. *J. Exp. Med.* **1994**, *179*, 1127–1135.

28. Dusinska, M.; Collins, A.R. The comet assay in human biomonitoring: Gene-environment interactions. *Mutagenesis* **2008**, *23*, 191–205.

29. Anderson, D.; Dhawan, A.; Laubenthal, J. The comet assay in human biomonitoring. *Methods Mol. Biol.* **2013**, *1044*, 347–362.

30. Zock, J.P.; Rodriguez-Trigo, G.; Pozo-Rodriguez, F.; Barberà, J.A.; Bouso, L.; Torralba, Y.; Antó, J.M.; Gómez, F.P.; Fuster, C.; Verea, H.; *et al.* Prolonged respiratory symptoms in clean-up workers of the *Prestige* oil spill. *Am. J. Respir. Crit. Care Med.* **2007**, *176*, 610–616.

31. Fenech, M. Cytokinesis-block micronucleus cytome assay. *Nat. Protoc.* **2007**, *2*, 1084–1104.

32. ISCN. *An International System for Human Cytogenetic Nomenclature*; Shafer, L.G., McGowan-Jordan, J., Schmid, M., Eds.; S. Karger: Basel, Switzerland, 2013.

33. Liang, K.Y.; Zeger, S.L. Longitudinal data analysis using generalized linear models. *Biometrika* **1986**, *73*, 13–22.

34. De la Chica, R.A; Ribas, I.; Giraldo, J.; Egozcue, J.; Fuster, C. Chromosomal instability in amniocytes from fetuses of mothers who smoke. *JAMA* **2005**, *293*, 1212–1222.

35. Thompson, S.L.; Compton, D.A. Chromosomes and cancer cells. *Chromosome Res.* **2011**, *19*, 433–444.

36. Luzhna, L.; Kathiria, P.; Kovalchuk, O. Micronuclei in genotoxicity assessment: From genetics to epigenetics and beyond. *Front. Genet.* **2013**, *4*, 131.

37. Ishioka, N.; Umeki, S.; Hirai, Y.; Akiyama, M.; Kodama, T.; Ohama, K.; Kyoizumi, S. Stimulated rapid expression *in vitro* for early detection of *in vivo* T-cell receptor mutations induced by radiation exposure. *Mutat. Res.* **1997**, *390*, 269–282.

38. Littlefield, L.G.; Joiner, E.E. Analysis of chromosome aberrations in lymphocytes of long-term heavy smokers. *Mutat. Res.* **1986**, *170*, 145–150.

Effect of Vegetation on the Late Miocene Ocean Circulation

Gerrit Lohmann [1,2,*], **Martin Butzin** [1,2] **and Torsten Bickert** [2]

[1] Alfred-Wegener-Institut Helmholtz Zentrum für Polar- und Meeresforschung, Bussestr. 24, 27570 Bremerhaven, Germany; E-Mail: Martin.Butzin@awi.de
[2] MARUM, Center for Marine Environmental Sciences, University of Bremen, P.O. Box 330440, 28334 Bremen, Germany; E-Mail: tbickert@marum.de

* Author to whom correspondence should be addressed; E-Mail: Gerrit.Lohmann@awi.de

Academic Editor: Nathalie Fagel

Abstract: We examine the role of the vegetation cover and the associated hydrological cycle on the deep ocean circulation during the Late Miocene (~10 million years ago). In our simulations, an open Central American gateway and exchange with fresh Pacific waters leads to a weak and shallow thermohaline circulation in the North Atlantic Ocean which is consistent with most other modeling studies for this time period. Here, we estimate the effect of a changed vegetation cover on the ocean general circulation using atmospheric circulation model simulations for the late Miocene climate with 353 ppmv CO_2 level. The Late Miocene land surface cover reduces the albedo, the net evaporation in the North Atlantic catchment is affected and the North Atlantic water becomes more saline leading to a more vigorous North Atlantic Deep Water circulation. These effects reveal potentially important feedbacks between the ocean circulation, the hydrological cycle and the land surface cover for Cenozoic climate evolution.

Keywords: ocean circulation; Late Miocene; hydrological cycle; Central American gateway

1. Introduction

The Eocene-Oligocene and the Mid-Miocene climate transitions are two major cooling steps in the Cenozoic climate evolution (Zachos *et al.*, 2001, [1]) from greenhouse to "icehouse" climate conditions. Ocean circulation changes and atmospheric pCO_2 variations are often cited as potential catalysts of these cooling events (DeConto and Pollard, 2003, [2]). Tectonic reorganizations of gateways may have altered the large-scale ocean circulation, which in turn may have resulted in ice growth and global cooling (Kennett, 1977, [3]; Zachos *et al.*, 2001, [1]). Carbon-13 proxy evidence (e.g., Wright and Miller, 1996, [4]; Billups, 2002, [5]) indicates pronounced ocean circulation changes in conjunction with the timing of tectonic events at critical ocean pathways like the Drake Passage, the Tasmanian Seaway, the Indonesian Seaway (Cane and Molnar, 2001, [6]; Lawver and Gahagan, 2003, [7]), the eastern Tethys (Flower and Kennett, 1994, [8]), and the Central American Seaway (e.g., Haug and Tiedemann, 1998, [9]). For a detailed review about the timing of the proxy records and their uncertainties, we refer to Mudelsee *et al.*, (2014, [10]).

Here, we examine the climate for the early Late Miocene, *i.e.*, the Tortonian (11–7 Ma, Ma: million years before present). The Tortonian is characterized by intensive Antarctic glaciation and the buildup of ice sheets in the North Atlantic realm. We focus on the spatial temperature distribution which is a principal problem in understanding Cenozoic climate change. In the case of the Miocene, elevated global-mean surface temperatures and weaker equator-to-pole temperature gradients are proposed (Greenwood and Wing, 1995, [11]; Crowley and Zachos, 2000, [12]; Pound *et al.*, 2012, [13]). While numerical simulations exhibit rising global-mean temperatures for increasing greenhouse gas concentrations, they do not capture the reconstructed reduction in the meridional temperature gradient (Barron, 1987, [14]; Huber and Sloan, 2001, [15]; Micheels *et al.*, 2011, [16]).

Some authors (Schmidt and Mysak, 1996, [17]; Hay *et al.*, 1997, [18]; Micheels *et al.*, 2011, [16]) have suggested that atmospheric heat transport may have played an important role in the temperature distribution. We would expect a warmer atmosphere to transport more latent heat toward the poles, helping to reduce meridional temperature gradients (polar amplification). However, despite the exponential increase of saturation vapor pressure with temperature, this water vapor feedback becomes less powerful as temperature rises (Caballero and Langen, 2005, [19]). Other possible mechanisms located in the atmosphere involve the atmospheric stationary wave response due to changing paleogeography and sea level.

On the other side, marine proxy data indicate that ocean gateway changes and major reorganizations of the global ocean circulation can play a crucial role for the climate evolution (e.g., Kennett, 1977, [3]; Wright *et al.*, 1992, [20]; Zachos *et al.*, 2001, [1]; Mudelsee *et al.*, 2014, [10]). Concerning the Tortonian, the still open Central American Seaway (CAS, *i.e.*, the Panama Strait) enabled the exchange of saline Atlantic water with comparatively fresher Pacific water (Montes *et al.*, 2015, [21]), and it has been shown that this leads to a weakening of the thermohaline circulation in the North Atlantic Ocean (e.g., Mikolajewicz *et al.*, 1993, [22]; Bice *et al.*, 2000, [23]). Therefore, the Atlantic thermohaline circulation is unlikely to be responsible for a warmer climate at higher latitudes; at least in the Northern Hemisphere.

The question of temperatures might be linked to other feedbacks in the climate system, such as changes in the hydrological cycle and vegetation cover. Paleontological and palynological data give evidence for drastic changes in vegetation and therefore climate during the Cenozoic (Retallack, 2001, [24]; Willis and Mc Elwain, 2002, [25]). For example, during the Eocene/Oligocene glaciation (~34 Ma), tropical rain forests virtually disappeared poleward of the northern and southern high-pressure zones (Retallack, 2001, [24]; Willis and Mc Elwain, 2002, [25]). Grasslands, which had begun to develop under dry conditions during the Eocene, covered larger areas in the Oligocene (after ~34 Ma). During the Mid-Miocene Climatic Optimum (~17 to 15 Ma), moist, warm forests expanded poleward of the subtropical high-pressure zones for a short period (Retallack, 2001, [24]; Willis and Mc Elwain, 2002, [25]). Following the global climatic deterioration after the Mid-Miocene Climatic Optimum tropical rain forests withdrew again to the equatorial zone. Grasslands and deserts expanded through much of the lower mid-latitudes (Morley, 2000, [26]; Bredenkamp *et al.*, 2002, [27]). C_4-type grasslands became widespread during the interval from about 8 to 5 Ma (Cerling *et al.*, 1997, [28]; Freeman and Colarusso, 2001, [29]). During the Miocene most of the climatically arranged vegetation belts developed: ranging from rain forest along the equator to polar desert at high latitudes. However, to date, little is known about the connection between continental vegetation change and climate change during the Cenozoic. It is still an open question whether the vegetation adapted to hydrological changes or whether it played an active role as a modifier of major climate transitions. In principle, vegetation can contribute to a polar amplification through modifying the local albedo (e.g., Dutton and Baron, 1997, [30]; Otto-Bliesner and Upchurch, 1997, [31]). In the light of these findings we investigate whether such a feedback could have been effective during the Late Miocene. In particular, we are interested in the sensitivity of the Miocene ocean circulation with respect to the vegetation cover and associated hydrological cycle in conjunction with the open Central American gateway.

2. Methods

In order to evaluate the Miocene land surface cover and its associated hydrological cycle, we use the output of an atmospheric circulation model under different assumptions for the land cover parameters to force a dynamical vegetation model and an ocean circulation model.

2.1. Atmospheric Circulation Model

For the Late Miocene climate simulations, we apply the atmosphere general circulation model ECHAM4 (Roeckner *et al.*, 1996, [32]). The prognostic variables are calculated in the spectral domain with a triangular truncation at wave number 30 (T30), which corresponds to a Gaussian longitude-latitude grid of approximately 3.75°. The vertical domain is represented by 19 hybrid sigma-pressure (terrain following) levels with the highest level at 10 hPa. The model is coupled to a 50 m slab ocean. This allows a prescription of the Miocene ocean heat transport consistent with proxy data (Steppuhn *et al.*, 2006, [33]). Furthermore, the orography is adapted to the Tortonian when the height of mountain ranges was generally reduced (references in Steppuhn *et al.*, 2006, [33]). For example, Greenland reaches only about a tenth of its recent elevation. In addition to the aforementioned boundary conditions, the atmospheric CO_2 is prescribed with the present-day level of 353 ppmv for all experiments. This value ranges in the spectrum commonly used for the Miocene (Freeman and Hayes, 1992, [34]; Cerling *et al.*, 1997, [28]; Pagani *et al.*, 1999, [35]; Pearson and Palmer, 2000, [36]; Demicco *et al.*, 2003, [37]; Pagani *et al.*, 2005, [38]; Kürschner *et al.*, 2008, [39]; Zhang *et al.*, 2013, [40]).

For the land surface, sensitivity experiments were performed which are described below. Each model simulation with the atmospheric general circulation model (AGCM) was run over 20 years. The model reaches an equilibrium state after 5 years, and the last 10 years are taken into account for further analysis. Vegetation is a fixed factor represented through the specification of different land surface parameters like albedo, roughness length, vegetation ratio, leaf area index and maximum soil water capacity. This model approach has been applied and validated with proxy data in an investigation of heat transport mechanisms for the Late Miocene (Micheels *et al.*, 2011, [16]) and warm climate during the Late Miocene (Knorr *et al.*, 2011, [41]). Depending on the surface and vegetation type, physical quantities like the background albedo or roughness length are changed and prescribed. The albedo is a function of the background albedo, and on climate variables like snow depth and ice (Roeckner *et al.*, 1996, [32]).

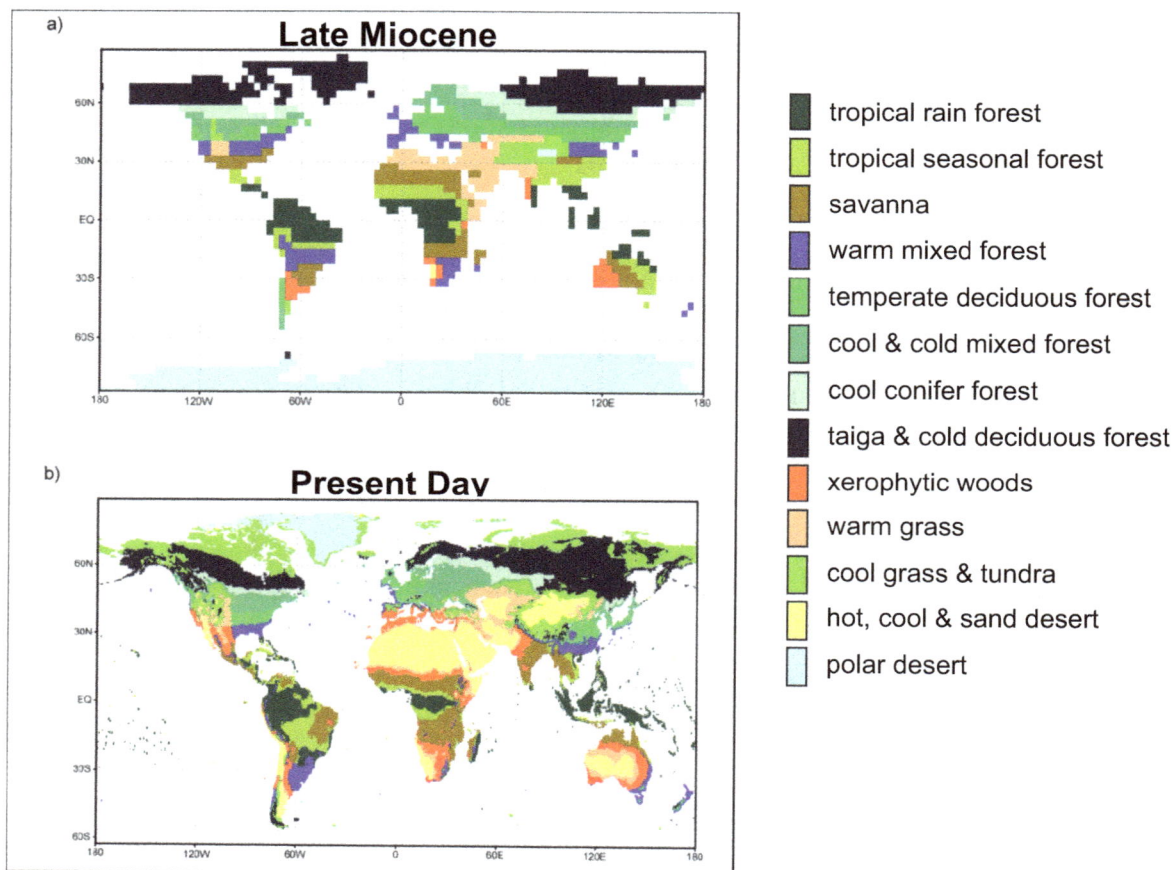

Figure 1. (**a**) The proxy-based reconstructed Tortonian vegetation; and (**b**) the present-day's vegetation (New *et al.*, 1999) [42]. These maps serve as an input into the atmospheric general circulation model (AGCM) experiments TVEG (**a**), TGEO (**b**) and CTRL (**b**). Warm grass corresponds to subdesertic Mediterranean-like open vegetation.

Based on the Late Miocene boundary conditions described above, we performed two Tortonian sensitivity experiments with respect to the vegetation. The first Tortonian simulation (TGEO) used the present-day vegetation, except that Greenland glaciers were replaced by tundra vegetation (Steppuhn *et al.*, 2006, [33]). The second Tortonian run (TVEG) uses a proxy-based reconstruction of the palaeovegetation (Figure 1) (Micheels, 2003, [43]; Micheels *et al.*, 2007, [44]). The palaeovegetation represents a generally larger forest cover than today and forest margins shift farther poleward. Grasslands and deserts/semi-deserts are reduced in the Late Miocene. For example, in North Africa the Late Miocene vegetation reconstruction shows warm grassland to savanna vegetation without any desert area. It is an on-going debate whether North Africa was humid until the Pliocene (e.g., Pickford, 2000, [45]; Micheels et al., 2009, [46]) or whether the Sahara desert first appeared at the end of the Miocene (~7 to 6 Ma) (e.g., Vignaud et al., 2002, [47]; Schuster et al., 2006, [20]). For the Tortonian (11 to 7 Ma), a non-desert situation in North Africa as an intermediate state in between the Early Miocene tropical forests (Wolfe, 1985, [48]; Dutton and Barron, 1997, [8]) and desert in the Pliocene (e.g., Schuster et al., 2006, [20]) is consistent with the fossil record. For TVEG, land surface parameters (albedo, leaf area index, vegetation and forest cover, and maximum soil water capacity) are changed for each grid point. A list of the experiments is given in Table 1.

Table 1. List of experiments.

	CTRL	TGEO	TVEG
Atmosphere: ECHAM4			
SST	modern	Steppuhn et al., (2006), [33]	Steppuhn et al., (2006), [33]
ocean heat transport	modern	Steppuhn et al., (2006), [33]	Steppuhn et al., (2006), [33]
CO_2	353 ppmv	353 ppmv	353 ppmv
land	modern	Greenland/Tibet	Greenland/Tibet
vegetation	modern	modern	Micheels et al., (2007), [36]
Ocean: LSG			
ocean gateways	modern	CAS 500 m	CAS 500 m
wind and hydrological cycle	from CTRL	from TGEO	from TVEG
salinity and temperature	calculated	calculated	calculated
Vegetation: LPJ			
vegetation	calculated	-	calculated

2.2. Dynamical Vegetation Model

The Lund-Potsdam-Jena dynamical vegetation model LPJ (Sitch et al., 2003, [49]) combines process-based descriptions of terrestrial ecosystem structure (vegetation composition, biomass and height) and function (energy absorption, carbon cycling). Vegetation composition is described by nine different plant functional types (PFTs), which are distinguished according to their physiological (C_3-, C_4-type photosynthesis), morphological (tree, grass) and phenological (deciduous, evergreen) attributes. The model is run on a horizontal $2° \times 2°$ grid, directly forced with the output of the AGCM experiments. The model is run on a grid cell basis with input of soil texture, monthly fields of temperature, precipitation, short- and long-wave radiation. The soil texture is taken from present values. Each grid cell is divided into fractions covered by the PFTs and bare ground. Both the presence and the covered fraction of PFTs within a grid cell depend on their specific environmental limits and on resource competition among the PFTs. Carbon isotope fractionation is included in the model (Kaplan et al., 2002, [50]; Scholze et al., 2003, [51]).

2.3. Ocean Circulation Model

Our ocean model is an updated version of the LSG circulation model developed by Maier-Reimer *et al.*, (1993, [52]). We implemented some significant improvements such as a new advection scheme for tracers (Schäfer-Neth and Paul, 2001, [53]; Prange *et al.*, 2002, [54]; Prange *et al.*, 2003, [55]) and an overflow parameterization for the bottom boundary layer (Lohmann, 1998, [56]; Lohmann and Schulz, 2000, [57]). The spatial resolution is $3.5° \times 3.5°$ in the horizontal and 22 levels in the vertical. We calibrated the model by simulating anthropogenic carbon-14 (Butzin *et al.*, 2005, [58]). The ocean is forced by 10-year averaged monthly fields of wind stress, surface air temperature, and freshwater flux (precipitation-evaporation), which serve as background climatology and originate from the simulations with the AGCM ECHAM4 described in Section 2.1.

In the ocean circulation experiments, we employ a hybrid coupled modeling approach, which allows an adjustment of surface temperatures and salinity to changes in the ocean circulation, based on an atmospheric energy balance model (Lohmann and Gerdes, 1998, [59]; Prange *et al.*, 2003, [55]). No flux correction is applied for present day and other climate conditions. This approach permits that sea surface temperatures (SST) can freely adjust to ocean circulation changes (e.g., see Prange *et al.*, 2003, [55]; Knorr and Lohmann, 2003, [60]; Butzin *et al.*, 2005, [58]). The hydrological cycle is closed by a runoff scheme that considers continental catchment areas and allows for variable land-sea distributions, which permits that sea surface salinities (SSS) can freely evolve. The total integration time of each experiment is 5000 years. For the late Miocene simulations, we assumed a 500 m deep and three gridpoints wide (between 9° N and 18° N) gateway between the Atlantic and Pacific Oceans.

3. Results

3.1. Albedo, Hydrological Cycle and Vegetation Cover

In order to check the consistency of the reconstructed vegetation distribution with the modeled climate in TVEG, we apply the dynamical vegetation model LPJ. We use the monthly output of the last 10 years of the CTRL and TVEG simulations, iterating these simulations 200 times to achieve an equilibrium of the dynamical vegetation model after 2000 model years. We build an average over the last 500 years and identify the spatial patterns of the PFTs for the Tortonian and present-day vegetation cover (Figure 2). For the late Miocene, simulated tropical trees are spread into subtropical Africa (North and South) and parts of Australia, whereas temperate trees are extended over Asia relative to present conditions. The extension of boreal forests far into the northern high latitudes during the Tortonian is in accordance with proxy data (Boulter and Manum, 1997, [61]). Grassland is extended into subtropical areas, over Greenland and over Alaska. The Sahara desert is smaller than today and consists of steppes and open grassland rather than sand desert which is consistent with fossil data (Schuster *et al.*, 2006, [62]; Micheels *et al.*, 2009, [46]).

When comparing TVEG with CTRL and TGEO, altering the land surface parameters to that appropriate for the Late Miocene causes the surface albedo to decrease for most regions on the globe (Figure 3). The strongest changes are found over northern Asia and North America, and over North Africa. The removal of inland ice on Greenland causes surface temperatures to rise and the associated reduction in the ice-albedo feedback mechanism reduces sea ice in the northern high latitudes.

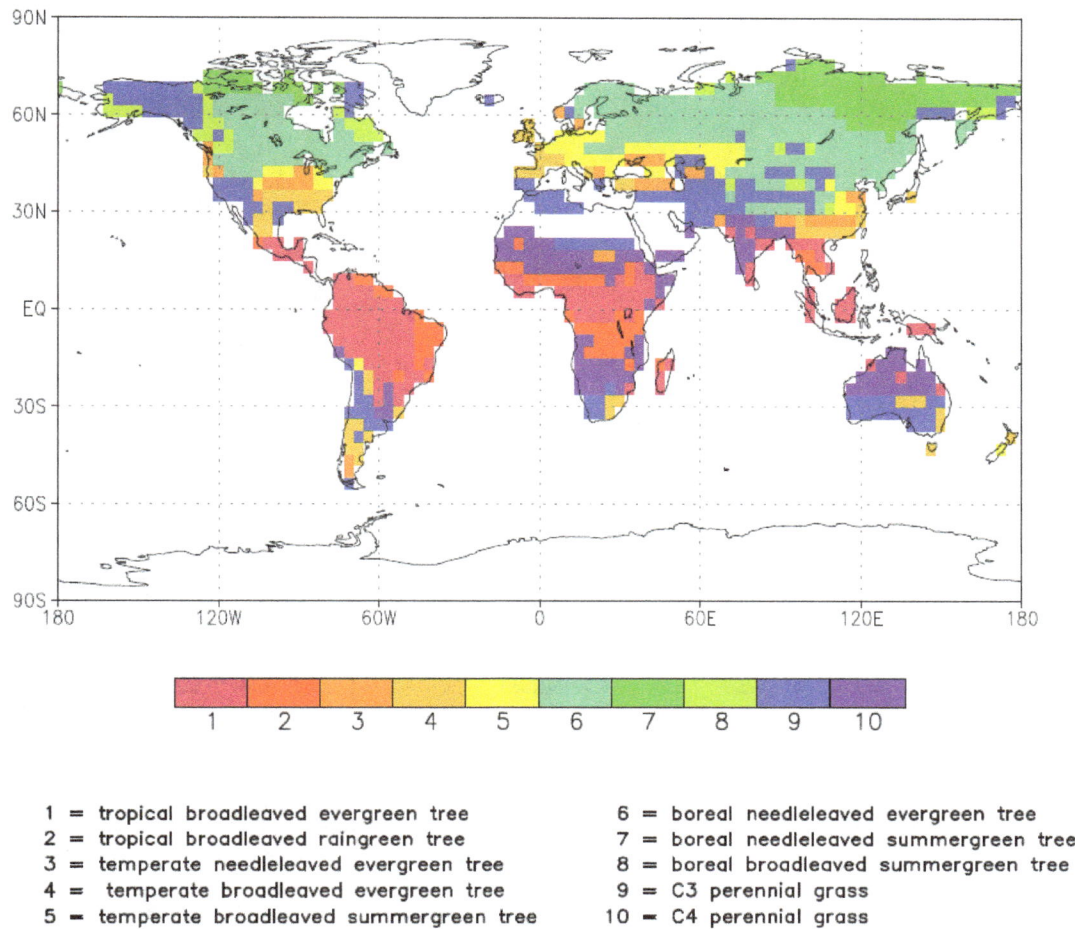

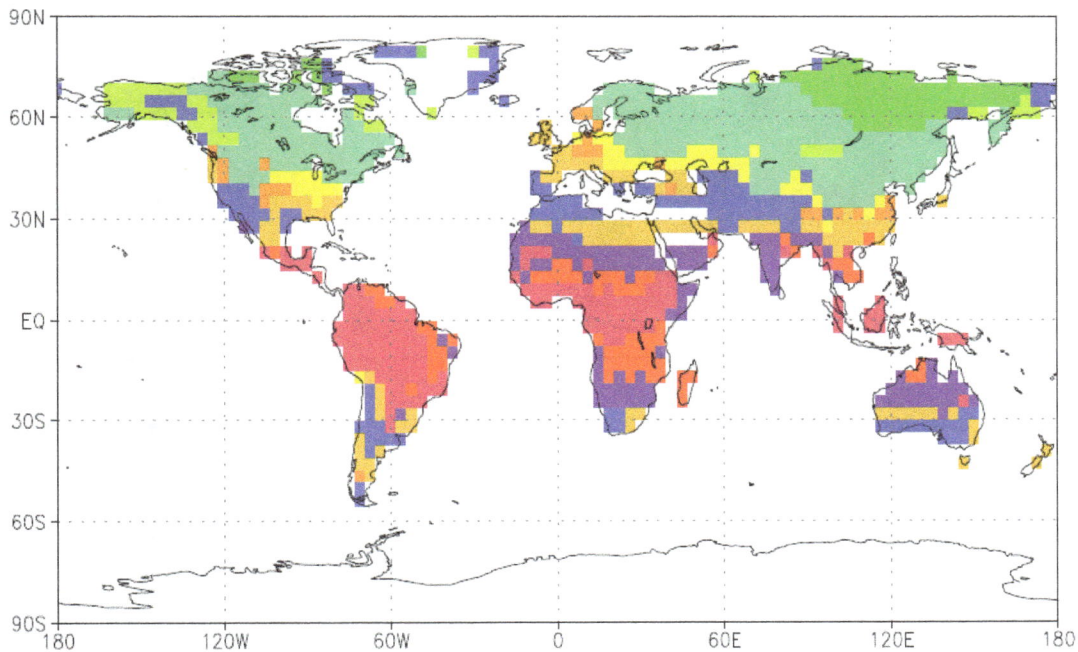

Figure 2. Dominant plant functional types as simulated by the LPJ dynamical vegetation model for present (upper panel) and Tortonian conditions (lower panel). The model has been run to equilibrium conditions (see text for details). White areas indicate regions with no vegetation.

a)

albedo anomalies [frac.] TVEG-CTRL

b)

albedo anomalies [frac.] TVEG-TGEO

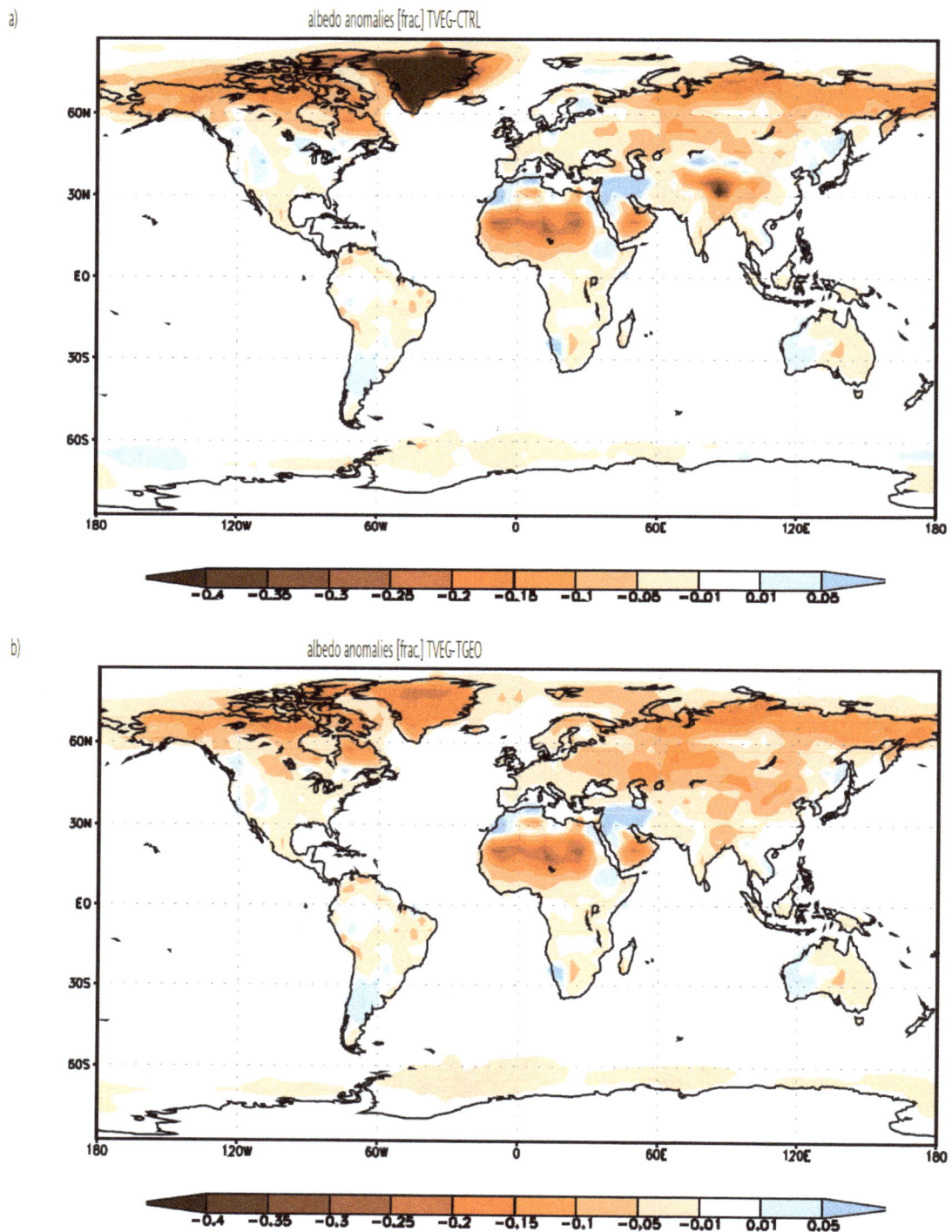

Figure 3. Changes in the mean annual albedo (frac.) for (**a**) TVEG minus CTRL; and for (**b**) TVEG minus TGEO.

Figure 4 shows the zonally integrated freshwater flux over the Atlantic catchment area (*cf.*, Lohmann, 2003, [63]). In the subtropics there is considerably more net evaporation for TVEG than for CTRL and TGEO. The integrated net evaporation between $10°$ and $50°$ N is 0.227 Sv (CTRL), 0.209 Sv (TGEO), 0.305 Sv (TVEG), respectively. (The unit 1 Sv corresponds to a mass transport of 10^9 kg s^{-1}, equivalent to a volume transport of 10^6 m^3 s^{-1} of liquid water.) Maximum transport is found for TVEG, which is linked to zonal winds and a water vapor export out of the Atlantic catchment area. The wind anomalies between TVEG and TGEO are shown in Figure 5. The wind anomalies are largely related to changes in surface temperature. Pronounced changes are seen over North Africa with enhanced monsoonal circulation in the boreal summer (Figure 5b). The zonal circulation towards the African continent and the Indian Ocean is enhanced in TVEG, associated with increased land-sea contrast between 2 °C and 6 °C and increased precipitation over the Sahel region and the Indian Ocean. In a sensitivity study, Micheels *et al.*, (2009, [46]) found that the Sahara region greening leads to a temperature rise of up to 6 °C and a significant increase in precipitation in the Sahel region and the Indian region (*cf.* Figure 2, Micheels *et al.*, 2009, [46]). The pronounced clockwise circulation over a center at 60° W, 30° N (Figure 5b) is most likely linked to the barotropic response to a relative cooling in this area. This circulation indicates an increased transport out of the Atlantic basin, especially during the boreal summer season when the moisture transport is maximal. Furthermore, we detect a strong change in the North Pacific Ocean which reaches its maximum in boreal winter (Figure 5a). A pronounced change can also be detected for the annual mean circulation (Figure 5c).

For the momentum flux it is interesting to note that the zonal wind stress in the Atlantic Ocean is only slightly affected for TVEG and lies between the CTRL and TGEO simulations (Figure 6). The meridional wind stress (Figure 7) is similar in the experiments TVEG and TGEO, except for a region around 50° N.

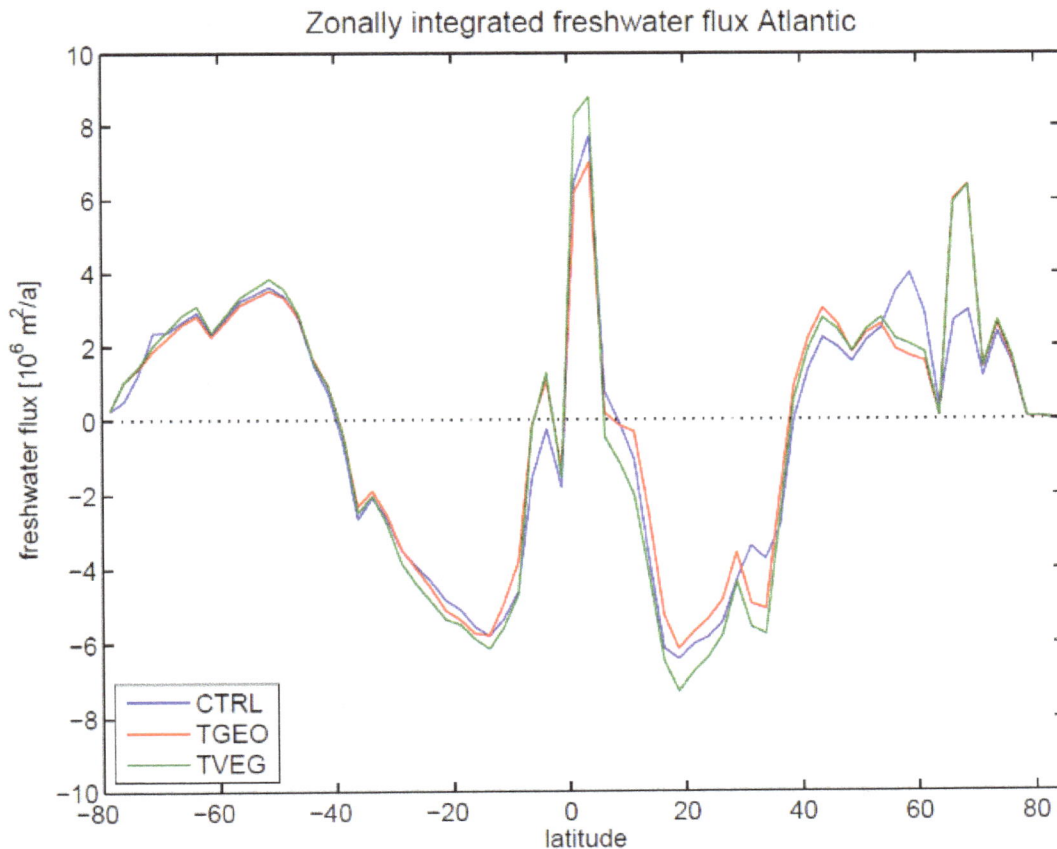

Figure 4. Annual mean net freshwater fluxes in the Atlantic Ocean for CTRL, TGEO, and TVEG. For the calculation of the net freshwater flux, the catchment areas in the ocean model have been taken into account (as in Lohmann, 2003, [63]). South of 35° S, the Atlantic Ocean is defined by the longitudes of the southern tips of South America and Africa, respectively.

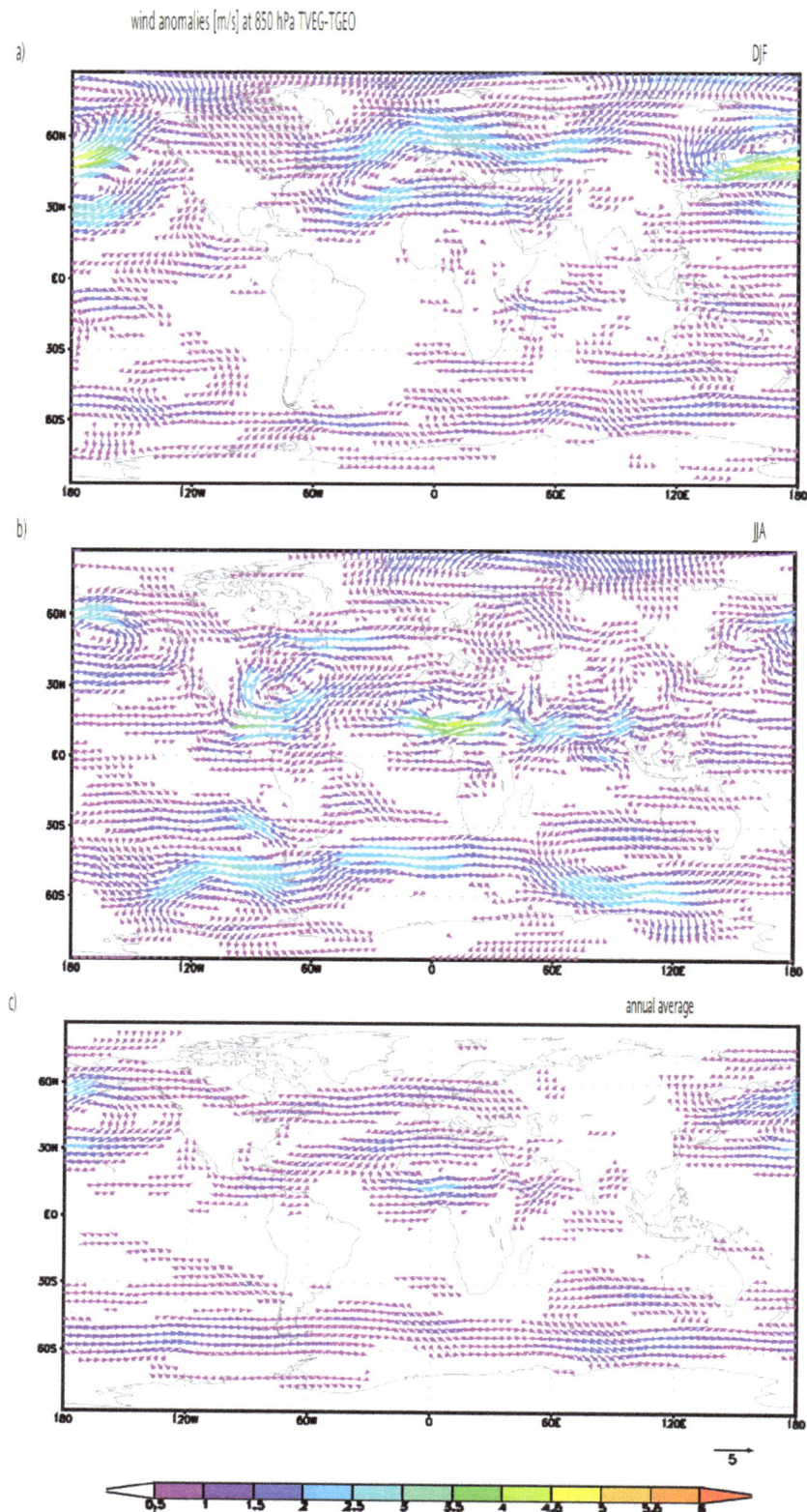

Figure 5. Changes in the wind field (m/s) at 850 hPa for TVEG minus TGEO. (**a**) Boreal winter: December-January-February; (**b**) boreal summer: June-July-August; (**c**) annual mean. The reference arrow represents 5 m/s. Vectors for differences of less than 0.5 m/s are not shown.

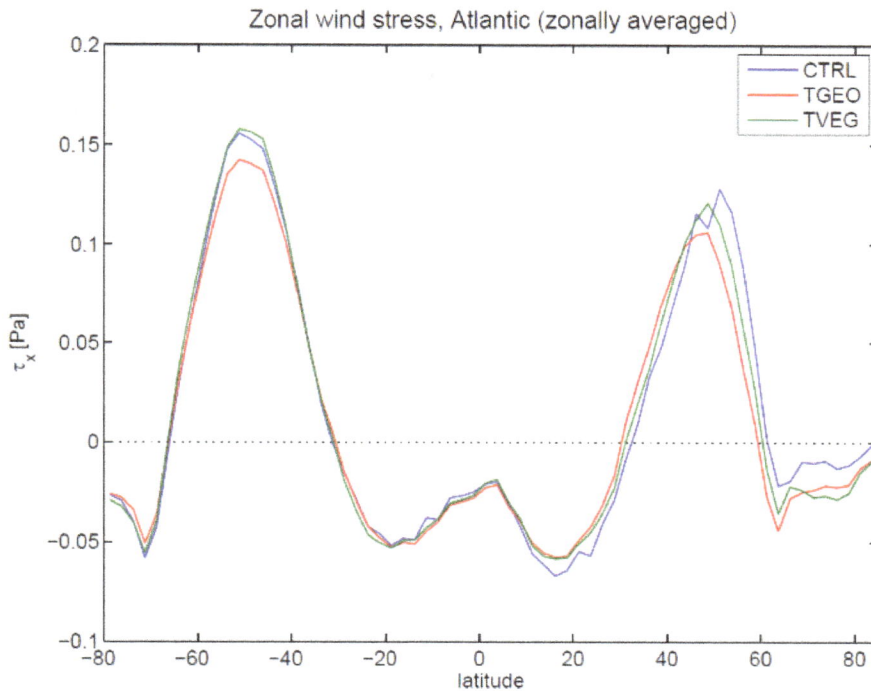

Figure 6. Zonal mean zonal wind stress in the Atlantic Ocean for CTRL, TGEO, and TVEG.

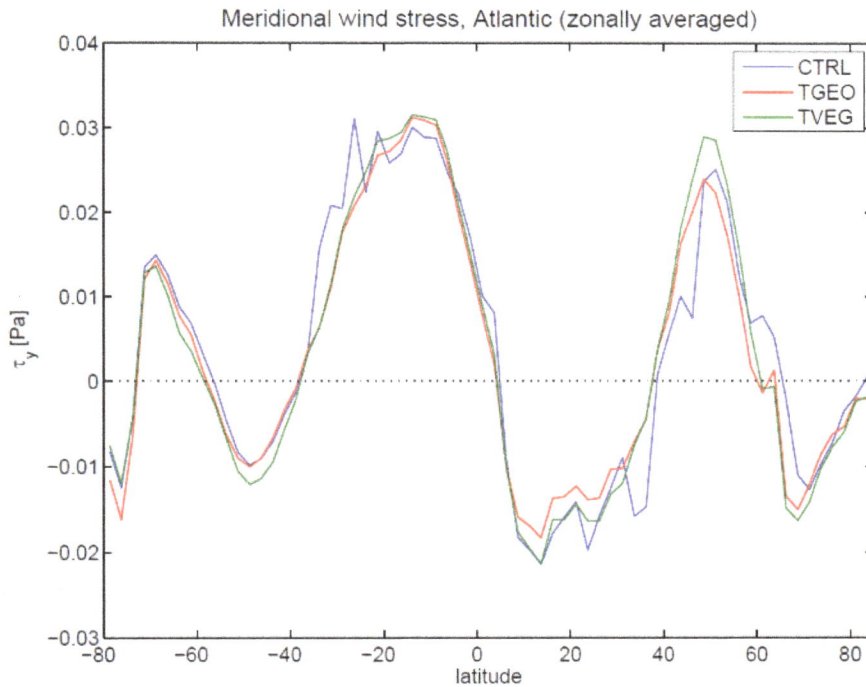

Figure 7. Same as Figure 6, but for the meridional wind stress.

3.2. Ocean Circulation

The control experiment for present-day conditions (Figure 8a) reasonably reflects the modern Atlantic Ocean circulation with a southward water export of 16 Sv at 30° S and a heat transport of 0.96 PW (1 PW = 10^{15} W) at 30° N, which is in the range of oceanographic observations (Schmitz, 1995, [64];

Macdonald and Wunsch, 1996, [65]). A comparison of the control run with the Tortonian experiments (TGEO, TVEG) reveals significant changes in the meridional overturning circulation (Figure 8b,c): The formation of deep water in the North Atlantic is strongly reduced (TGEO) when the Central American Seaway (CAS) is open (Figure 8). The meridional circulation is only 3 Sv and represents a "mini-conveyor belt" circulation with an ocean heat transport at 30° N of 0.19 PW (Figure 8b). In experiment TVEG, the circulation strength is similar to present-day conditions (14 Sv export at 30° S, 0.83 PW at 30° N), but slightly shallower than today. The reason might be the increased flow of bottom water from the Antarctic (Figure 8c).

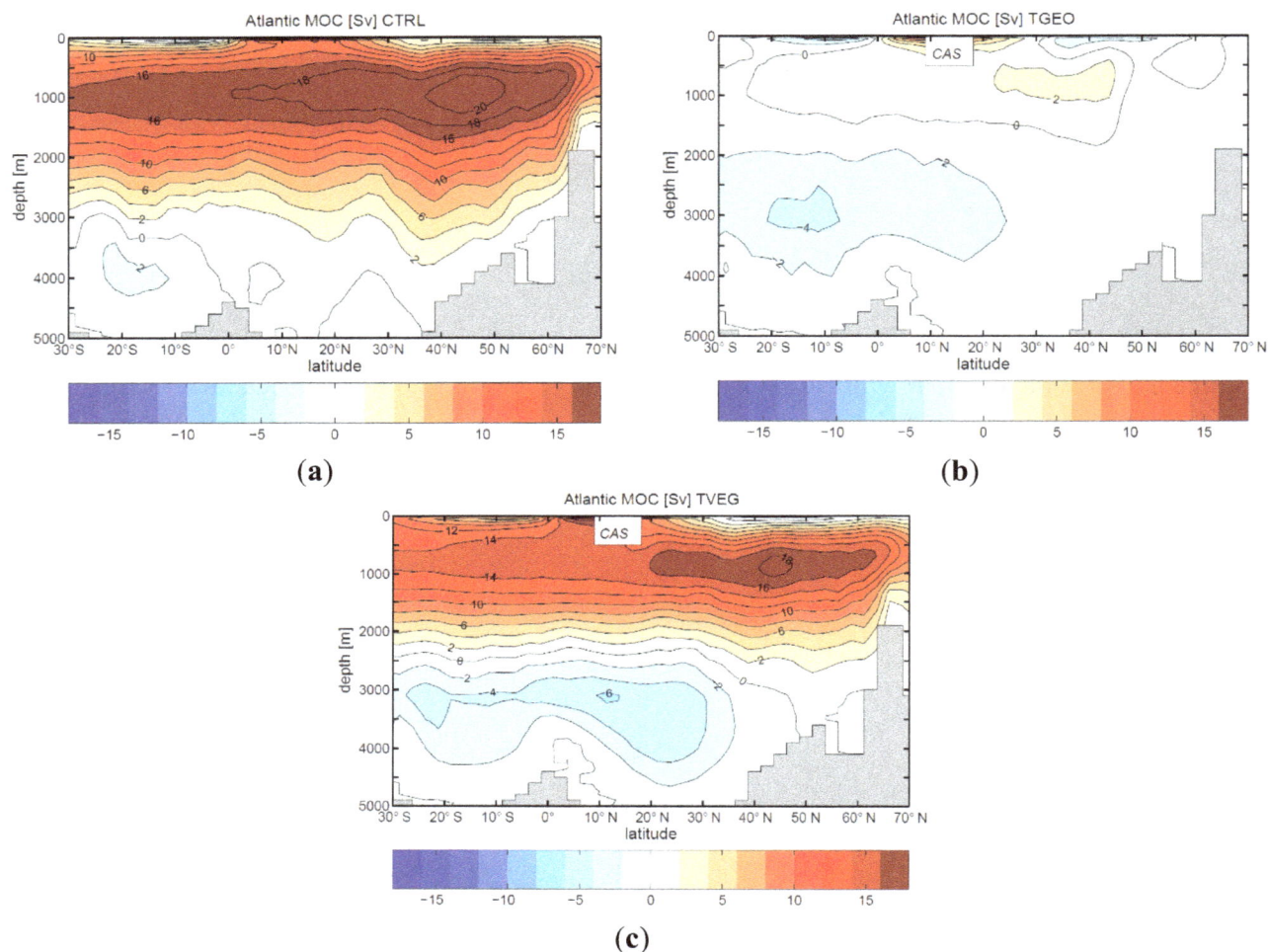

(a)

(b)

(c)

Figure 8. Atlantic meridional overturning circulation (Sv = 10^6 m³/s) for present-day (**a**), and the late Miocene configuration with open Central American Seaway (CAS); (**b**) with present vegetation cover (TGEO); and (**c**) with reconstructed vegetation cover (TVEG). Note that the Atlantic meridional overturning stream function is only defined if the ocean basin is bounded either side by land. We calculate the stream function from the bottom of the ocean to the top. Therefore, the open seaway is blanked out in the panels b and c.

A detailed analysis of the flow patterns in various depths of the CAS shows an export of surface water from the Atlantic to the Pacific Ocean (Figure 9a). The import of thermocline and intermediate layer water from the Pacific to the Atlantic Ocean is responsible for a reversal of the Northeast Brazil Current (Figure 9a). The subsurface transport for depths of more than 200 m is 27 Sv and 29 Sv for TGEO and

TVEG, respectively. The net flux of Pacific water through the CAS into the Atlantic leads to thermocline water of relative low salinity. This freshening inhibits deep-water formation in the North Atlantic. In TGEO, the surface winds and atmosphere-ocean net freshwater flux in the North Atlantic are not able to overcome this freshening by the open CAS (Figure 8b). In contrast to TVEG, a stronger northward flow (Figure 9b) and increase in net evaporation are sufficient to push the ocean circulation into a present-day-like circulation mode (Figure 8c). Both the increased ocean circulation with a northward shift of the Arctic sea ice, and a local warming associated to the land surface quantities, induce an anomalous warming between TVEG and TGEO of up to 8 °C (Figure 9b). Note that all experiments were carried out for pCO_2 = 353 ppmv. Figure 10a displays the zonal mean sea-surface temperature in the Atlantic Ocean. North of 30° S, the surface temperature in TVEG is increased compared to CTRL and TGEO in the zonal mean. Interestingly, the strongest temperature change between TVEG and CTRL (more than 3 °C) is in the subsurface of the Atlantic Ocean (Figure 10b). This increase is strongest between 40° S and 20° N at 500 m depth. Pronounced warming is detected for a region around Iceland and in the northern North Pacific (Figure 9a).

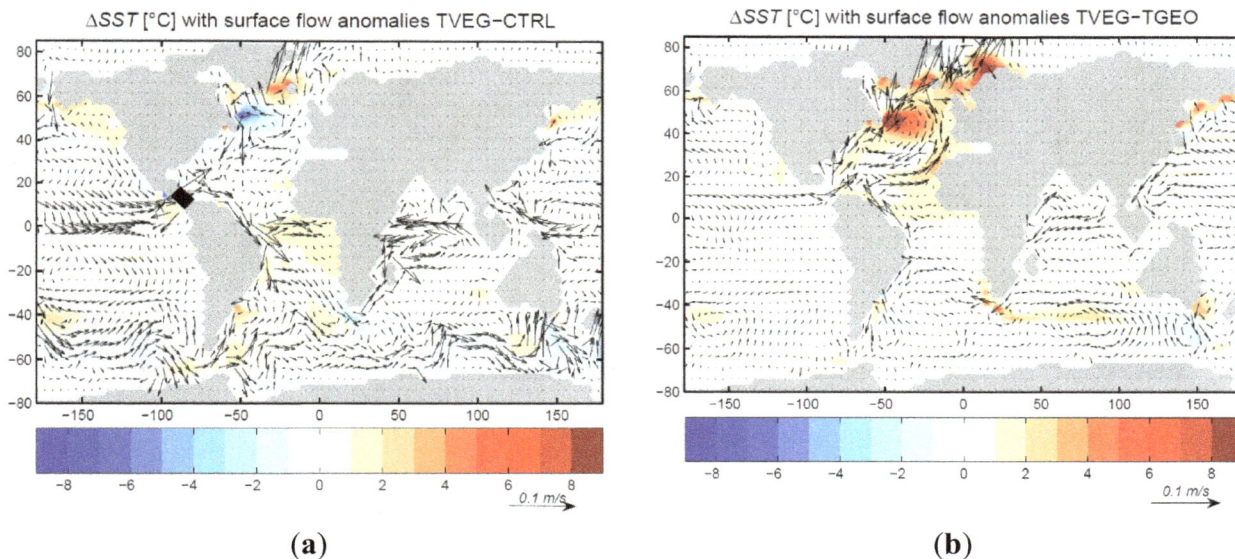

Figure 9. Modeled sea surface temperature anomalies [°C] and surface flow [m/s]. (**a**) Difference between TVEG and CTRL; (**b**) difference between TVEG and TGEO.

In TGEO, the drop in meridional ocean circulation is accompanied by a reduction in sea surface salinity in the North Atlantic (Figure 11a). Due to the exchange of surface water in the CAS, the surface water in the tropical Pacific becomes more saline. In contrast to TGEO, the stronger ocean circulation for TVEG is related to the considerably higher sea surface salinities in the North Atlantic Ocean (Figure 11b). The increased water vapor transport (as seen in the increase in 850 hPa winds) provides a source for net negative freshwater flux in the North Atlantic basin (a net evaporation anomaly) affecting sea surface salinity. The strong increase in North Atlantic upper 500 m salinity is clearly emphasized in the Atlantic zonal-mean salinity distribution (Figure 12a). The surface and subsurface warming of TVEG relative to TGEO is strongest in the subtropics and polar latitudes (Figure 12b). At northern polar latitudes, the warming is associated to strong poleward surface currents (Figure 10b), sea ice retreat, and meridional heat transport.

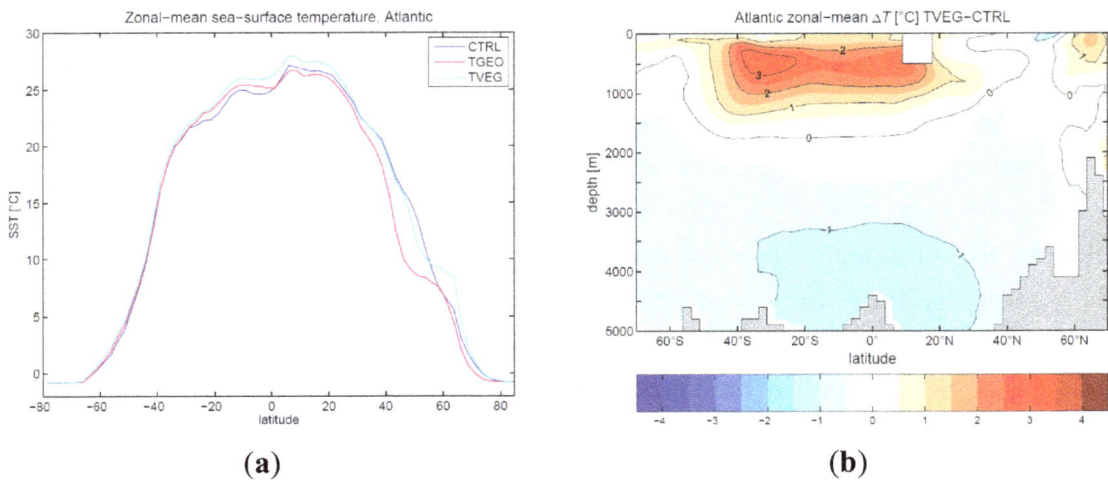

Figure 10. (**a**) Zonal mean sea surface temperature [°C] for CTRL, TGEO, and TVEG; (**b**) zonal mean Atlantic temperature anomaly TVEG-CTRL [°C].

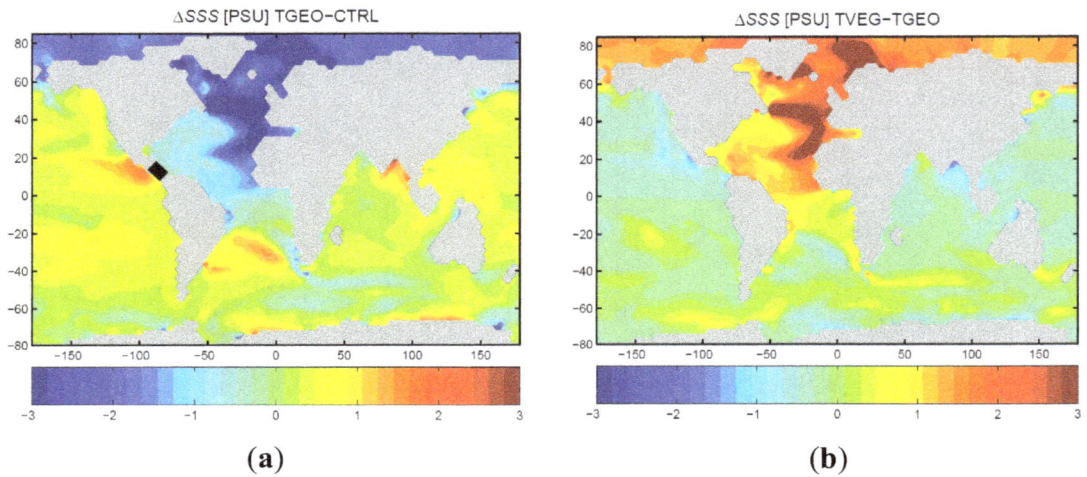

Figure 11. Modeled sea surface salinity anomalies [PSU]. (**a**) Difference between TGEO and CTRL; (**b**) difference between TVEG and TGEO.

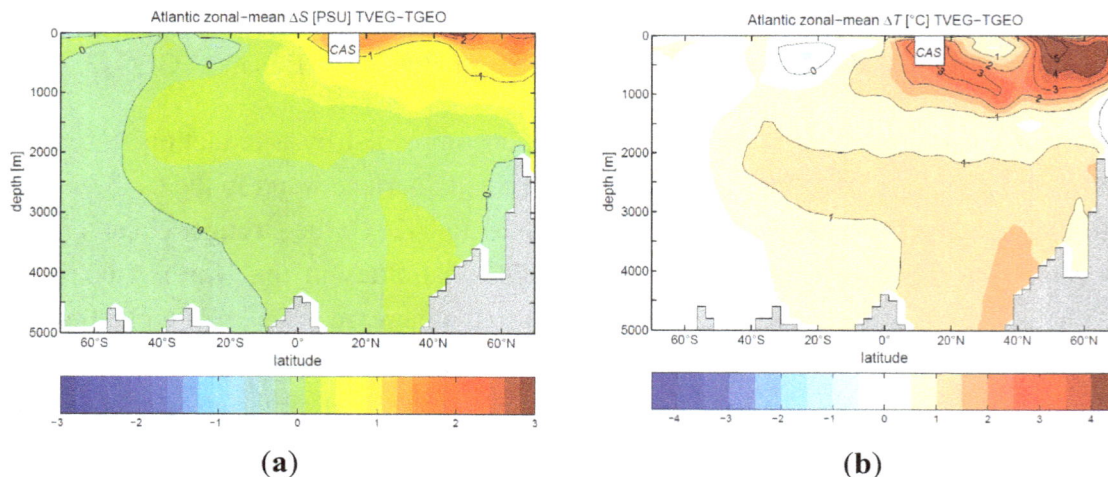

Figure 12. Zonal mean difference between TVEG and TGEO in the Atlantic Ocean: (**a**) salinity; (**b**) temperature.

4. Discussion

We have evaluated the vegetation effect on the surface albedo and ocean circulation during the late Miocene. We find that both effects contribute to a relatively warm late Miocene climate in the North Atlantic realm. Such warming has been reported from terrestrial proxy evidence (e.g., Wolfe, 1994, [66]). The albedo is strongly reduced through the presence of vegetation (greening) and through the vegetation-snow-albedo feedback (reduction of the snow-albedo feedback in the presence of vegetation). Several modeling studies found that paleovegetation-climate feedback may have significantly contributed to the Cenozoic climate evolution (Dutton and Barron, 1997, [30]; Knorr et al., 2011, [41]; Bradshaw et al., 2015, [67]). However, the possible synergy of vegetation and climate can vary with different background conditions. Bradshaw et al., (2015, [67]) suggest that climate sensitivity to CO_2 forcing is directly affected by the palaeogeographic configuration and that the inferred climate sensitivity is higher for the late Miocene than we might expect for future climate because of the differences in vegetation distribution (e.g., tropical forest changes) in conjunction with differences in ocean circulation and sea ice. Most models indicate a weaker overturning rate of the deep ocean circulation during the Miocene (Herold et al., 2012, [68] and references therein) which is accompanied with reduced high-latitude salinities and lowered northern North Atlantic temperatures.

Associated with the changes in the atmospheric circulation, we detect enhanced water vapor export out of the Atlantic catchment area for the Tortonian in the presence of vegetation. The enhanced water vapor transport is caused by an increase in the zonal moisture transport associated with the Atlantic trade winds. Triggered by high salinities at northern high latitudes, North Atlantic Deep Water is formed and the sea ice edge is moved poleward, in general agreement with proxy data (e.g., Wolf and Thiede, 1991, [69]). A stabilizing effect of the ocean circulation has been proposed for tropical water vapor transport during glacials (Lohmann and Lorenz, 2000, [70]) which may be responsible for an additional sea surface salinity contrast between the Atlantic and Pacific/Indian Oceans (Broecker, 1992, [71]; Zaucker and Broecker, 1992, [72]; Zaucker et al., 1994, [73]), for times when the Atlantic overturning is weaker (Lohmann, 2003, [63]), as well as for El Niño conditions (Schmittner et al., 2000, [74]; Soden, 2000, [75]; Latif et al., 2000, [76]). As pointed out by Steppuhn et al., (2006, [33]), there is a significant warming of more than 2 °C at the eastern margin of the Pacific Ocean associated with a decreased upwelling in this area. This is linked to a weaker equatorial Walker circulation and a Tortonian permanent El-Niño state. Fedorov et al., (2006, [77]) proposed that a permanent El-Niño state might have contributed to the Pliocene glaciation and Cenozoic climate evolution. This aspect will be analyzed in a subsequent study using a fully coupled atmosphere-ocean circulation model for the late Miocene. In our experiments, we changed the wind stress and the hydrological cycle in our hybrid-coupled model approach. Sensitivity experiments indicate that the direct wind effect turns out to be of secondary importance in our simulations (not shown).

New proxy compilations (LaRiviere et al., 2012, [78]; Goldner et al., 2014, [79]; references therein) indicate a substantial increase of several degrees in SST. Our simulations suggest a pronounced warming in the northern North Atlantic and North Pacific Oceans and a moderate warming in the tropical Atlantic Ocean and some regions in the Southern Hemisphere. Compared to LaRiviere et al., (2012, [78]), the warming in the northern North Pacific Ocean is not as strong as in reconstructions. It has been speculated that other mechanisms such as a long-term trend of thermocline shoaling and feedbacks may have been responsible for the warmth of the late Miocene (LaRiviere *et al.*, 2012, [78]). A detailed model-data comparison is however not our intension here. Goldner *et al.*, (2014, [79]) illustrate that the Mid-Miocene climate optimum is difficult to be simulated by a state-of-the-art climate model with CO_2 concentrations reconstructed for the Miocene, and concluded that either the models are not sensitive enough or that additional forcings remain missing to explain half of the anomalous warmth and pronounced polar amplification. In addition, it is possible that other mechanisms not included in the present generation of GCMs also might have had an important impact on Tortonian climate, such as high-latitude radiative warming by polar stratospheric clouds (Sloan and Pollard, 1998, [80]), increased ocean heat transport driven by tropical cyclone-induced mixing (e.g., Emanuel, 2002, [81]), or increased levels of greenhouse gases such as methane. Methane can be estimated through stable carbon isotopes (biological processes preferentially incorporate carbon-12) and areas of wetlands as calculated from the land surface scheme including the vegetation distribution.

In our approach we used a "one-way interaction" model. Sensitivity experiments using fully coupled models (*i.e.*, two-way interaction between vegetation and ocean) would provide more comprehensive information on the interaction between vegetation and ocean circulation in the Late Miocene. Furthermore, we have to take into account that the uncertainty in the CO_2 reconstructions is at least 150 ppm (larger than the glacial-interglacial variations), therefore we are unable to state that the CO_2 concentrations did not vary (*cf.* Kürschner *et al.*, 2008, [39]; Zhang *et al.*, 2013, [40]; and references therein). Therefore, fully interactive atmosphere-ocean-ice-biogeochemistry models might be necessary in order to examine the Earth system feedbacks for the Miocene.

To further investigate major developments during the Miocene, a combined approach between modeling and establishing proxy records from selected key locations is needed. Model results on changing patterns of heat transport can be validated by temperature reconstructions (Mg/Ca, alkenones, TEX86), both from the deep (benthic fauna) and the shallow (planktonic) ocean (Lear *et al.*, 2003, [82]; Billups *et al.*, 2002, [83]; Sluijs et al, 2006, [84]). Interestingly, the deep water in the North Atlantic Ocean indicates a warm signature during the late Miocene (Lear *et al.*, 2003, [82]) which can be interpreted as an already established meridional overturning circulation at this time. Before and after the Tortonian North Atlantic temperatures were much lower and indicate a reduced ocean circulation. In a next step, we should test the extent to which temporal shifts in growing season or vertical shifts in depth habitat can reduce model-data and data-data misfits and uncertainties (*cf.* Lohmann *et al.*, (2013, [85]) for a similar analysis of Holocene sea surface temperatures). Figure 10b indicates large vertical temperature changes between the experiments which are interesting in terms of differing temperature reconstuctions. For example, some proxies might reflect subsurface signature more explicitly than the other proxies and they show a different equator-to-pole gradient. Similarly, a quantitative terrestrial model-data comparison for the late Miocene has been undertaken by Bradshaw *et al.*, (2012, [86]), which incorporated a conservative estimate of uncertainties associated with both the model output and the data reconstructions.

Major changes in ocean circulation can be further traced by using water mass characteristic proxies like Cd/Ca, Nd isotopes, and carbon-13 (Frank *et al.*, 1999, [87]; Frank *et al.*, 2002, [88]; Delaney and Boyle, 1987, [89]; Butzin *et al.*, 2011, [90]). The combination of temperature reconstructions with oxygen-18 gives evidence on changes in salinity and may provide indications on the high salinities in the northern North Atlantic and associated strong ocean circulation.

It can be speculated that the overturning effect which is described here is relevant for other time slices as well. Zhang *et al.*, (2013) [91] analyzed the models participating in Pliocene Modeling Intercomparison Project (PlioMIP) and showed that the simulated mid-Pliocene Atlantic northward heat transport is similar to the pre-industrial. The consistent change in Atlantic deep circulation is a moderate shoaling of the overturning cell in the Atlantic, but there might be also model-dependent factors like the Arctic Ocean sea surface salinity (Stepanek and Lohmann, 2012, [92]) which potentially stem from vegetation-ocean feedbacks for such climate states. Further studies are required to examine the possible feedbacks for such climate states, which might be crucial for the high-latitude climate during the Cenozoic.

5. Conclusions

The Cenozoic climate evolution includes significant changes in the oceanic transports which are ultimately linked to the paleotopography and opening and closing of passages. Our study aims at exploring the effect of land-cover changes as a non-CO_2 forcing that might have happened during the late Miocene. When the CAS permitted exchange of fresh Pacific water with saline Atlantic water the density in the North Atlantic Ocean is reduced, thus affecting the large-scale ocean circulation. We find that a modified land surface cover can compensate this gateway effect showing an almost present oceanic northward heat transport. Essential for this mechanism is the increased interbasin fresh water transport in conjunction with a positive feedback of enhanced thermohaline-driven ocean circulation.

Our results indicate that the reorganization of vegetation zones, topographical changes and changes in the global ocean circulation could have played a dominant role in the major Cenozoic climate transitions. Consequently, it is of utmost importance not only to understand the behavior of these individual systems in better detail but also to investigate the full dynamics, feedbacks, and synergisms of the coupled system (including the carbon cycle). The high-latitude warming likely involves many different nonlinear feedbacks to the imposed changes (Lyle *et al.*, 2008, [93]; Knorr *et al.*, 2011, [41]; Knorr and Lohmann, 2014, [94]). Kürschner *et al.*, (2008, [57]) point to a long-term coupling between atmospheric CO_2 and climate, but indicate large fluctuations in greenhouse gas concentrations during the Miocene affecting the evolution of terrestrial ecosystems.

The Cenozoic is a proper test bed for large structural changes in a similar, but warmer climate with major regional heterogeneities. In the late Miocene, the reconstructed elevated global-mean surface temperatures and weaker equator-to-pole temperature gradients are underestimated in our model (as in all models which we are aware of). However, comparing proxy data with models, one should furthermore test the assumptions in the recorder systems (e.g., temporal shifts in recording season, shifts in depth habitat, recorder specific assumptions) by using proxy models and statistical analysis. As a logical next step, we will apply such tools to examine the large-scale temperature gradients under external forcings during the Miocene.

Acknowledgments

Arne Micheels and Volker Mosbrugger are acknowledged for discussions on an earlier manuscript and for providing some of the forcing data. Thanks go to Stefanie Klebe for proofreading. Silke Schubert and Marco Scholze are acknowledeged for help in performing the LPJ runs, and Andreas Manschke is acknowledged for help in model data processing. The work has been supported by the Helmholtz programme PACES, the DFG research group UCCC, and the excellence cluster marum. The authors thank three anonymous referees for their constructive comments.

Author Contributions

GL designed the study and wrote the manuscript with significant contributions of MB and TB. MB conducted the ocean model simulations. All authors contributed to interpretation and the preparation of the final manuscript.

Conflicts of Interest

The authors declare no conflict of interest.

References

1. Zachos, J.; Pagani, M.; Sloan, L.; Thomas, E.; Billups, K. Trends, rhythms, and aberations in global climate 65 Ma to Present. *Science* **2001**, *29*, 686–693.

2. DeConto, R.M.; Pollard, D. Rapid Cenozoic glaciation of Antarctica induced by declining atmospheric CO2. *Nature* **2003**, *421*, 245–249.

3. Kennett, J.P. Cenozoic evolution of Antarctic glaciation, the Circum Antarctic Ocean, and their impact on global paleoceanography. *J. Geophys. Res.* **1977**, *82*, 3843–3860.

4. Wright, J.D.; Miller, K.G. Control of North Atlantic Deep Water circulation by the Greenland-Scotland Ridge. *Paleoceanography* **1996**, *11*, 157–169.

5. Billups, K. Late Miocene through early Pliocene deep water circulation and climate change viewed from the sub-Antarctic South Atlantic. *Palaeogeogr. Palaeoclimatol. Palaeoecol.* **2002**, *185*, 287–307.

6. Cane, M.A.; Molnar, P. Closing of the Indonesian seaway as a precursor to east African aridification around 3–4 million years ago. *Nature* **2001**, *411*, 157–161.

7. Lawver, L.A.; Gahagan, L.M. Evolution of Cenozoic seaways in the circum-Antarctic region. *Palaeogeogr. Palaeoclimatol. Palaeoecol.* **2003**, *198*, 11–37.

8. Flower, B.P.; Kennett, J.P. The middle Miocene climatic transition: East Antarctic ice sheet development, deep ocean circulation and global carbon cycling. *Palaeogeogr. Palaeoclimatol. Palaeoecol.* **1994**, *108*, 537–555.

9. Haug, G.; Tiedemann, R. Effect of the formation of the Isthmus of Panama on Atlantic Ocean thermohaline circulation. *Nature* **1998**, *393*, 673–676.

10. Mudelsee, M.; Bickert, T.; Lear, C.H.; Lohmann, G. Cenozoic climate changes: A review based on time series analysis of marine benthic $\delta 18O$ records. *Rev. Geophys.* **2014**, *52*, 1–42.

11. Greenwood, D.R.; Wing, S.L. Eocene continental climates and latitudinal temperature gradients. *Geology* **1995**, *23*, 1044–1048.

12. Crowley, T.J.; Zachos, J.C. Comparison of zonal temperature profiles for past warm periods. In *Warm Climates in Earth History*; Huber, B., MacLeod, K.G., Wing, S.C., Eds.; Cambridge University Press: New York, NY, USA, 2000; pp. 50–76.

13. Pound, M.; Haywood, A.M.; Salzman, U.; Riding, J.B. Global vegetation dynamics and latitudinal temperature gradients during the Mid to LateMiocene (15.97–5.33 Ma). *Earth Sci. Rev.* **2012**, *112*, 1–22.

14. Barron, E.J. Eocene equator-to-pole surface ocean temperature: A significant climate problem? *Paleoceanography* **1987**, *2*, 729–739.

15. Huber, M.; Sloan, L. Heat transport, deep waters and thermal gradients: Coupled climate simulation of an Eocene greenhouse climate. *Geophys. Res. Lett.* **2001**, *28*, 3481–3484.

16. Micheels, A.; Bruch, A.A.; Eronen, J.; Fortelius, M.; Harzhauser, M.; Utescher, T.; Mosbrugger, V. Analysis of heat transport mechanisms from a late Miocene model experiment with a fully-coupled atmosphere-ocean general circulation model. *Palaeogeogr. Palaeoclimatol. Palaeoecol.* **2011**, *304*, 337–350.

17. Schmidt, G.A.; Mysak, L.A. Can increased poleward oceanic heat flux explain the warm Cretaceous climate? *Paleoceanography* **1996**, *11*, 579–593.

18. Hay, W.W.; DeConto, R.M.; Wold, C.N. Climate: Is the past key to the future? *Geol. Rundsch.* **1997**, *86*, 471–491.

19. Caballero, R.; Langen, P. The dynamic range of poleward energy transport in an atmospheric general circulation model. *Geophys. Res. Lett.* **2005**, *32*, doi:10.1029/2004GL021581.

20. Wright, J.D.; Miller, K.G.; Fairbanks, R.G. Miocene stable isotopes: Implications for deepwater circulation and climate. *Paleoceanography* **1992**, *7*, 357–389.

21. Montes, C.; Cardona, A.; Jaramillo, C.; Pardo, A.; Silva, J.C.; Valencia, V.; Pérez-Angel, L.C.; Ayala, C.; Rodriguez-Parra, L.A.; Ramirez, V.; *et al.* Middle Miocene closure of the Central American Seaway. *Science* **2015**, *348*, 226–229.

22. Mikolajewicz, U.; Maier-Reimer, E.; Crowley, T.J.; Kim, K.J. Effect of Drake and Panamanian gateways on the circulation of an ocean model. *Paleoceanography* **1993**, *8*, 409–426.

23. Bice, K.L.; Scotese, C.R.; Seidov, D.; Barron, E.J. Quantifying the role of geographic change in Cenozoic ocean heat transport using uncoupled atmosphere and ocean models. *Palaeogeogr. Palaeoclimatol. Palaeoecol.* **2000**, *161*, 295–310.

24. Retallack, G. Cenozoic expansion of grasslands and climatic cooling. *J. Geol.* **2001**, *109*, 407–426.

25. Willis, K.J.; McElwain, J.C. *The Evolution of Plants*; Oxford University Press: Oxford, UK, 2002; p. 378.

26. Morley, R.J. *Origin and Evolution of Tropical Rain Forests*; Wiley: Chichester, UK, 2000; p. 362.

27. Bredenkamp, G.J.; Spada, F.; Kazmieczak, E. On the origin of northern and southern hemisphere grasslands. *Plant Ecol.* **2002**, *163*, 209–229.

28. Cerling, T.E.; Harris, J.M.; MacFadden, B.J.; Leakey, M.G.; Quade, J.; Eisenmann, V.; Ehleringer, J.R. Global vegetation change through the Miocene/Pliocene boundary. *Nature* **1997**, *389*, 153–159.

29. Freeman, K.H.; Colarusso, L.A. Molecular and isotopic records of C4 grassland expansion in the late Miocene. *Geochim. Cosmochim. Acta* **2001**, *65*, 1439–1454.

30. Dutton, J.F.; Barron, E.J. Miocene to present vegetation changes: A possible piece of the Cenozoic puzzle. *Geology* **1997**, *25*, 39–41.

31. Otto-Bliesner, B.; Upchurch, G.R. Vegetation-induced warming of high-latitude regions during the Late Cretaceous period. *Nature* **1997**, *385*, 804–807.

32. Roeckner, E.; Arpe, K.; Bengtsson, L.; Christoph, M.; Claussen, M.; Dümenil, L.; Esch, M.; Giorgetta, M.; Schlese, U.; Schulzweida, U. *The Atmospheric General Circulation Model ECHAM-4: Model Description and Simulation of Present-Day Climate, Max-Planck-Institut für Meteorologie*; MPI Report No. 218; Max-Planck-Institute: Hamburg, Germany, 1996.

33. Steppuhn, A.; Micheels, A.; Geiger, G.; Mosbrugger, V. Reconstructing the Late Miocene Climate and oceanic heat flux using the AGCM ECHAM4 coupled to a mixed-layer ocean model with adjusted flux correction, *Palaeogeogr. Palaeoclimatol. Palaeoecol.* **2006**, *238*, 399–423.

34. Freeman, H.H.; Hayes, J.M. Fractionation of carbon isotopes by phytoplankton and estimates of ancient CO2 levels. *Global Biogeochem. Cycles* **1992**, *6*, 185–198.

35. Pagani, M.; Arthur, M.A.; Freeman, K.H. Miocene evolution of atmospheric carbon dioxide. *Paleoceanography* **1999**, *14*, 273–292.

36. Pearson, P.N.; Palmer, M.R. Atmospheric carbon dioxide concentrations over the past 600 million years. *Nature* **2000**, *406*, 695–699.

37. Demicco, R.V.; Lowenstein, T.K.; Hardie, L.A. Atmospheric pCO2 since 60 Ma from records of seawater pH, calcium and primary carbonate mineralogy. *Geology* **2003**, *31*, 793–796.

38. Pagani, M.; Zachos, J.C.; Freeman, K.H.; Tipple, B.; Bohaty, S. Marked decline in atmospheric carbon dioxide concentrations during the Paleogene. *Science* **2005**, *309*, 600–603.

39. Kürschner, W.M.; Kvaček, Z.; Dilcher, D.L. The impact of Miocene atmospheric carbon dioxide fluctuations on climate and the evolution of terrestrial ecosystems. *PNAS* **2008**, *105*, 449–453.

40. Zhang, Y.G.; Pagani, M.; Liu, Z.; Bohaty, S.M.; Deconto, R. A 40-million-year history of atmospheric CO2. *Philos. Trans. A Math. Phys. Eng. Sci.* **2013**, *371*, doi: 10.1098/rsta.2013.0096.

41. Knorr, G.; Butzin, M.; Micheels, A.; Lohmann, G. A Warm Miocene Climate at Low Atmospheric CO2 levels. *Geophys. Res. Lett.* **2011**, doi:10.1029/2011GL048873.

42. New, M.; Hulme, M.; Jones, P. Representing Twentieth-Century Space–Time Climate Variability. Part I: Development of a 1961–90 Mean Monthly Terrestrial Climatology. *J. Climate* **1999**, *12*, 829–856.

43. Micheels, A. Late Miocene climate modelling with ECHAM4/ML—The effects of the palaeovegetation on the Tortonian climate. Ph.D. Thesis, University of Tübingen, Tübingen, Germany, 2003; p. 110.

44. Micheels, A.; Bruch, A.A.; Uhl, D.; Utescher, T.; Mosbrugger, V. A Late Miocene climate model simulation with ECHAM4/ML and its quantitative validation with terrestrial proxy data. *Palaeogeogr. Palaeoclimatol. Palaeoecol.* **2007**, *253*, 251–270.

45. Pickford, M. Crocodiles from the Beglia Formation, Middle/Late Miocene Boundary, Tunisia, and their significance for Saharan palaeoclimatology. *Ann. Paleontol.* **2000**, *86*, 59–67.

46. Micheels, A.; Eronen, J.; Mosbrugger, V. The Late Miocene climate response to a modern Sahara desert. *Global Planet. Chang.* **2009**, *67*, 193–204.

47. Vignaud, P.; Duringer, P.; Mackaye, H.T.; Likius, A.; Blondel, C.; Boisserie, J.-R.; de Bonis, L.; Eisenmann, V.; Etienne, M.E.; Geraads, D.; *et al.* Geology and paleontology of the Upper Miocene Toros-Menalla hominidlocality, Chad. *Nature* **2002**, *418*, 152–155.

48. Wolfe, J.A. Distribution of major vegetational types during the Tertiary. In *The Carbon Cycle and Atmospheric CO2 Natural Variations Archean to Present*; Sundquist, E.T., Broecker, W.S., Eds.; American Geophysical Union: Washington, DC, USA, 1985; pp. 357–375.

49. Sitch, S.; Smith, B.; Prentice, I.C.; Arneth, A.; Bondeau, A.; Cramer, W.; Kaplan, J.O.; Levis, S.; Lucht, W.; Sykes, M.T.; *et al.* Evaluation of ecosystem dynamics, plant geography and terrestrial carbon cycling in the LPJ dynamic global vegetation model. *Global Chang. Biol.* **2003**, *9*, 161–185.

50. Kaplan, J.O.; Prentice, I.C.; Knorr, W.; Valdes, P.J. Modelling the dynamics of terrestrial carbon storage since the Last Glacial Maximum. *Geophys. Res. Lett.* **2002**, *29*, 2074–2078.

51. Scholze, M.; Knorr, W.; Heimann, M. Modelling terrestrial vegetation dynamics and carbon cycling for an abrupt climate change event. *Holocene* **2003**, *13*, 327–333.

52. Maier-Reimer, E.; Mikolajewicz, U.; Hasselmann, K. Mean circulation of the Hamburg LSG OGCM and its sensitivity to the thermohaline surface forcing. *J. Phys. Oceanogr.* **1993**, *23*, 731–757.

53. Schäfer-Neth, C.; Paul, A. Circulation of the glacial Atlantic: A synthesis of global and regional modeling. In *The northern North Atlantic: A changing environment*; Schäfer, P., Ritzrau, W., Schlüter, M., Thiede, J., Eds.; Springer: Berlin, Germany, 2001; pp. 446–462.

54. Prange, M.; Romanova, V.; Lohmann, G. The glacial thermohaline circulation: Stable or unstable? *Geophys. Res. Lett.* **2002**, *29*, doi:10.1029/2002GL015337.

55. Prange, M.; Lohmann, G.; Paul, A. Influence of vertical mixing on the thermohaline hysteresis: Analyses of an OGCM. *J. Phys. Oceanogr.* **2003**, *33*, 1707–1721.

56. Lohmann, G. The influence of a near-bottom transport parameterization on the sensitivity of the thermohaline circulation. *J. Phys. Oceanogr.* **1998**, *28*, 2095–2103.

57. Lohmann, G.; Schulz, M. Reconciling Bølling warmth with peak deglacial meltwater discharge. *Paleoceanography* **2000**, *15*, 537–540.

58. Butzin, M.; Prange, M.; Lohmann, G. Radiocarbon simulations for the glacial ocean: The effects of wind stress, Southern Ocean sea ice and Heinrich events. *Earth Planet. Sci. Lett.* **2005**, *235*, 45–61.

59. Lohmann, G.; Gerdes, R. Sea ice effects on the sensitivity of the thermohaline circulation in simplified atmosphere-ocean-sea ice models. *J. Clim.* **1998**, *11*, 2789–2803.

60. Knorr, G.; Lohmann, G. Southern Ocean origin for resumption of Atlantic thermohaline circulation during deglaciation. *Nature* **2003**, *424*, 532–536.

61. Boulter, M.C.; Manum, S.B. A lost continent in temperate Arctic. *Endevour* **1997**, *21*, 105–108.

62. Schuster, M.; Duringer, P.; Ghienne, J.-F.; Vignaud, P.; Mackaye, H.T.; Likius, A.; Brunet, M. The Age of the Sahara Desert. *Science* **2006**, *311*, doi:10.1126/science.1120161.

63. Lohmann, G. Atmospheric and oceanic freshwater transport during weak Atlantic overturning circulation. *Tellus A* **2003**, *55*, 438–449.

64. Schmitz, W.J. On the interbasin scale thermohaline circulation. *Rev. Geophys.* **1995**, *33*, 151–173.

65. Macdonald, A.M.; Wunsch, C. An estimate of global ocean circulation and heat fluxes. *Nature* **1996**, *382*, 436–439.

66. Wolfe, J.A. An analysis of Neogene climates in Beringia. *Palaeogeogr. Palaeoclimatol. Palaeoecol.* **1994**, *108*, 207–216.

67. Bradshaw, C.D.; Lunt, D.J.; Flecker, R.; Davies-Barnard, T. Disentangling the roles of late Miocene palaeogeography and vegetation—Implications for climate sensitivity. *Palaeogeogr. Palaeoclimatol. Palaeoecol.* **2015**, *417*, 17–34.

68. Herold, N.; Huber, M.; Müller, R.D.; Seton, M. Modeling the Miocene climatic optimum: Ocean circulation. *Paleoceanography* **2012**, *27*, doi:10.1029/2010PA002041.

69. Wolf, T.C.W.; Thiede, J. History of terrigenous sedimentation during the past 10 m.y. in the North Atlantic (ODP Legs 104 and 105 and DSDP Leg 81). *Mar. Geol.* **1991**, *101*, 83–102.

70. Lohmann, G.; Lorenz, S. On the hydrological cycle under paleoclimatic conditions as derived from AGCM simulations. *J. Geophys. Res.* **2000**, *105*, 417–436.

71. Broecker, W.S. The salinity contrast between the Atlantic and Pacific during glacial time. *Paleoceanography* **1992**, *4*, 207–212.

72. Zaucker, F.; Broecker, W.S. The influence of atmospheric moisture transport on fresh water balance of the Atlantic drainage basin: General circulation model simulations and observations. *J. Geophys. Res.* **1992**, *97*, 2765–2773.

73. Zaucker, F.; Stocker, T.F.; Broecker, W.S. Atmospheric freshwater fluxes and their effect on the global thermohaline circulation. *J. Geophys. Res.* **1994**, *99*, 12443–12457.

74. Schmittner A.; Appenzeller, C.; Stocker, T.F. Enhanced Atlantic freshwater export during El Niño. *Geophys. Res. Lett.* **2000**, *27*, 1163–1166.

75. Soden, B.J. The sensitivity of the tropical hydrological cycle to ENSO. *J. Clim.* **2000**, *13*, 538–549.

76. Latif, M.; Roeckner, E.; Mikolajewicz, U.; Voss, R. Tropical stabilization of the thermohaline circulation in a greenhouse warming simulation. *J. Clim.* **2000**, *13*, 1809–1813.

77. Fedorov, V.; Dekens, P.S.; McCarthy, M.; Ravelo, A.C.; deMenocal, P.B.; Barreiro, M.; Pacanowski, R.C.; Philander, S.G. The Pliocene paradox (Mechanisms for a permanent El Niño). *Science* **2006**, *312*, 1485–1489.

78. LaRiviere, J.P.; Ravelo, A.C.; Crimmins, A.; Dekens, P.S.; Ford, H.L.; Lyle, M.; Wara, M.W. Late Miocene decoupling of oceanic warmth and atmospheric carbon dioxide forcing. *Nature* **2012**, *486*, 97–100.

79. Goldner, A.; Herold, N.; Huber, M. The challenge of simulating the warmth of the mid-Miocene climatic optimum in CESM1. *Clim. Past* **2014**, *10*, 523–536.

80. Sloan, L.C.; Pollard, D. Polar stratospheric clouds: A high latitude warming mechanism in an ancient greenhouse world. *Geophys. Res. Lett.* **1998**, *25*, 3517–3520.

81. Emanuel, K. A simple model for multiple climate regimes. *J. Geophys. Res.* **2002**, *107*, doi:10.1029/2001/JD001002.

82. Lear, C.H.; Rosenthal, Y.; Wright, J.D. The closing of a seaway: Ocean water masses and global climate change. *Earth Planet. Sci. Lett.* **2003**, *210*, 425–436.

83. Billups, K.; Channell, J.E.T.; Zachos, J. Late Oligocene to early Miocene geochronology and paleoceanography. *Paleoceanography* **2002**, *17*, doi:10.1029/2000PA000568.

84. Sluijs, A.; Schouten, S.; Pagani, M.; Woltering, M.; Brinkhuis, H.; Sinninghe Damste, J.S.; Dickens, G.R.; Huber, M.; Reichart, G.-J.; Stein, R.; *et al.* Subtropical Arctic Ocean temperatures during the Palaeocene/Eocene thermal maximum. *Science* **2006**, *441*, 610–613.

85. Lohmann, G.; Pfeiffer, M.; Laepple, T.; Leduc, G.; Kim, J.-H. A model-data comparison of the Holocene global sea surface temperature evolution. *Clim. Past* **2013**, *9*, 1807–1839.

86. Bradshaw, C.D.; Lunt, D.J.; Flecker, R.; Salzmann, U.; Pound, M.J.; Haywood, A.M.; Eronen, J.T. The relative roles of CO_2 and palaeogeography in determining late Miocene climate: Results from a terrestrial model-data comparison. *Clim. Past* **2012**, *8*, 1257–1285.

87. Frank, M.; O'Nions, R.K.; Hein, J.R.; Banakar, V.K. 60 Ma records of major elements and Pb-Nd isotopes from hydrogenous ferromanganese crusts: Reconstruction of seawater paleochemistry. *Geochim. Cosmochim. Acta* **1999**, *63*, 1689–1708.

88. Frank, M.; Whiteley, N.; Kasten, S.; Hein, J.R.; O'Nions, R.K. North Atlantic Deep Water export to the Southern Ocean over the past 14 Myr: Evidence from Nd and Pb isotopes in ferromanganese crusts. *Paleoceanography* **2002**, *17*, doi:10.1029/2000PA000606.

89. Delaney, M.; Boyle, E.A. Cd/Ca in Late Miocene benthic foraminifera and changes in the global organic carbon budget. *Nature* **1987**, *330*, 156–159.

90. Butzin, M.; Lohmann, G.; Bickert, T. Miocene ocean circulation inferred from marine carbon cycle modeling combined with benthic isotope records. *Paleoceanography* **2011**, *26*, doi:10.1029/2009PA001901.

91. Zhang, Z.-S.; Nisancioglu, K.H.; Chandler, M.A.; Haywood, A.M.; Otto-Bliesner, B.L.; Ramstein, G.; Stepanek, C.; Abe-Ouchi, A.; Chan, W.-L.; Bragg, F.J.; *et al.* Mid-Pliocene Atlantic Meridional Overturning Circulation not unlike modern? *Clim. Past* **2013**, *9*, 1495–1504.

92. Stepanek, C.; Lohmann, G. Modelling mid-Pliocene climate with COSMOS. *Geosci. Model Dev.* **2012**, *5*, 1221–1243.

93. Lyle, M.; Barron, J.; Bralower J.; Huber, M., Lyle, A.O.; Ravelo, A.C.; Rea, R.K.; Wilson, P. Pacific Ocean and cenozoic evolution of climate. *Rev. Geophys.* **2008**, *46*, doi:10.1029/2005RG000190.

94. Knorr, G.; Lohmann, G. A warming climate during the Antarctic ice sheet growth at the Middle Miocene transition. *Nat. Geosci.* **2014**, *7*, 376–381.

8

A Novel Mooring Tether for Highly-Dynamic Offshore Applications; Mitigating Peak and Fatigue Loads via Selectable Axial Stiffness

Tessa Gordelier [†], David Parish [†,*], Philipp R. Thies and Lars Johanning

University of Exeter, Treliever Road, Penryn, Cornwall TR10 9FE, UK;
E-Mails: tjg206@exeter.ac.uk (T.G.); p.r.thies@exeter.ac.uk (P.R.T.); l.johanning@exeter.ac.uk (L.J.)

[†] These authors contributed equally to this work.

[*] Author to whom correspondence should be addressed; E-Mail: d.n.parish@exeter.ac.uk

Academic Editors: Bjoern Elsaesser and Umesh A. Korde

Abstract: Highly-dynamic floating bodies such as wave energy convertors require mooring lines with particular mechanical properties; the mooring system must achieve adequate station keeping whilst controlling mooring tensions within acceptable limits. Optimised compliant mooring systems can meet these requirements but where compliance is achieved through system architecture, the complexity of the system increases together with the mooring footprint. This work introduces the "Exeter Tether", a novel fibre rope mooring tether providing advantages over conventional fibre ropes. The tether concept aims to provide a significantly lower axial stiffness by de-coupling this attribute from the minimum breaking load of the line. A benefit of reduced axial stiffness is the reduction of mooring system stiffness providing a reduction of peak and fatigue loads, without increasing mooring system complexity. Reducing these loads improves system reliability and allows a reduction in mass of both the mooring system and the floating body, thus reducing costs. The principles behind the novel tether design are presented here, along with an outline of eight prototype tether variants. Results from the proof of concept study are given together with preliminary findings from sea trials conducted in Falmouth Bay. Results demonstrate that the Exeter Tether can be configured to achieve a significantly lower axial stiffness than conventional fibre rope and that the stiffness is selectable within limits for a given breaking strength. Strain values greater than 0.35 are achieved at 30% of line breaking strength; this represents more than a threefold increase of the strain achievable with a conventional rope of the same material. The tether was subjected to six months of sea trials to establish any threats to its own reliability and to inform future design enhancements in this respect.

Keywords: elastomeric mooring; compliant mooring; peak load; fatigue load; mooring load; tether; wave energy; reliability; axial stiffness; DMaC; SWMTF

1. Introduction

The mooring system is one of the most critical sub-systems for a floating offshore installation. In particular, marine renewable energy developers seek to install devices in highly-dynamic environments governed by wave and tidal conditions. The requirements and design issues are extensively described by [1–4]. Importantly, mooring systems must satisfy the following requirements:

1. Survivability under extreme load conditions.
2. Long-term reliability.
3. Provision of required compliance so as to minimise peak loads.
4. Minimise the mooring spread footprint.

As a consequence, items 1 and 2 typically require a high Minimum Breaking Load (MBL), to allow sufficiently high factors of safety (FOS) to warrant long-term reliability. For conventional mooring systems both requirements conflict with objectives three and four, and *vice versa*. The cost of conventional mooring line material (e.g., chain, steel wire and polyester) is directly proportional to the rated MBL [4]. As a consequence any peak loads, such as those experienced during storm events, have a direct impact on the mooring cost. The dilemma for floating offshore installations is that the capital cost of the mooring system is driven by extreme (peak load) conditions, whilst the revenue is generated under normal operating conditions. If peak loads can be mitigated the cost of mooring systems and associated structural elements, as well as deployment and installation costs, can be significantly reduced. The design challenge is to find a feasible combination of all four objectives listed above. Objectives 1–3 are directly affected by the compliance and associated MBL of the mooring lines. In some cases system architecture can assist with achieving objectives 1–3 (e.g., by introducing multi-catenary mooring configurations); however, this comes at the expense of objective 4[4].

Wave buoys typically feature a highly elastic mooring configuration using rubber materials [5]. This satisfies the design requirements for wave buoys to follow the orbital wave motion (item three above), whilst absorbing some of the wave and tide-induced forces to increase system reliability (item two above).

At the other end of the spectrum are taut mooring systems for tension-leg platforms (TLPs) using steel wire, which are one of the proposed solutions for floating offshore wind installations [6,7]. The compliance of these systems is minimal and very high MBLs are required to satisfy item 1 (survivability).

Whilst the different mentioned mooring solutions match the requirements for wave buoys and TLPs well, a combination of soft and stiff response elements would constitute an optimised mooring solution as it would allow a reconciliation of the diverging design objectives. A number of systems have been proposed to combine these characteristics, among which are the Seaflex buoy mooring system [8] and the TfI mooring tether [9,10]. The development and proof of concept for a third innovative mooring design, the Exeter Tether, is the subject of this paper.

Recent studies [9,11] have indicated that mooring systems with this non-linear, combined response offer significant potential for peak load reductions. A numerical study [12] specifically performed for the Exeter Tether described in this paper has demonstrated peak load reduction by a factor of three. The non-linear coupled modelling and simulations compared the non-linear mooring load response against real load data during high energy wave conditions for a conventional nylon rope.

Conventional fibre ropes display hysteresis when subjected to cyclic loading, a proportion of the work done during the extension being transferred to heat [13]. This is an important consideration for both conventional ropes and the Exeter Tether (which also displays hysteretic behaviour) as the hysteresis will contribute to the damping of the system. The hysteretic damping characteristics of the Exeter Tether are not detailed within this paper but will be addressed as further work.

As for all innovative systems the performance characteristics and long-term behaviour require careful consideration, research and demonstration. This paper addresses these aspects for the Exeter Tether by introducing the design philosophy behind the tether and presenting results from a proof of concept study conducted with eight tether prototypes.

2. Technical Background

2.1. Axial Stiffness, Maximum Strain, and Minimum Breaking Load

Three important properties of a mooring line that strongly influence its performance [4] are:

2.1.1. Axial Stiffness in Tension

This parameter describes the extension of a line in relation to its original length, when it is subjected to a given tensile load. A line with high stiffness (low compliance), for instance steel wire or steel chain, will not yield much when a load is applied [14]. This high stiffness can lead to excessively high "snatch" loads being generated within the mooring system which are transmitted into the floating structure [4].

Axial stiffness is defined as *load/strain*, or the gradient of the *load vs. strain* plot line.

2.1.2. Minimum Breaking Load

The minimum breaking load (MBL) under tension is specified for any rope, chain or similar structural tie. This value can be considered to be the least value at which a rope, chain or other will fail completely. Some permanent damage or change might occur at a lower load.

2.1.3. Maximum Limit of Axial Strain

This defines the maximum extension that a line can achieve before breaking at MBL. Conventional fibre ropes can achieve a maximum strain of around 0.40 (nylon, three-strand laid construction, new rope) [13]. The ability to achieve high values of strain can be useful where large displacements must be allowed e.g., when tide height varies significantly in relation to the water depth.

In conventional fibre ropes the axial stiffness and MBL are strongly associated parameters. Consequently there is little capability to vary the stiffness of any particular rope, which is governed by the MBL. Some selection of stiffness for a given MBL is possible by means of the following:

- Material selection: fibre rope for offshore mooring might be of polyester, nylon, high modulus polyethylene or other polymer construction. The different polymer yarns exhibit differing extension and recovery properties; nylon has the lowest stiffness [13].
- Construction geometry: fibre rope for offshore use can be constructed such that the main load carrying sub-ropes run either parallel to the rope itself (parallel lay), are helically wound within the rope (three-strand laid), or those that approximate a helical form, such as braided or plaited ropes. Ropes with parallel lay sub-ropes will exhibit higher stiffness than ropes where the load is carried helically [15].

The lowest axial stiffness for any given MBL of conventional fibre rope will therefore be achieved with a nylon rope. However, the advantageous maximum strain of 0.4 is only available at loads approaching the MBL of the rope. If a factor of safety is applied, to allow for uncertainties in the load case and degradations to the rope, the axial stiffness increases with the increase to MBL.

2.2. The Exeter Tether

The Exeter Tether [16,17] is a tether assembly comprising a hollow braided fibre rope, an elastomeric core assembly and at least one anti-friction membrane between the rope and the core (see Figure 1).

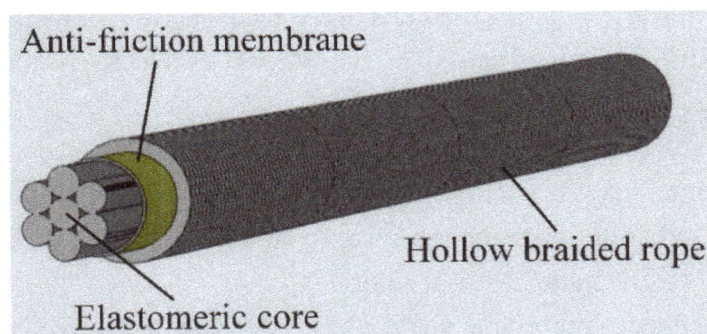

Figure 1. Representation of the Exeter Tether assembly.

The tension load exerted onto the tether is carried predominantly by the hollow rope which is terminated with an eye splice at each end. As the braided rope extends, its diameter contracts according to the pitch angle of the braid. The primary function of the elastomeric core is to resist this diametric change and in so doing, to resist the extension of the braided rope. The degree to which the rope's diametral contraction (and, hence, extension) is resisted, will depend upon the compressibility of the elastomeric core. This can be altered by design changes to the elastomer material hardness and to the cross sectional form of the core assembly. A core structure with seven round sections as shown will provide easy initial compression of the core whilst the elastomer cords deform towards a more solid cross section. This will result in an initial soft extension response of the tether which will stiffen markedly as the elastomer core assembly becomes solid. From this point on, further extension is achieved via the Poisson's diminution of the elastomer material, the extension of the polyester rope strands and a degree of embedment of the rope strands into the surface of the core. With a braid angle in excess of 50° (from the axis), the hollow rope has a mechanical advantage in compressing the core during the initial extension. However, as the braid angle decreases with the tether extension, so does the mechanical advantage over the core. This gradual change acts to reinforce the two stage extension property of the tether attributed to the core structure. Braid angle change and diametral contraction are demonstrated in Figure 2.

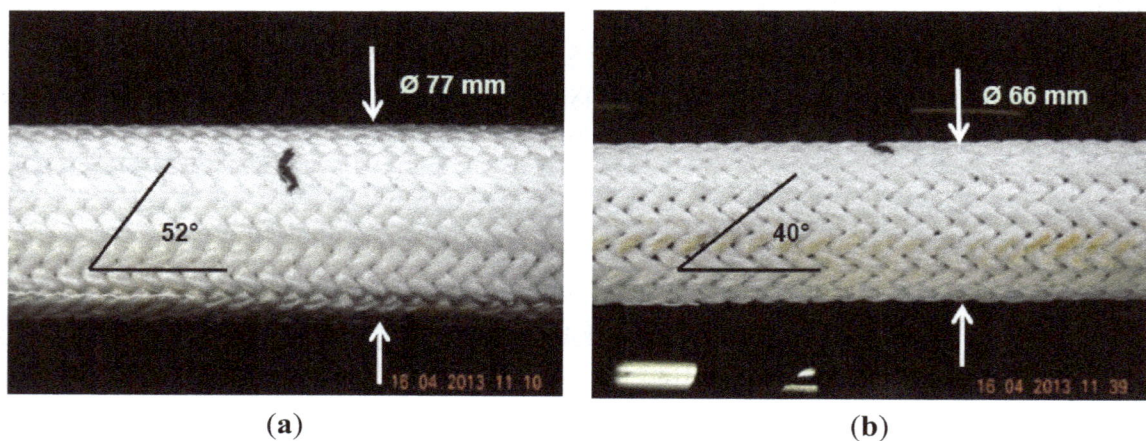

Figure 2. (**a**) P1-2 tether at zero extension; and (**b**) P1-2 tether at 30% extension.

The core assembly extends as the tether extends and therefore the elastomer cords carry a tensile load themselves. However, due to the low Young's modulus of the core's elastomer material, this contribution to load carrying is less than 10% of the load carried by the tether [12]. The remaining balance of the load is carried by the hollow rope.

Importantly, design changes to the core's material and structure can be made independently of the strength of the hollow rope such that the extension properties of the tether are not coupled to its MBL. This allows the selection, at tether design stage, of lower axial stiffness and a higher strain limit whilst specifying the MBL to allow an adequate factor of safety.

Scaling of the tether between prototype stages has been successfully achieved through the maintenance of geometric proportions in the cross section of the tether. This includes the cross sectional area of the load carrying rope strands which scale the linear density and hence the MBL of the rope accordingly. The diameter of the braided rope strands does not need to follow the geometric scaling directly and the number of strands comprising the braid is not critical. The solidity of the pack of the rope strands in relation to the cross sectional area of the hollow rope is however important and this is maintained as a constant when scaling the tether.

3. Proof of Concept Prototypes

Prototype tethers were constructed for the proof of concept study and are referred to as the P1 series prototypes. The elastomer cores, together with their anti-friction membranes, were assembled by University of Exeter (UoE). These core assemblies were then taken to Lankhorst Ropes' manufacturing facility in Maia, Portugal, where the rope was braided onto the cores and the eye splices were made. The completed tether assemblies were then shipped back to UoE for test work and analysis.

A total of 12 tether variants were manufactured in the P1 series; this paper presents extension performance results for five of these tethers P1-2, P1-3, P1-4, P1-5, and P1-6. Also presented are durability findings from P1-8 and P1-16 as well as breaking strength results from P1-17. These eight tether variants are described in Table 1.

Table 1. Construction of the P1 series prototypes presented in this paper.

Tether Identity	Specified Elastomer Hardness (Shore A)	Measured Elastomer Hardness (Shore A)	Elastomer Cord Diameter (mm)	Anti—Friction Membrane Material	Working Tether Length (m)
P1-2	50	54	25	PVC	2.5
P1-3	60	59	25	PVC	2.5
P1-4	70	71	25	PVC	2.5
P1-5	80	70	25	PVC	2.5
P1-6	90	81	25	PVC	2.5
P1-8	70	-	25	Dacron	2.5
P1-16	70	-	20	PVC	2.0
P1-17	70	71	25	PVC	2.0

Note: P1-16 and P1-17 are shorter in working and overall length so that the breaking load of the tether is achievable within the 1 m stroke of DMaC.

3.1. Core Construction

The core architecture as detailed in Figure 1 comprised a seven strand bundle of Ø25 or Ø20 mm section cords. The elastomer material used for the P1 series is ethylene propylene diene monomer (EPDM). The variants of the tether introduced in this paper are constructed from EPDM with measured durometer hardness values of between 54 and 81 Shore A.

3.2. Anti-Friction Membrane

The anti-friction membrane serves two purposes: initially the membrane binds the elastomer core assembly together providing some limited structural integrity prior to over-braiding with hollow rope; in service the membrane offers a lower friction surface for the rope strands to move across. For the P1 tethers this membrane was achieved with a helically wound tape of 50 mm in width, the windings overlapping by approximately one third of the tape width. Two alternative materials are used, adhesive PVC tape of 0.13 mm thickness and Dacron sail tape of 0.2 mm thickness.

3.3. Hollow Rope

The material selected for the hollow rope was polyester. The construction was a 1 × 1 braid of 48 strands (24 in each helix direction) with a strand diameter of 4.5 mm. The braiding machine was set to produce a 202.5 mm pitch helix for each strand. The resulting outer diameter of the hollow rope was 60 mm which increased to approximately 80 mm when braiding onto the standard core (constructed of 25 mm cords) which was fed into the rear of the machine. The hollow rope was terminated at both ends using a form of Lankhorst Rope's A3 eye splice.

3.4. Tether Assemblies

The construction and identification of the P1 series of prototypes described in this paper is detailed in Table 1. The dimensions of the tether are detailed in Figure 3.

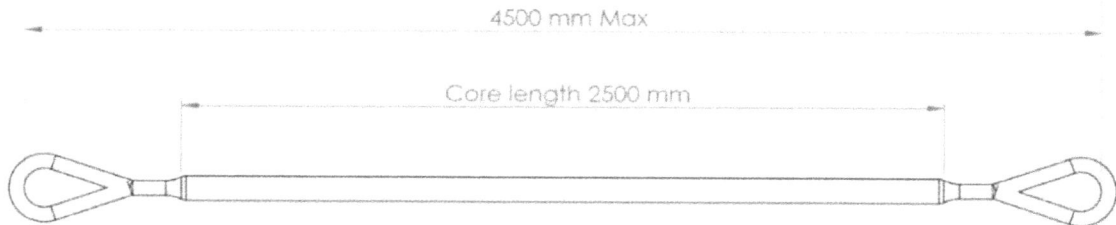

Figure 3. Working length (core length) and overall length of the P1 tethers (except P1-16/17).

4. Methods

4.1. Dynamic Marine Component Test Facility

4.1.1. Facility Overview

The Dynamic Marine Component test facility (DMaC) is based in Falmouth Docks and is owned and operated by the University of Exeter. It is a large horizontal test machine that has a linear actuator and a two-degrees-of-freedom headstock. Further specifications of DMaC and examples of other component tests are detailed in [18–20]. For the tether test work the headstock is not utilised and the linear actuator is used to provide displacement of up to 1000 mm and tension of up to 222 kN. The linear actuator follows a prescribed time series for either displacement or for tension and in both cases has full feedback control of the driving parameter.

The test piece can be submerged in fresh water which is essential for the tether test work in order that the assembly is properly lubricated. For the tether test work, an interchangeable headstock platen was manufactured that provided 800 mm of pre-tension travel via an M64 thread. The "top hat" form of this platen also increases the effective test bed length of DMaC by 300 mm. Figure 4a shows DMaC with a tether assembly fitted ready to test (submerged in water) and Figure 4b shows the pre-tension adjuster providing maximum pre-tension. The pre-tension adjuster is important because it allows the slack to be removed from the test piece without using any of the 1000 mm linear stroke available from the hydraulic ram.

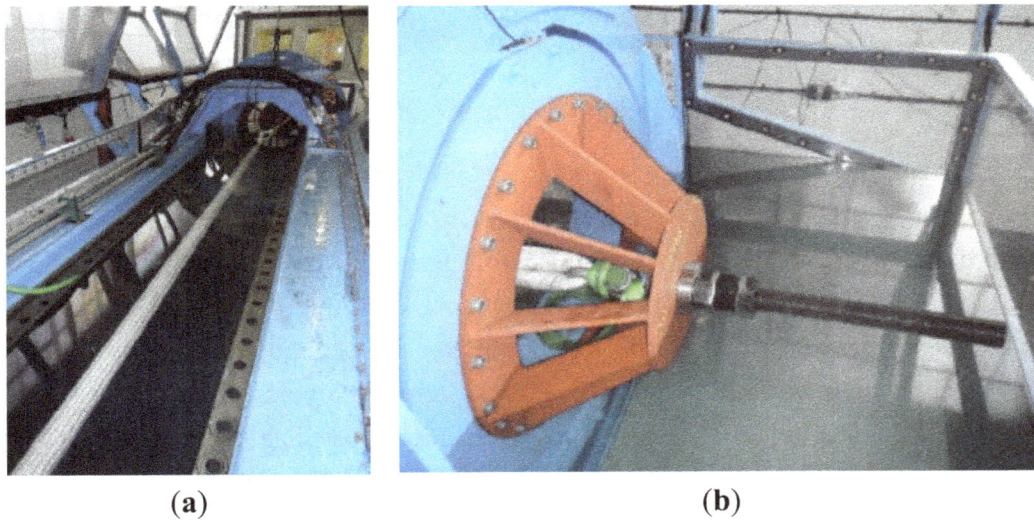

(a)   **(b)**

Figure 4. (**a**) DMaC with a tether fitted and full of water; and (**b**) the pre-tension adjuster and "top hat".

4.1.2. Calibration

DMaC was calibrated using a reference five tonne load cell which itself has calibration traceable to national standards. The results of the final DMaC calibration run are given in Figure 5.

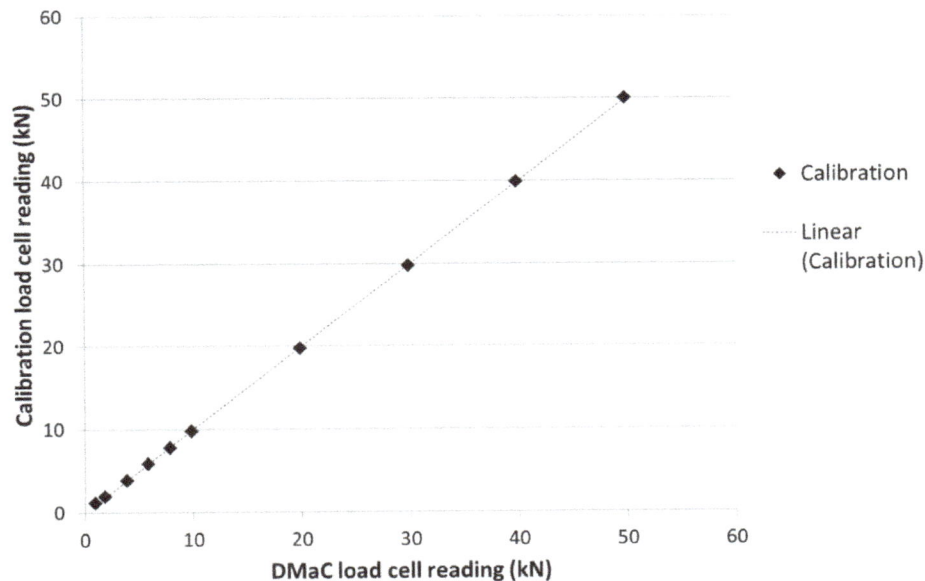

Figure 5. Final calibration run of DMaC using 5 T reference load cell.

A line of best fit has been fitted to the data points in the final calibration run which achieves a R^2 value of 0.9999 (the square of the Pearson product moment correlation coefficient). The equation for the line in the final calibration run here is:

$$y = 1.0016x - 0.0752 \qquad (1)$$

Applying this linear relationship to the range of loads investigated in the test work reported below provides a maximum error of ±0.07 kN.

4.2. Test Considerations

The following considerations inform the general test methodology for the tether extension tests:

- The Mullins effect, or stress softening, occurs in most thermoplastic elastomers [21]. The effect manifests itself as a reduction in the stress required to achieve a particular strain value once that strain value has been achieved on a single occasion; it occurs in both tension and compression. Cantournet *et al.* [22] state that the behaviour of these elastomers does not strictly follow the idealised case and that the effect is still significant for the second strain cycle. They add that stability is soon reached with only negligible effect after 5–10 cycles.
- The Exeter Tether is specifically designed for use underwater. The elastomeric core assembly requires the lubrication and cooling afforded to it by submersion in water.
- Conventional fibre ropes require repeated strain cycles before the stability is adequate for representative test results. Ten cycles are recommended although individual test plans differ [13]. During these "bedding in" cycles, individual fibre components of the rope adopt their optimal position. The hollow braided rope of the P1 prototypes is likely to require less bedding in than ropes of a more complex construction.
- The tether is a novel and unknown technology. The test methodology must strike a balance between rigorous test work and the preservation of the prototypes which had unknown durability at the onset of the test work.
- The tether is primarily designed for use as a mooring tether for floating marine renewable energy converters. The scale of the P1 prototypes is suited to small devices moored in the nearshore environment.

4.3. General Test Methodology for Tether Extension Tests

The considerations listed above resulted in the following general practises for extension tests:

- All extension test work was conducted with the tethers fully submerged in fresh water with the exception of the eye splice extension test and the breaking load test.
- All of the P1 tethers were subjected to conditioning tests to achieve "bedding in" prior to any primary data tests or durability tests. Primary data is the data which is subsequently presented or used in further analysis. The conditioning tests are summarised in Table 2 together with the primary data and durability tests.
- Cyclical primary data tests have five extension cycles, the fifth cycle providing the primary data.
- Extension cycles follow a sine wave, having a frequency of 0.125 Hz (8 s period).

Table 2. Summary of the conditioning, primary data, and durability extension tests.

Test I.D.	Test Type	Driving Parameter	Pre-Tension (kN)	Peak Tension (kN)	Displacement Limits (m)	Period (s)	Cycles (No.)
ETT_03	conditioning	force	1	10		8	10
ETT_04	conditioning	force	2	20		8	10
ETT_05	conditioning	force	2	40		8	5

Table 2. *Cont.*

ETT_06	conditioning	force	2	60		8	5
ETT_08	primary data	displacement	1.55	-	0.5–0.9	8	5
ETT_19	primary data	displacement	variable	-	0–0.99	8	5
TETT_26	durability	force	2.2	110		8	1000
TETT_27	durability	force	2.2	132		8	1000
TETT_28	durability	force	2.2	154		8	1000

4.4. DMaC Extension Tests

4.4.1. Eye Splice Extension Tests

The extension data output by DMaC relates to the extension of the entire tether rather than the working length. It is therefore necessary to quantify the axial stiffness of the eye splice terminations so that the extension of the eye splices can be subtracted from the total extension data to reveal the extension experienced by the working length of the tether.

Tests were performed on tethers P1-3 and P1-6 (following conditioning tests) using the displacement driven test ETT_08 (see Figure 6). A draw wire linear transducer was used to measure the extension between the connection shackle and the closest end of the working tether length. These tests were performed without submersion to eliminate the risk of water ingress and damage to the transducer.

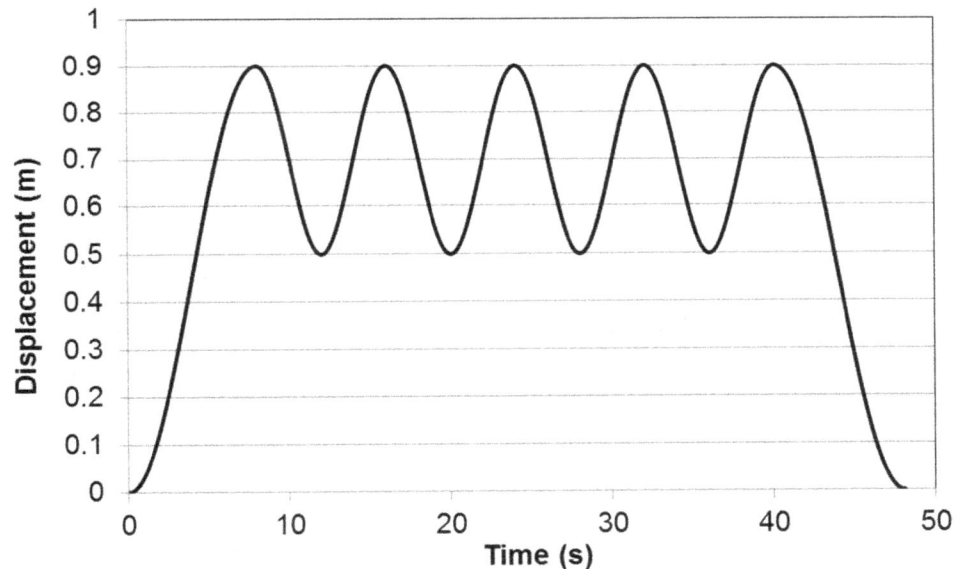

Figure 6. The displacement (extension) drive data for test ETT_08.

4.4.2. Breaking Load Test

Due to the compliance of the P1 series tethers, the maximum tension load that could be realised using all of the pre-tension and the full hydraulic stroke length of DMaC fell well short of parting a tether. It was necessary to manufacture a reduced length tether, P1-17, see Table 1.

Displacement mode test ETT_08 was used to break the tether following the conditioning tests. The test was conducted without submerging the tether but using a hose to thoroughly wet the assembly.

4.4.3. Performance Tests Referenced to a Tension Load Datum

A tension load datum might refer to the static pre-tension of a mooring line when the floating body is at calm and the tide height is at a minimum. Following completion of the conditioning tests the pre-tension was set to 1550 N and the tether was left at this tension for a prolonged period (overnight) to stabilise. At the end of this stabilisation period the pre-tension was reset to the tension datum (1550 N) if any drift had occurred. A displacement mode test (the displacement time series drives the linear actuator) was then conducted according to test script ETT_08. The drive data for this test is given in graphical form as Figure 6.

4.4.4. Performance Tests Referenced to a Displacement Datum

Referencing to a displacement datum allows for easier comparisons between the P1 tethers and conventional rope through comparison of calculated strain values.

The ETT_19 displacement driven test was developed to utilise a greater displacement range on DMaC (0–990 mm). The displacement followed a sine wave form with a period of 8 s and five cycles. Tethers P1-2 and P1-6 were tested according to ETT_19. The test was conducted four times on each tether. Incremental increases in the test pre-tension were made up to a maximum possible pre-tension resulting from the full uptake of the adjuster thread.

4.4.5. Fatigue Endurance Test

The current design stage of the tether aims to demonstrate and investigate the functional performance characteristics and does not yet address weaknesses related to fatigue and durability. However, this test stage was included to gain an early understanding of any critical weakness that might exist with the concept. To this end, a "Thousand Cycle Load Limit" (TCLL) test was conducted on a single tether.

The TCLL test was developed by the Oil Company's International Marine Forum (OCIMF) to quantify mooring hawser response to tension—tension fatigue (cycling between lower and higher tension values) [23]. Here, the basic concepts of the TCLL test have been adapted to make it more appropriate for the P1 series tether and DMaC. These adaptations are associated with the frequency of cycling, the rate of increase of strain and the wetting of the test piece.

The tests load the tether cyclically for 1000 cycles per test at increasing load steps, starting from 50% MBL, and increasing in 10% increments until failure. Short periods of static load are permitted between each test step, with the load always maintained above 1% MBL. An adapted tether (P1-16) was prepared in order that the required loads could be achieved within the 1000 mm stroke available from DMaC. The MBL for the tether was allocated a value of 222 kN based on the result of the breaking load test detailed in Section 4.4.2 Test scripts were prepared according to these test requirements and are listed in Table 2.

4.5. Sea Trials: South West Mooring Test Facility (SWMTF)

Qualitative testing was employed to review the effects of prolonged operation in the marine environment on the tether. The use of conventional polyester rope as the predominant load carrier allows for predictable design for anticipated loads and raises minimal durability concerns. However, due to the novel nature of other components of the tether, key durability considerations of interest from this sea trial are:

- Biofouling: the extent marine growth can infiltrate the rope, anti-friction membrane, and core of the tether is of interest. In the short term the change in surface friction and additional weight could affect tether operation, and in the long term the introduction of rough particles could lead to friction wear and eventually failure of the tether [24,25].
- Anti-friction membrane durability: the effect of the environment and continuous tether operation on the membrane was reviewed.
- Core integrity: the effect of the environment and continuous compression loading on the elastomer core was reviewed.

The SWMTF operated by the UoE is located in Falmouth Bay and enables mooring systems and components to be exposed to representative sea conditions, as described in the literature [20,26]. The 3.25 tonne displacement SWMTF buoy has a three-limbed compliant mooring system, each limb comprising a chain catenary and a fibre rope upper element. The major components of the standard mooring limb are as follows (listed from the seabed upwards):

1. 1.1 tonne drag embedment anchor
2. DN32 stud link chain × 5 m
3. DN24 open link chain × 36 m
4. Diameter 44 mm double braid nylon rope × 20 m

Four tethers were attached to each other end on end using cow hitches to form a single 19 m long tether. This line of tethers was substituted for the 20 m nylon rope tail (detailed in point four above) from the southern limb of the SWMTF mooring system (Figure 7). The tethers were ordered as shown in Table 3 with P1-8 being positioned at the top where marine growth and wave action is more intensive. Tether P1-8, having a mid-range nominal core hardness of 70 Shore A and Dacron helical tape anti-friction membrane, was selected as the primary subject for this trial.

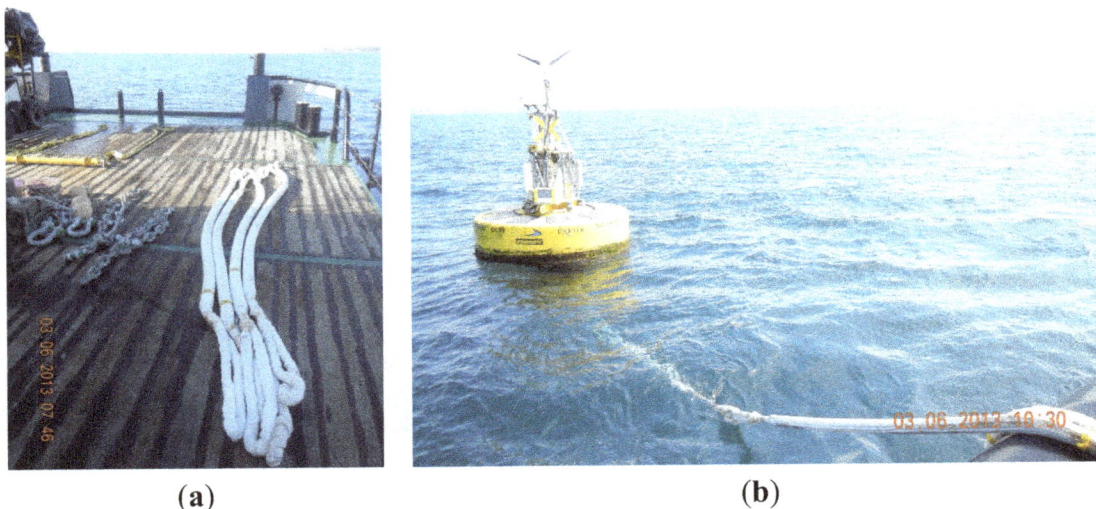

(a)   (b)

Figure 7. (a) Four tethers for deployment; (b) tethers being deployed at SWMTF.

Table 3. The tethers deployed at SWMTF and their position in the southern limb.

Position	Lower (1st)	2nd	3rd	Upper (4th)
Tether	P1-10	P1-12	P1-3	P1-8

Deployment of the tethers at the SWMTF was between 3 June 2013 and 26 November 2013 after which they were recovered for examination.

4.6. Durometer Hardness Testing

The core elastomer material was supplied as five extruded round section lengths of 25 mm diameter having specified durometer hardness values of 50, 60, 70, 80, and 90 Shore A. A sample of 18 mm in length was cut from the middle part of each extrusion. The test end of each sample piece was polished using a wet 240 grit micro-section polishing wheel to produce a uniform flat surface. A Mitutoyo Hardmatic HH-331(A) durometer was used to take three readings for each test piece. Care was taken to distribute the three tests around the face of each test piece so as to avoid misrepresentation caused by slow material recovery after penetration of the indenter. Test indentations were made approximately 8 mm from the edge of the test face.

5. Results and Discussion

5.1. DMaC Extension Tests

5.1.1. Eye Splice Extension Tests

Figure 8 shows the extension of a P1-3 eye splice recorded by the linear transducer over the five cycles of the ETT_08 test. The final sine wave is selected from the data set and the gradient from the cycle load up data (as shown by dotted line in Figure 8) is identified as a reference value.

This test and data analysis was repeated for P1-6 and the results are given in Table 4. The mean value of 1965.93 kN/m was inverted to 5.09×10^{-4} m/kN and then doubled to 1.02×10^{-3} m/kN to approximate the total eye splice extension of a P1 series tether under load up conditions.

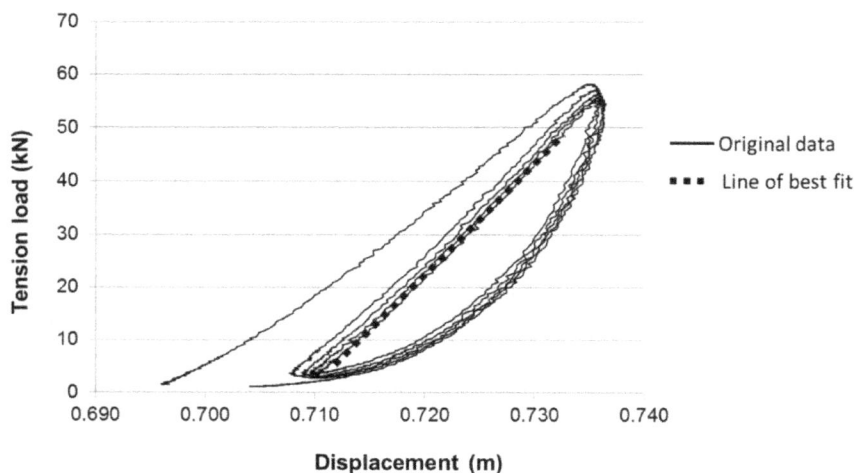

Figure 8. P1-3 eye splice extension during test ETT_08.

Table 4. Results of the eye splice extension tests (where R^2 value is the square of the Pearson product moment correlation coefficient).

Tether I.D.	Straight Line Gradient (kN/m)	R^2 Value
P1-3 (single end)	2065.68	0.999
P1-6 (single end)	1866.18	0.999
Mean (single end)	1965.93	-

5.1.2. Breaking Load Test

Figure 9 shows the load extension curve obtained from the breaking load test performed on tether P1-17. The maximum load recorded was 222 kN; this value is used to represent the MBL of the P1 tether series.

Failure of P1-17 occurred where the braided hollow rope is converted into two sub-ropes which then enter into the eye splice (Figure 10).

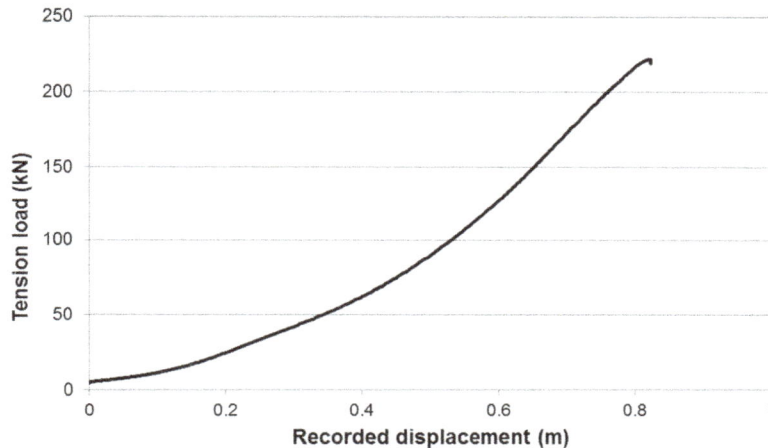

Figure 9. Load/displacement graph showing failure of P1-17 at 222 kN.

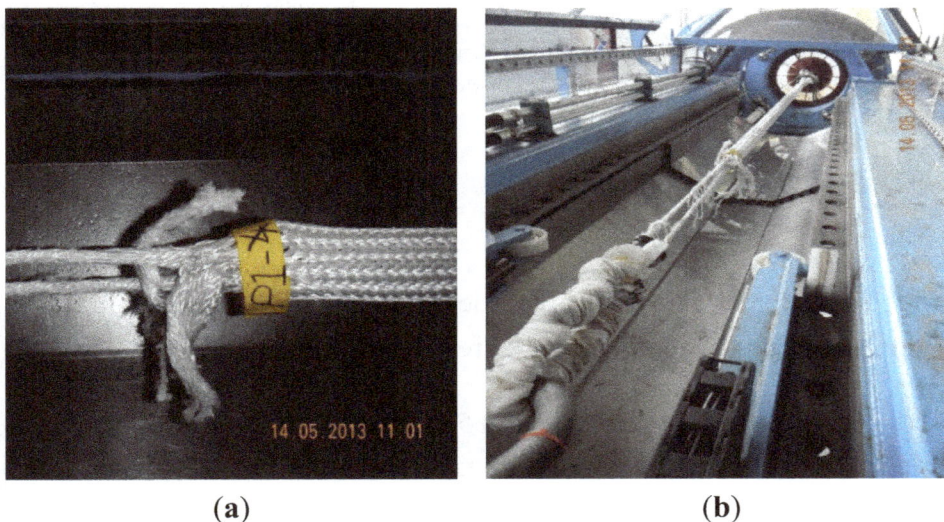

(a)	**(b)**

Figure 10. (a) Breaking load test failure detail; **(b)** breaking load test tether following failure in DMaC test facility.

5.1.3. Performance Tests Referenced to a Tension Load Datum

The final cycle (fifth cycle) load up data is identified. For each data time step, the incremental increase in tension is used to calculate the extension of the eye splices by applying the value 1.02×10^{-3} m/kN derived in Section 5.1.1. The eye splice extension is then subtracted from each extension value recorded by DMaC to provide data corresponding to the extension of the working part of the tether. The extension is normalised against the original working length and expressed as a percentage. The tension load is normalised against the MBL of 222 kN and expressed as a percentage. Figure 11a shows the outcome of these tests in graphical form.

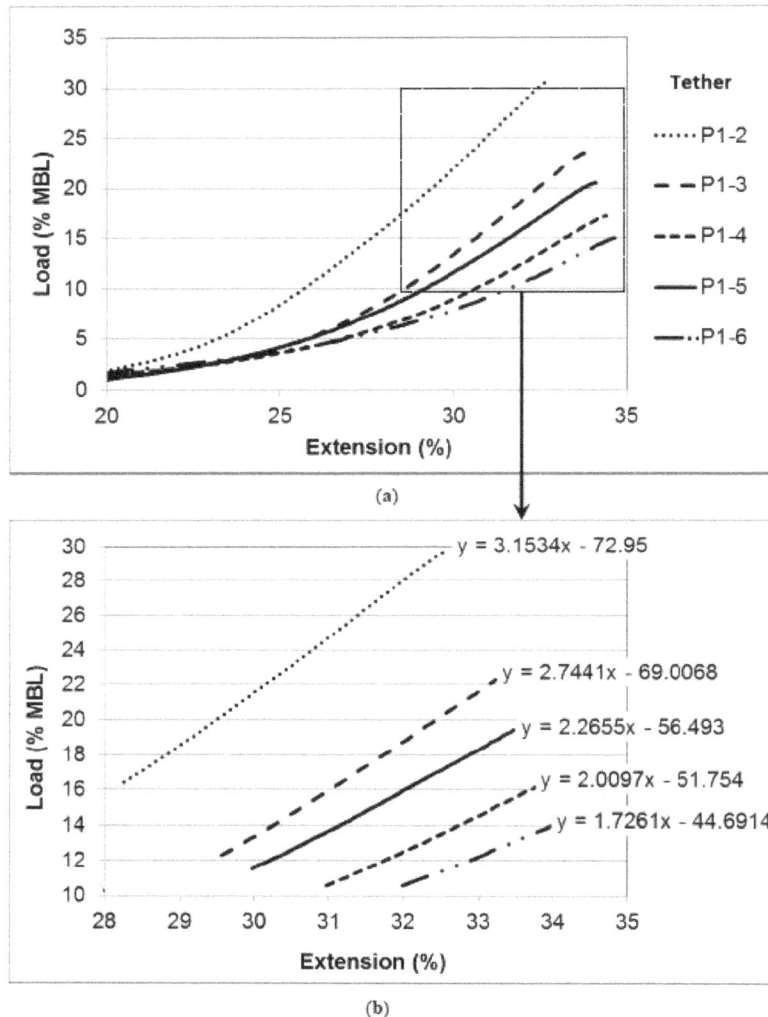

(a)

(b)

Figure 11. (a) P1 series tether extension properties from a 1550 N pre-tension datum; and **(b)** data clipped to achieve a R^2 0.9995 linear regression.

The five tethers exhibit a similar load response at 0.5 m displacement (see displacement test script, Figure 6). The divergence of the five plot lines thereafter demonstrates the differences in axial stiffness through the range of tethers. It is clear that in all cases the tether approximates a linear relationship between load and extension beyond a certain tension load. Figure 11b shows further analysis of this behaviour. In this figure data points have been clipped from both ends of each plot line to expose the near linear portion:

- The final 20 data points have been clipped from each data set to remove a small portion of non-linear behaviour at the top end of the load up cycle. This non-linear behaviour is caused by the viscous, time dependant properties of the elastomers as the displacement sine wave causes the stroke velocity to tend towards zero.
- A much greater amount of data has been removed from the lower end of the plot lines. Sufficient data was removed so that an R^2 value (the square of the Pearson product moment correlation coefficient) of 0.9995 was achieved with the remaining data.

The equation for the near linear line was then detailed (Figure 11b); the critical value being the gradient, as this represents the tether axial stiffness, the crucial property under investigation.

The equation of the best fit straight line is shown in Figure 11b and the gradients are repeated in Table 5. The tethers are ranked according to their gradient and it is apparent that there is a relationship between the durometer hardness of the elastomer and the gradient (durometer hardness results are detailed in Section 5.3). It should be noted that the stiffest tether is achieved with the softest core material and the most compliant tether with the hardest core material. This counter-intuitive result will be discussed in Section 6.

Table 5. Tabulated results of the linear regressions shown in Figure 10.

Hardness (Shore A)	Tether	Gradient	Stiffness Ranking
54	P1-2	3.1534	1
59	P1-3	2.7441	2
70	P1-5	2.2655	3
71	P1-4	2.0097	4
81	P1-6	1.7261	5

5.1.4. Performance Tests Referenced to a Displacement Datum

The final cycle (fifth cycle) load up data is identified. In these tests a waterproof linear transducer recorded extension of the eye splice at one end of the tether during every test. For each displacement data point, the single eye splice extension result was doubled to approximate the total eye splice extension. This value was then subtracted from the total tether extension recorded by DMaC to derive the extension of the working part of the tether.

The tethers yield significantly upon initial loading, taking on a temporary extension "set". For this analysis, the extension results are referenced to a "dynamic zero load length" that better represents the free length of the tether during cyclic loading. This zero load length is derived from a simple static load *versus* extension graph for each tether. The equation of the best fit straight line is then applied to the load recorded at the first data point to derive the corresponding extension.

The tension load for the tethers is referenced against the MBL of 222 kN based on the breaking load test (Section 5.1.2). This allows load to be plotted as % of MBL for direct comparison to other ropes. Figure 12 details the extension properties for the tethers tested alongside a reference rope: a double braid polyester rope in the as new condition (data obtained from Lankhorst Ropes). The P1 series prototypes exhibit two phases of extension with an intermediate transition phase. The first phase has very low axial stiffness, the second phase being markedly stiffer. This closely matches the expectations of the mechanisms described in Section 2.2.

The comparison of P1-2 and P1-6 to the reference rope demonstrates the reduced stiffness achieved with the tether design, with P1-6 achieving a strain in excess of 35% whilst remaining under 30% MBL.

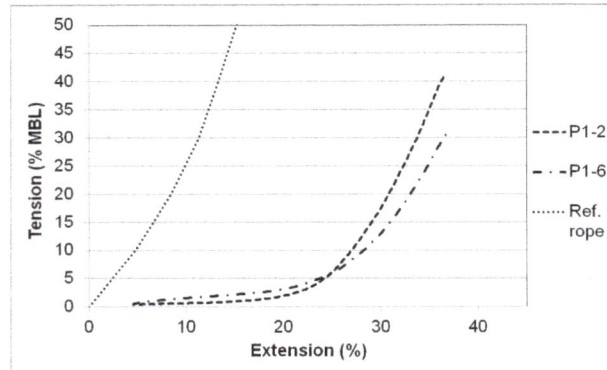

Figure 12. Normalised extension properties of Exeter Tether prototypes shown alongside a double braid polyester reference rope.

5.1.5. Fatigue Endurance Test

Tether P1-16 failed during test TETT 28 (a load range of 1%–70% MBL or 2.2–154 kN) at the 176th cycle. The calculation for the thousand cycle load level is detailed in [23]:

$$TCLL = 100\% - \frac{6.91(100\% - TLL)}{Ln\ CTF} = 59.91\% \tag{2}$$

where TLL = test load level at which cycles to failure was determined

CTF = cycles to failure at test load level

6.91 = natural logarithm of 1000

Further work is required to fully quantify the TCLL value in relation to other mooring options. Some publications suggest a TCLL of just 52% for polypropylene ropes [27]; however improvements in rope technology are now producing ropes with TCLL values approaching 80% [27,28]. For an early proof of concept prototype 60% is an acceptable TCLL with clear potential for improvement in subsequent prototypes.

Figure 13. Failed tether under fatigue cycle loading.

The failure (parting of strands) occurred at the point where the rope increases in diameter to envelop the core bundle. The edge of the core bundle caused fretting which is likely to have promoted this failure, shown in Figure 13.

Minor changes to the geometry of the core, such as a more gradual slope from the full diameter bundle to the empty rope should reduce the fretting in this area and lead to an improved TCLL value. Other variations on membrane could also be trialled to reduce friction at this point. As previously mentioned the P1 series tethers were designed as a proof of concept and durability was not a main objective at this stage.

5.2. Sea Trials: South West Mooring Test Facility (SWMTF)

Significant marine growth developed on the upper tethers (Figure 14a). After cleaning with a bristle brush and fresh water, there was no external evidence of fretting or degradation other than a noticeable colour change to the rope (Figure 14b). Visual inspection and microscope investigation were used to assess the tether components. The results from these qualitative sea trials are intended as a preliminary durability assessment to inform a series of quantitative laboratory tests.

(a)   **(b)**

Figure 14. (a) Upper tethers with mussel growth; **(b)** P1-8 tether after cleaning.

P1-8 was opened up to review the effect of the sea trial on the internal components of the tether.

- Biofouling: evidence of marine growth infiltrating all parts of the tether was clear. As detailed in Figure 15 the significant growth on the outer rope had penetrated through the rope and was visible on the anti-friction membrane with some debris evident on the core strands. Although much of the growth had been crushed by the action of the tether (Figure 16a), further investigation with a microscope revealed byssal filaments on both the rope inner surface (Figure 16b) and on the anti-friction membrane (Figure 16c) suggesting mussels had fully infiltrated the rope and were attempting to attach to the membrane itself. The potential effect of this marine growth can be seen in Figure 16d, which details the wear caused by the infiltration of the marine growth on the yarns of the rope. During longer deployments wear of this nature could ultimately lead to failure of the rope.
- Anti-friction membrane durability: in addition to the marine growth, considerable degradation had occurred to the Dacron membrane as shown in Figure 17a,b. The edges of the membrane

have frayed significantly and "pressure tears" are evident within the membrane. These rough edges create a higher friction surface which could lead to increased levels of wear to the load carrying rope during tether operation. The early stage of this may be detailed in Figure 18 where evidence of the filaments of the rough membrane can be seen in an area of worn rope.

- Core integrity: minor compression deformations were visible on the outer surface of the core elastomer strands as shown in Figure 19. The repeated high pressure exerted by the rope on the core during tether operation is likely to have caused this imprint. Although the permanent deformation of the core will not lead to a critical failure of the tether, it requires further investigation as it may lead to altered operating parameters and therefore reduced mooring load mitigation.

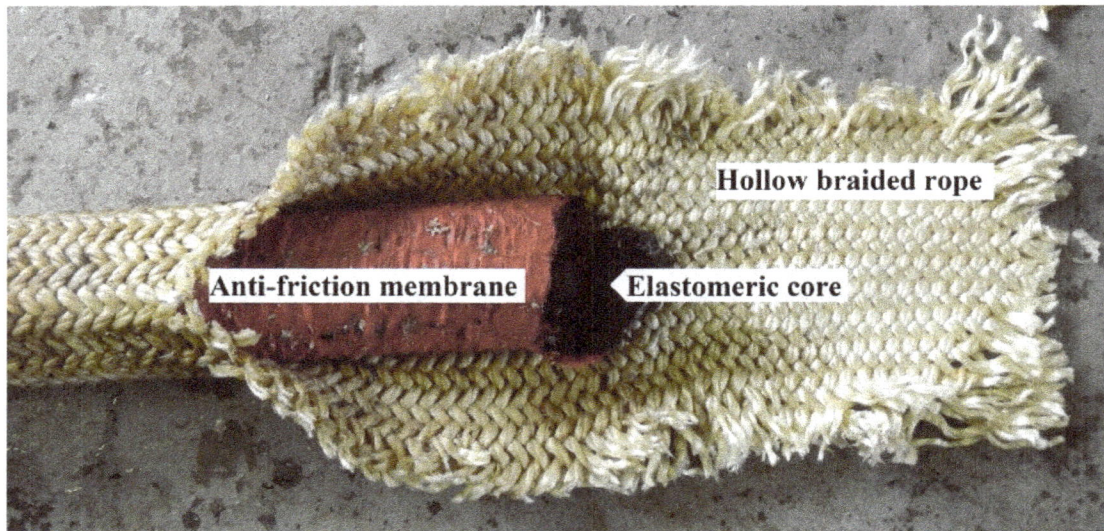

Figure 15. Internal view of P1-8 following sea trials, debris from marine growth is evident on the anti-friction membrane.

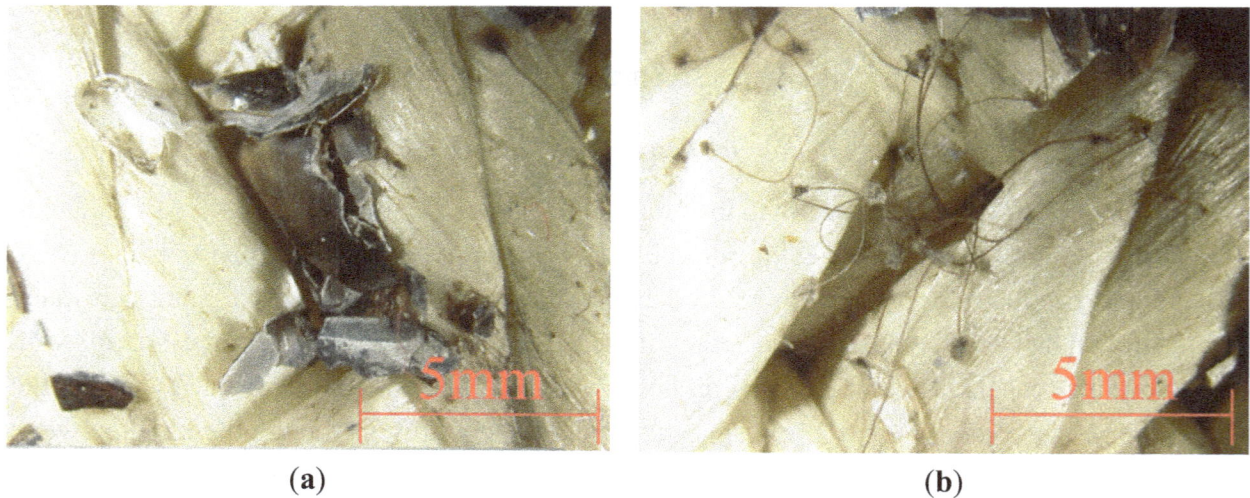

(a)   (b)

Figure 16. *Cont.*

(c)

(d)

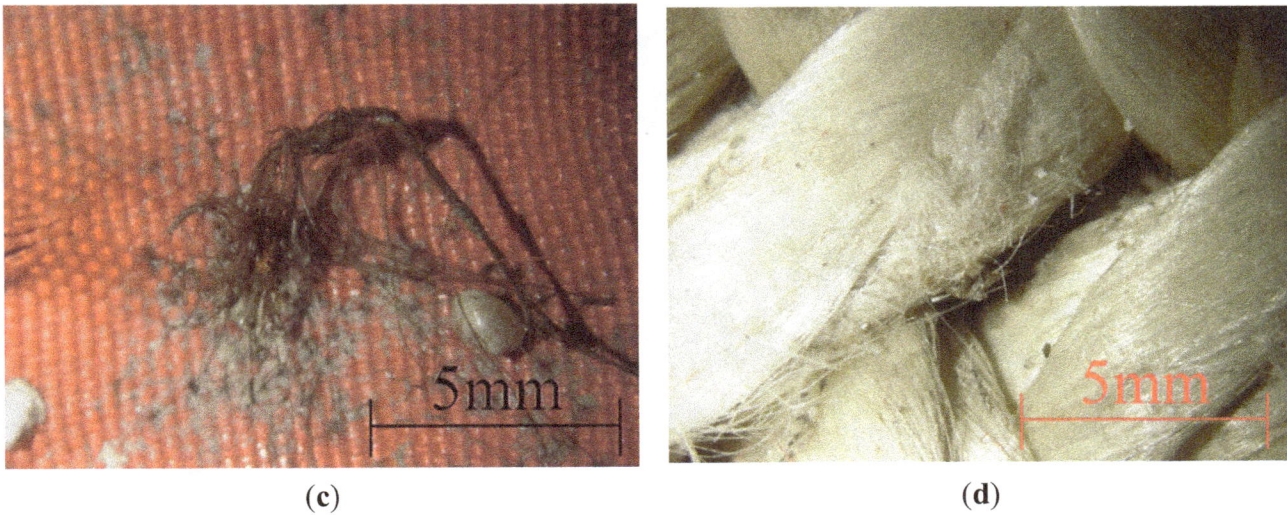

Figure 16. (a) Crushed mussel growth; **(b)** byssal filaments on inner surface of rope; **(c)** byssal filaments on the anti-friction membrane; and **(d)** evidence of wear of rope yarns.

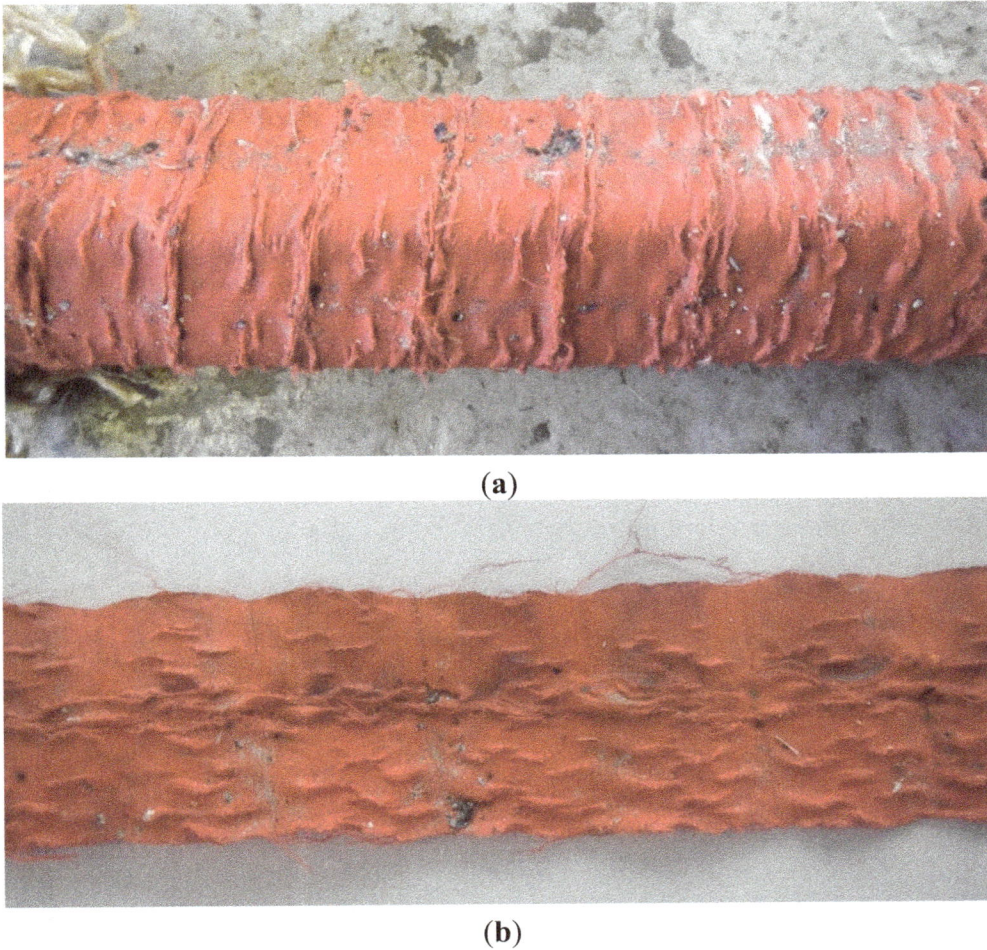

(a)

(b)

Figure 17. (a) Degradation of outer surface of anti-friction membrane; and **(b)** degradation of inner surface of anti-friction membrane.

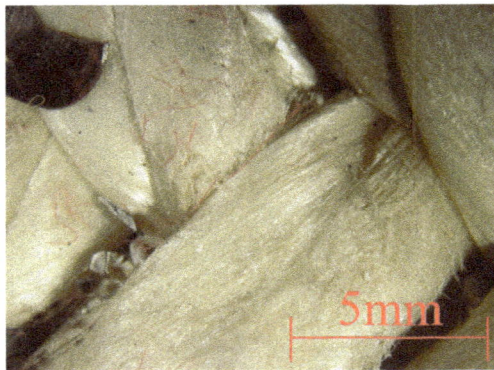

Figure 18. Red strands from anti-friction membrane evident in areas exhibiting wear of rope yarns.

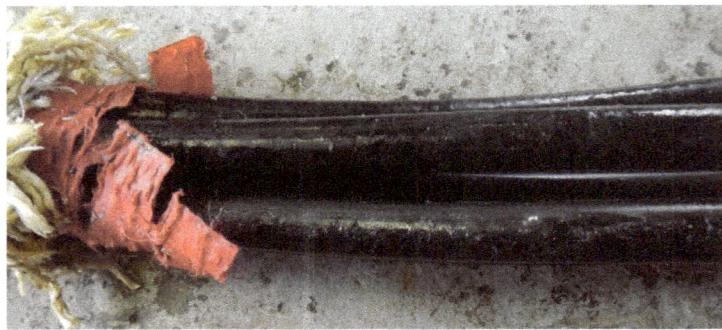

Figure 19. Minor compression deformations on the outer surface of the tether core strands.

5.3. Durometer Hardness Tests

Durometer hardness readings and mean results are given in Table 6.

From the results detailed, it is clear that the EPDM used for both P1-5 and P1-6 was not as specified. Results detailed in Section 5.1.3 relating tether performance to hardness values have, therefore, been presented with reference to the mean hardness values detailed in Table 6.

Table 6. Durometer test results.

Target Tether I.D.	Specified Hardness (Shore A)	Hardness Readings (Shore A)			Mean Hardness (Shore A)
P1-2	50	54	54	54	54.0
P1-3	60	59	59	59	59.0
P1-4	70	70	71	71	70.7
P1-5	80	70	70	70	70.0
P1-6	90	81	80	81	80.7

6. Conclusions and Further Work

The results described here prove the working concept of the Exeter Tether. The tether successfully de-couples the extension properties from the MBL of the load carrier. Thus, the tether allows the selection, between certain limits, of axial stiffness for a given MBL. The tether is shown to have satisfactory load carrying capability and durability for this prototype stage of its development.

During tests, the tether displayed two phases of extension according to the design principles, each phase having a distinct axial stiffness. The two phases are separated by a smooth transition phase. The initial phase is one that provides soft extension properties up to a load limit of around 5% of MBL. The second phase of extension displays a markedly stiffer behaviour but remains less stiff than a double braid rope of the same material.

The stiffness of the second phase is shown to be inversely related to the durometer hardness of the core material. It is possible that this effect is due to bulging of the core material into the space between the braided rope strands; this material then provides direct resistance to the braid angle decrease during tether extension. The softer materials provide a greater resistance due to increased penetration into the braid. The analysis of this effect needs to be progressed further before this theory can be confirmed.

Strain values in excess of 0.35 (35% extension) are shown to be achievable whilst remaining below 30% MBL. This is more than three times greater than the strain value achievable at 30% MBL for a conventional double braid rope of the same material.

Results from the first sea trials of the P1 prototypes are promising and indicate that the tether can survive nearly six months of exposure to the marine environment with realistic loading conditions and show minimal wear to the principal load carrier. However, the sea trials have also highlighted three aspects of the tether that require further investigation. Firstly, the development of an outer protective jacket is necessary to prevent the ingress of marine growth into the inner components of the tether. Secondly, the anti-friction membrane should be developed to better withstand the operating conditions of the tether. Several alternative anti-friction membranes have been assembled and will be trialled under controlled conditions in DMaC prior to another round of sea testing. Finally, following evidence of permanent deformation of the tether core, detailed quantitative testing has been conducted on the elastomer, investigating both the effects of the marine environment and continuous fatigue load cycles on key material properties. These test results will be detailed in future publications.

The results reported here demonstrate that the Exeter Tether has the potential to mitigate the conflict between axial stiffness and MBL discussed in the introduction. This will enable mooring designers to achieve a more compliant mooring design thus reducing peak and fatigue loads and subsequently reducing the costs of all structural elements within the system. Further work is on-going to quantify the reductions in peak line loads possible through use of the tether.

Acknowledgments

The authors gratefully acknowledge the support of Lankhorst Ropes who collaborated with the University of Exeter to conduct the proof of concept study. The authors would also like to acknowledge the support of the UK Centre for Marine Energy Research (UKCMER) through the SuperGen programme funded by the EPSRC. Support from IFREMER's Materials in a Marine Environment Laboratory is also gratefully acknowledged, funded through the MARINET Programme.

Author Contributions

T.G. and D.P. contributed equally to this paper, and performed all experiments in collaboration. D.P. conceived the Exeter Tether and developed the P1 series. D.P. led on tether functionality experimentation and data processing. T.G. led on tether durability experimentation and data processing.

P.T. wrote Section 1 and L.J. provided scientific advice and supervision throughout. All authors discussed the results and implications, and commented on the manuscript at all stages.

Conflicts of Interest

The authors declare no conflict of interest.

References

1. Johanning, L.; Wolfram, J. Challenging tasks on moorings for floating wecs. In Proceedings of the International Symposium on Fluid Machinery for Wave and Tidal Energy (IMechE), London, UK, 19 October 2005.
2. Johanning, L.; Smith, G.; Wolfram, J. Towards design standards for wec moorings. In Proceedings of the 6th European Wave Tidal Energy Conference, Glasgow, UK, 29 September–2 October 2005.
3. Johanning, L.; Smith, G.H.; Wolfram, J. Measurements of static and dynamic mooring line damping and their importance for floating wec devices. *Ocean Eng.* **2007**, *34*, 1918–1934.
4. Harris, R.E.; Johanning, L.; Wolfram, J. Mooring systems for wave energy converters: A review of design issues and choices. In Proceedings of the 3rd International Conference on Marine Renewable Energy, Blyth, UK, 6–9 July 2004.
5. Joosten, H. Elastic mooring of navigation buoys. *Int. Ocean Syst.* **2006**, *10*, 16.
6. Musial, W.; Butterfield, S.; Boone, A. Feasibility of floating platform systems for wind turbines. In Proceedings of the 23rd ASME Wind Energy Symposium, Reno, NV, USA, 5–8 January 2004.
7. Bachynski, E.E.; Moan, T. Design considerations for tension leg platform wind turbines. *Mar. Struct.* **2012**, *29*, 89–114.
8. Bengtsson, N.; Ekström, V. *Seaflex. The Buoy Mooring System: Increase Life Cycle and Decrease Cost for Navigation Buoys*; Technical Report for Navigations Teknik AB: Vinslöv, Sweden, 2010.
9. McEvoy, P. Combined elastomeric and thermoplasit mooring tethers. In Proceedings of the 4th International Conference on Ocean Energy, Dublin, Ireland, 17–19 October 2012.
10. Thies, P.R.; Johanning, L.; McEvoy, P. A novel mooring tether for peak load mitigation: Performance and service simulation testing. *Int. J. Mar. Energy* **2014**, *7*, 43–56.
11. Thies, P.R.; Crowley, S.; Johanning, L.; Micklethwaite, W.; Ye, H.; Tang, D.; Cui, L.; Li, X. Novel mooring design options for high-intensity typhoon conditions—An investigation for wave energy in china. In Proceedings of the Royal Institution of Naval Archiects (RINA) Structural Load & Fatigue on Floating Structures, London, UK, 25–26 February 2015.
12. Parish, D. A Novel Mooring Tether for Highly Dynamic Offshore Applications. University of Exeter: Exeter, UK, 2015.
13. McKenna, H.A.; Hearle, J.W.S.; O'Hear, N. *Handbook of Fibre Rope Technology*; Woodhead Publishing in Textiles: Cambridge, UK, 2000.
14. Oil Companies International Marine Forum. *Effective Mooring: Your Guide to Mooring Equipment and Operations*, 2nd ed.; Witherby's Publishing: Livingston, UK, 2005.
15. Weller, S.; Davies, P.; Johanning, L. The influence of load history on synthetic rope. In Proceedings of the European Wave and Tidal Energy Conference, Aalborg, Denmark, 2–5 September 2013.

16. Parish, D.; Johanning, L. Mooring Limb. International Patent Application Publication Number: Wo 2011/089545 A1, 28 July 2011.

17. Parish, D.; Johanning, L. Mooring limb. US Patent Application Publisher No. US 2012/0298028 A1, 29 November 2012.

18. Johanning, L.; Thies, P.R.; Smith, G.H. Component test facilities for marine renewable energy converters. In Proceedings of the Marine and Offshore Renewable Energy Conference, London, UK, 21–23 April 2010.

19. Thies, P.R.; Johanning, L. Development of a marine component testing facility for marine energy converters. In Proceedings of the 3rd International Conference on Ocean Energy, Bilbao, Spain, 6–8 October 2010.

20. Weller, S.D.; Thies, P.R.; Gordelier, T.; Parish, D.; Harnois, V.; Johanning, L. Navigating the valley of death: Reducing reliability uncertainties for marine renewable energy. In Proceedings of the ASRSNet International Conference on Offshore Renewable Energy, Glasgow, UK, 15–17 September 2014.

21. Mark, J.E.; Erman, B.; Roland, M. *The Science and Technology of Rubber*; Academic Press: Waltham, MA, US, 2013.

22. Cantournet, S.; Desmorat, R.; Besson, J. Mullins effect and cyclic stress softening of filled elastomers by internal sliding and friction thermodynamics model. *Int. J. Solids Struct.* **2009**, *46*, 2255–2264.

23. Oil Companies International Marine Forum. *Guidelines for the Purchasing and Testing of Spm Hawsers*; Witherby's Publishing: Livingston, UK, 2000.

24. *Polyurethane Elastomer Coating*; Offshore division, Lankhorst Ropes: Póvoa de Varzim, Portugal, 2010.

25. Ayres, R. *Characterizing Polyester Rope Mooring Installation Damage*; Inc. report to Minerals Management Service: Washington, DC, USA, 2001.

26. Johanning, L.; Spargo, A.; Parish, D. Large scale mooring test facility: A technical note. In Proceedings of the 2nd International Conference on Ocean Energy (ICOE), Brest, France, 15–17 October 2008.

27. Baaj, H.P. *Maritime Ropes Briefing 1/2011*; Lankhorst Ropes: Sneek, The Netherlands, 2011.

28. High Performance Ropes Euroflex (Catalouge Extract); Lankhorst Ropes: Sneek, The Netherlands, 2008.

9

Time Evolution of Man-Made Harbor Modifications in San Diego: Effects on Tsunamis

Aggeliki Barberopoulou [1,2,*], Mark R. Legg [3,†] and Edison Gica [4,5,†]

[1] National Observatory of Athens, Institute of Geodynamics, Lofos Nymphon, 11810 Athens, Greece

[2] AIR Worldwide, 131 Dartmouth Street, Boston, MA 02116, USA

[3] Legg Geophysical, 16541 Gothard St # 107, Huntington Beach, CA 92647, USA;
E-Mail: mrlegg@verizon.net

[4] Pacific Marine Environmental Laboratory, Center for Tsunami Research, National Oceanic and Atmospheric Administration, 7600 Sand Point Way NE, Seattle, WA 98115, USA;
E-Mail: edison.gica@noaa.gov

[5] Joint Institute for the Study of the Atmosphere and Ocean, University of Washington, 3737 Brooklyn Ave NE, Box 355672, Seattle, WA 98195-5672, USA

[†] These authors contributed equally to this work.

[*] Author to whom correspondence should be addressed; E-Mail: abarberopoulou@air-worldwide.com

Academic Editors: Valentin Heller, Billy Edge and Umesh A. Korde

Abstract: San Diego, one of the largest ports on the U.S. West Coast and home to the largest U.S. Navy base, is exposed to various local and distant tsunami sources. During the first half of the twentieth century, extensive modifications to the port included but were not limited to dredging, expansion of land near the airport and previous tidal flats, as well as creation of jetties. Using historical nautical charts and available Digital Elevation Models, this study gives an overview of changes to San Diego harbor in the last 150+ years due to human intervention and examines the effects of these changes on tsunamis. Two distant and two local scenarios were selected to demonstrate the impact of modified nearshore topography and bathymetry to incoming tsunamis. Inundation pattern, flow depths, and flooded localities vary greatly from year to year in the four scenarios. Specifically, flooded areas shift from the inner harbor to outer locations. Currents induced by the distant tsunamis intensify with modifications and shift from locations primarily outside the harbor to locations inside. A new characteristic in tsunami dynamics associated with port modifications is the introduction of high current spots. Numerical results also show that the introduction of high currents could threaten navigation, vessels, and facilities at narrow openings and also along the harbor "throat"—therefore, at an increased number of locations. Modifications in the port show that changes could have a negative but also a positive impact through constraint of flooding outside of the harbor and shifting of high currents to locations of minimal impact. The results of this study may be used as a first step toward future harbor design plans to reduce tsunami damages.

Keywords: tsunami; San Diego; port modifications

1. Introduction

Harbor modifications and redevelopment plans (see current plans for the Port of San Diego in [1]) are quite common to enhance or increase space in ports and, thus, to improve operations. In fact, harbor modifications at San Diego Bay decreased the water area by filling shoals around the main channels to add more land, including the area where Lindbergh Field—now known as San Diego Airport—was built. Space was developed for mooring boats, including private yachts and Navy vessels, but the reduction of the water surface area increased tsunami amplitudes and currents by constricting channels between Bay margins. Deepening of channels by dredging to improve navigability also enables greater penetration of tsunami energy into the harbor. Similar problems occurred along major rivers, e.g., the Mississippi River, when construction of levees restricted the channel width, negating the purpose of natural flood plains to distribute and absorb the excess water volume during major floods (e.g., severe flooding in the Midwest and South in 1973 [2]).

Ports are designed to be protected against vertical water fluctuations, but the subject of tsunami currents—a newly recognized problem—is not typically considered in the design of harbor modifications. Tsunami currents form a common threat to ports, as seen in previous tsunamis (22 May 1960, Chile; 27 March 1964, Alaska; 27 February 2010, Chile; 11 March 2011, Japan; e.g., [3,4]); usually, they are not accompanied by significant (>1 m) tsunami amplitudes. Tsunami amplitudes are related to the tsunamigenic potential of a source, but tsunami duration and tsunami currents—even if they relate to relatively small amplitude waves—are also important. The long duration of major trans-Pacific tsunamis (from Mw8+ subduction events) creates a persistent hazard that may produce the greatest damage many hours after the initial tsunami arrival.

In order to understand how anthropogenic changes to ports affect tsunamis, we studied the time evolution of man-made modifications in San Diego Bay since the late nineteenth century and its influence on tsunami amplitudes and currents. San Diego Bay is one of the largest harbors on the U.S. West Coast, and it has been significantly transformed in the twentieth century.

Most of the changes in San Diego Bay have included dredging, enlargement of the North Island/Coronado land area, and creation of new marinas by enhancing already existing sand bars and shoals with fill and creating breakwaters. As done along major rivers, dredging increases the channel depth, but addition of levees, breakwaters, and other shore protection structures combined with narrowing of the channel results in the increase of flow depths and current speeds during inundation (flooding or surge). Tsunami currents can cause damage to structures (e.g., by erosion or scouring; [5]).

We examined the effects of harbor modifications on tsunami hazard by simulating major tsunami events from both local and distant sources. The changes primarily occurred during the first half of the twentieth century. Four specific years were selected as representative of major changes in San Diego Bay: 1892, 1938, 1945 and present day. Post-1945, the bay continued to undergo changes but maintained a similar appearance and, therefore, the bathymetry/topography in and around the bay is represented by a Digital Elevation Model (DEM) showing the harbor shape as we know it today (dated 2014). To reflect these changes, historical nautical charts were used to show the state of the bay during that period.

2. San Diego Bay

San Diego Harbor is located on the southwestern corner of the United States. It has a long crescent-like shape of approximately 24 km in length and 1–3 km width. The harbor is the fourth largest in California and is home to a U.S. Navy base offering shelter to several aircraft carriers and other naval vessels[1]. Naval Air Station North Island (NASNI; [6]) is located at the north end of the Coronado peninsula in San Diego Bay. It is the home port of several aircraft carriers of the U.S. Navy, and is part of the largest aerospace-industrial complex in the U.S. Navy, Naval Base Coronado, California ([6]). A cruise ship terminal is located near downtown San Diego, a regular ferry service runs from Coronado to San Diego, and several marinas, for both private and commercial boats, exist throughout the harbor. Although it is naturally well protected from large wave amplitudes, the same does not apply for tsunami currents.

3. Time Evolution of San Diego Bay

San Diego Bay is one of the best natural harbors in the world, protected by the Point Loma Peninsula at the western harbor entrance, and by North Island, Coronado, and the Silver Strand enclosing the inner harbor. Coronado is an affluent resort town located in San Diego County, California, just over 8 km (5.2 mi) SSW of downtown San Diego, with a population of 24,697 ([7]). Coronado Island lies on a peninsula connected to the mainland by a 16-km (10 mi) isthmus called the Silver Strand (locally, The Strand; [8]). North Island was an uninhabited sand flat in the nineteenth century (Figure 1A). It was also referred to as North Coronado Island, because it was separated from South Coronado (now the city of Coronado) by a shallow bay, known as the Spanish Bight, which was later filled in during World War II (1945; see Figures 1 and 2). South Coronado, which is not an island but the terminus of a peninsula known as the Silver Strand, became the town of Coronado.

---

[1]   There are also some submarines, an explosives loading area, and mooring facilities and docks for many other naval vessels throughout the bay ([6]).

We present nautical charts that portray the changes that have occurred in San Diego Bay since the nineteenth century. Because modifications to the bay happened quite frequently, especially during the first half of the twentieth century, we summarized major changes over selected years. A small set of original nautical charts are presented next. The major changes primarily included modifications to Coronado Island (also referred to as The Island, The North Island, and Coronado, or North Beach Island and Coronado separated by Spanish Bight[2] in older charts) and the area around the Dutch Flat (see Figure 1). We use Coronado Island to refer to the entire island of Coronado and North Island after the Spanish Bight was filled in at the post-1945 configuration, and North Island to the pre-1945 configuration to distinguish the northern part of Coronado island separated by Spanish Bight in older charts (see Figure 1).

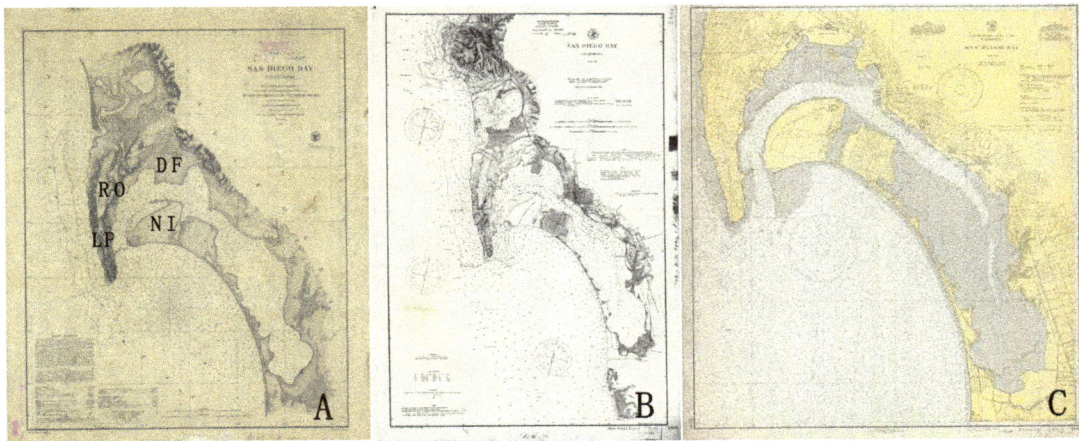

Figure 1. Maps of San Diego Bay. Depths are in feet. (**A**) Nautical chart of 1857. Approximate locations of Dutch Flat (DF), North Island (NI), Roseville (RO), and La Playa (LP) are shown in bold capitals; (**B**) Nautical chart of 1892. The major difference in this chart compared to (A) is the increase of depth at the entrance of the bay, east and northeast of Point Loma (between Roseville and LaPlaya) and between Ballast Point and North Island as a result of dredging. The change appears at least five-fold when compared to the earlier map of 1857; (**C**) Nautical chart of 1938. The major modifications include filling west and north of the North Island (compared to maps A and B) to unify with existing sand dunes and sand spits into North Island and Coronado, with a more regular shape and increased land space. The maps were downloaded from the historical map and chart collection of the Office of Coast Survey, NOAA [9].

3.1. Pre-Nineteenth Century to 1895

During this period, as shown in Figure 1B, the bay remained largely unmodified, with some dredging performed in the late nineteenth century. The depths inside the bay were fairly shallow at the

---

2   Spanish Bight is a graben associated with the Rose Canyon fault zone, e.g., [10]. San Diego Bay lies within the releasing stepover of the Rose Canyon fault zone, which created the bay and uplift of Point Loma. The Silver Strand is a narrow sand spit that connects North Island/Coronado to the mainland at Imperial Beach to the south. San Diego Bay was a natural lagoon formed behind this barrier beach ([11]).

time (characteristic depths 15–54 ft or 4.5–16.5 m near the entrance; Figure 1; [9]). The major characteristics of the bay were the irregularly shaped North Island[3] and the width of the bay around the Dutch Flat (North East of the Bay entrance; see Figure 1) to the northeast. The first tide gauge for San Diego was installed in 1854 at La Playa (near Roseville). It was removed in 1872 and in 1906, it was relocated to the south side of Navy Pier in downtown San Diego. The Dutch Flat area was part of the fan delta complex (depositional area) of the San Diego River, which was diverted to the west through Mission Bay (originally called False Bay) via levee and jetty construction after major floods in the mid-nineteenth century (*ca.* 1862; [12,13]).

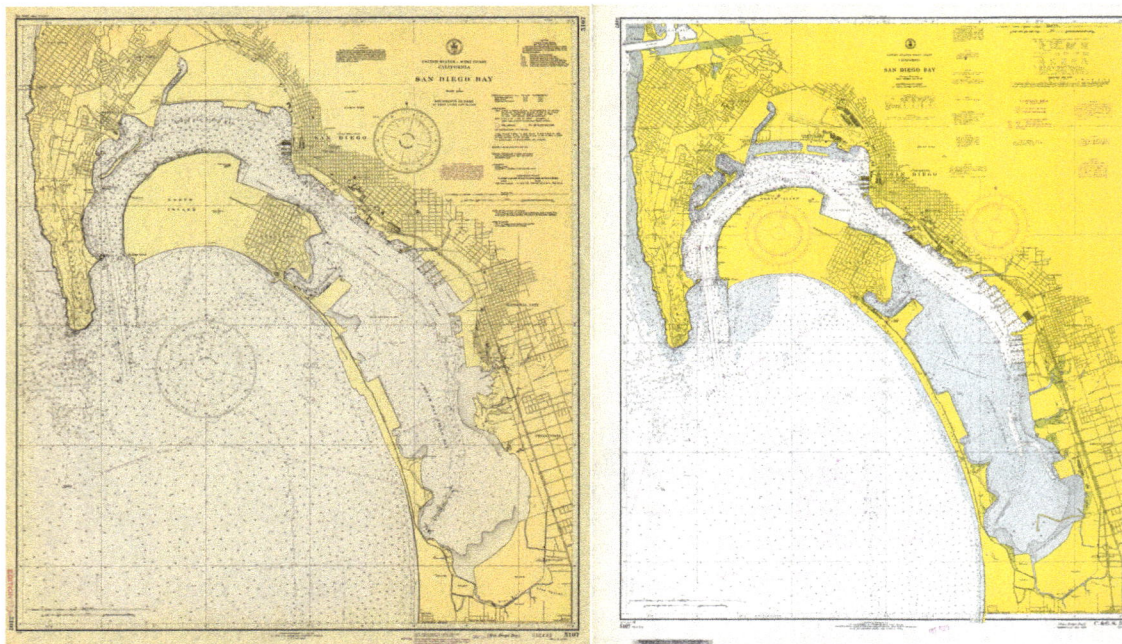

Figure 2. (**Left**) Nautical chart of San Diego Bay of 1945; (**Right**) Nautical chart of San Diego Bay of 1966. The maps were downloaded from the historical map and chart collection of the Office of Coast Survey, National Oceanic and Atmospheric Administration (NOAA; [9]).

3.2. Post-1895 and Early Twentieth Century

Most of the modifications for the enhancement of space in the bay happened during this period and prior to the 1950s. The modifications included dredging at and around the entrance to the bay and expansion around the Dutch Flat, probably to enable better use of the Naval Training Station and the U.S. Marine Corps base near the last location[4]. In addition to the previously noted modifications, sand

---

[3]   North Island was the source of fresh water for the lighthouse keeper on Point Loma during the nineteenth century.

[4]   Originally, this area was the fan delta for the San Diego River, which was flooded in the 1861/62 "Noachian Deluge"—a severe flood event. Levees (restraining wall on 1892, no 5106 chart) were built to redirect the river flow to False Bay (Mission Bay), which finally succeeded after a couple more floods broke through to San Diego Bay. Dutch Flat was expanded, and adjoining basins were subsequently filled as Lindbergh Field (the municipal airport) grew and adjacent land was developed. Finally, Shelter Island and Harbor Island were constructed to provide additional protection for marinas and land for hotels, restaurants, *etc.* (e.g., Figure 3).

spits and sand dunes close to North Island were connected through filling, for the creation of the more regularly shaped North Island[5] in the 1930s (Figure 1C).

3.3 Circa 1945 (1945 to 1960s)

By 1945, San Diego Bay had changed considerably, looking very close to how we know it today. The major changes to the harbor configuration resulted in narrow channels between islands and mainland shorelines, specifically from the harbor entrance to New San Diego (downtown San Diego). The area between San Diego and Coronado has always been narrow and the bathymetry fairly shallow, although it has increased over the years from 2.5 m (8 ft) in 1857 to about 13 m (40 ft) today; North Island and Coronado Island may also be uplifts flanking the Spanish Bight and Coronado fault segments of the Rose Canyon fault zone [10]. In fact, uplift exists along the west side of the north-trending Coronado Fault, which passes from the east side of Coronado Island to the mainland shore along the Pacific Highway (note how straight this section of coast was in the nineteenth century maps; Figure 1; west of New San Diego).

The Spanish Bight was a narrow channel that separated Coronado from North Island up to the early 1940s (Figure 1). The development of North Island by the U.S. Navy prior to and during World War II led to the filling of the Bight in 1943, combining the land areas into a single body. The Navy still operates NASNI on Coronado. On the southern side of the town is Naval Amphibious Base Coronado, a training center for Navy SEALs. Both facilities are part of the larger Naval Base Coronado complex [14].

Sand bars (shoals) to the north of the bay were further enhanced to create new marina locations (between Quarantine Station, now known as La Playa[6], and Roseville) in North San Diego Bay (Figures 1 and 2). By 1966, Shelter Island and Harbor Island were created by dredging and filling the area east of Roseville and Dutch Flat, providing more space for marinas (Figure 3). These islands are actually connected to the mainland by causeways on landfill. The depths in the bay are more uniform and larger than before. The area around Dutch Flat, Fisherman's Point and the municipal airport were filled, with the exception of a channel east of Roseville (Figure 2A). This channel provided boat access to the Marine Corps base. Small, elongated boat basins now exist between Shelter Island, Harbor Island and the mainland.

---

[5]   Spanish Bight was mostly filled to connect North Island to Coronado, and additional dredging created local bays, such as Glorietta Bay, for marinas and naval vessels. Piers were constructed at San Diego for commercial boats and naval shipyards to the south at National City where major shipyards were built. The southern end of the bay had salt evaporation ponds, although a marina was also built for the City of Chula Vista. The Silver Strand was enhanced to reduce flooding during storms, and marinas were constructed in some locations (other locations were restricted to military operations including an amphibious base and radio station).

[6]   La Playa is the location of the original tide gauge for San Diego. It was first installed in 1854. A strong local earthquake, 27 May 1862 M~6, created a wave in the bay that was observed by the tide gauge engineer (Andrew Cassidy) who was repairing the pier for the tide gauge at the time of the earthquake. He observed a runup of about 1–1.2 m (3–4 ft) on the beach at La Playa but the water returned to its normal level. We do not know if this runup was a vertical or horizontal distance [12,13]. The Rose Canyon fault was likely the source of the earthquake but data are inadequate to determine the epicenter with certainty.

Figure 3. View of Shelter Island in 1937, the 1950s and the 1990s, left to right. Photos courtesy of Unified Port of San Diego [1]. An example of extreme transformation showing how the sand spit, or "mud island" as it was referred to, now offers home to numerous vessels. View is southwest to northeast. The entrance to Shelter Island has also been subject to high currents in past tsunamis (e.g., 22 May 1960, Chile and 11 March 2011, Japan; [4]).

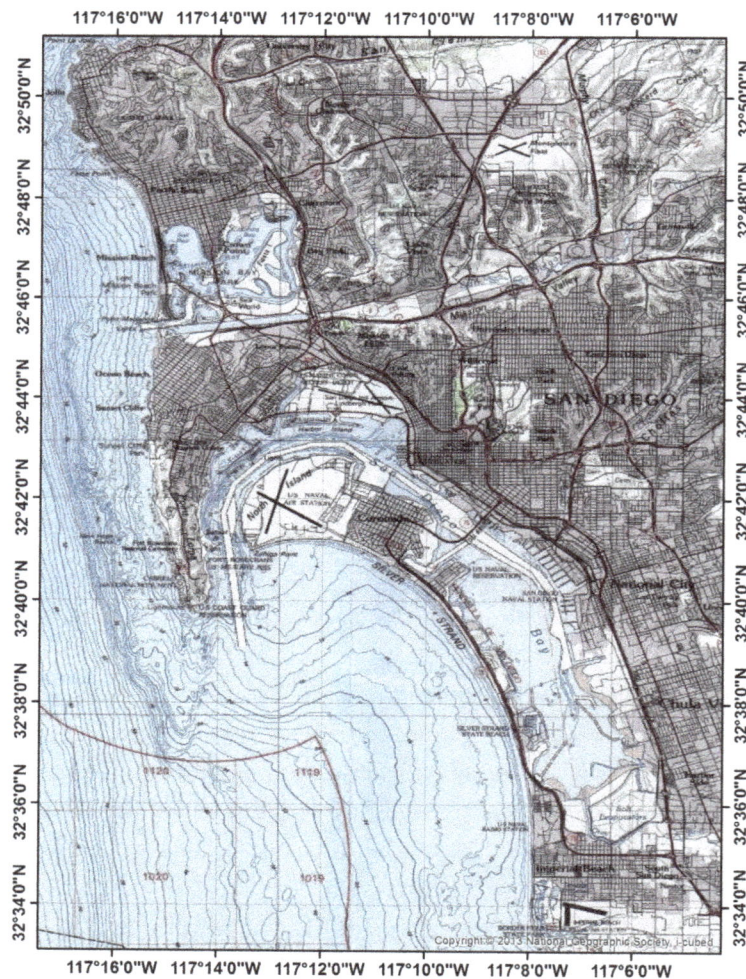

Figure 4. Map of San Diego Bay, showing the state of the harbor today with locations that are referenced frequently throughout the text.

4. Numerical Grids of San Diego

Four different DEMs for San Diego were used to depict the evolution of the port due to man-made modifications. Three were digitized from historical charts of the historical map and chart collection of the Office of Coast Survey, NOAA [9] for the purpose of our numerical simulations. The present day (2014) DEM originated from the National Geophysical Data Center (now National Centers for Environmental Information, or NCEI) [15]. The quality of the paper chart (*i.e.*, detail) and the years representing the major changes in the port were the primary factors that affected our choice of charts. The years representing the major changes to the port are: 1892, 1938, 1945, and present day for the current harbor configuration (Figure 4).

All numerical grids (1892, 1938, 1945, and present day) share two outer grids (usually referred to as A and B grids) that form a nested system. The A grid, or outermost grid, was created from interpolated ETOPO1 data to 30 arc-sec (Figure 5; ~ 900 m [16]). The B grid was also created from interpolated ETOPO1 data to 12 arc-sec [16], while the C grid uses a 3 arc-sec resolution, subsampled from a 1/3 arc-sec resolution dataset (~ 90 m; Figure 5; [17]).

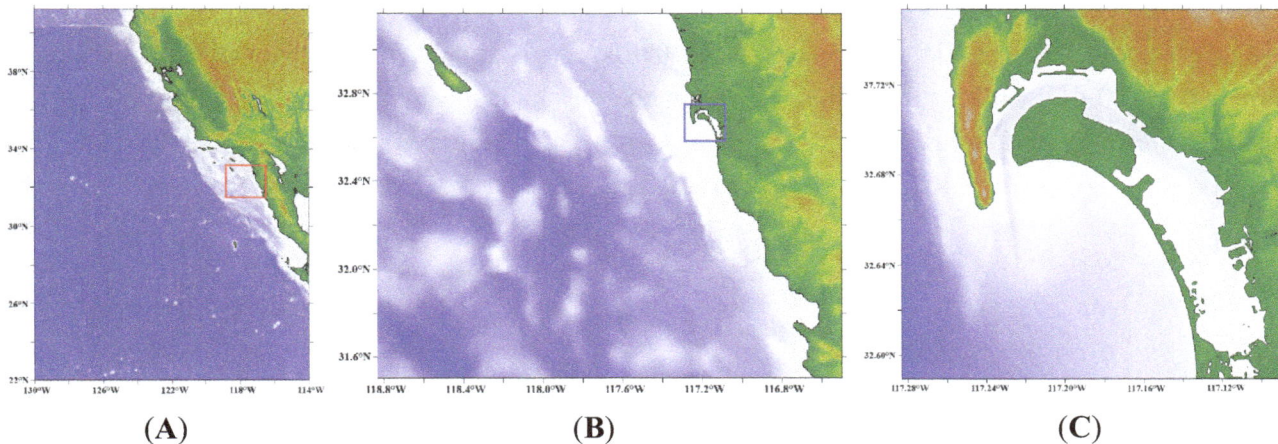

(A) (B) (C)

Figure 5. Extent of numerical grids **A (left)**, **B (middle**; shown as red rectangle within the A grid) and **C (right**; shown in blue colour within the B grid). C grid here only shows the current configuration but the extent of all four C grids (1892, 1938, 1945 and 2014) is the same.

Digitization of old nautical charts is challenging, especially in cases where not only the coastline and bathymetry around a port have changed, but also the topography has changed due to urbanization. Nautical charts are meant for navigation, and therefore, elevation information provided in them is limited. Elevation data for past years are also generally difficult to find so a number of assumptions were made for the creation of the DEMs. Elevation data of San Diego for the high elevation areas to the north of San Diego toward Mission Bay were assumed to have changed little in the last 100 years. This may be a reasonable assumption since construction of the restraining wall (levee) to redirect the San Diego River helped to maintain topography in the low area between Mission Bay and San Diego Bay. This allowed us to use topography from the United States Geological Survey (USGS) with datum

NAD83[7] (North American Vertical Datum 1988, or NAVD88; [18]). Flat areas close to shores usually have large uncertainties and were manually digitized to match available contour data (Figure 6). The coastline was digitized from the charts in addition to bathymetric points (Figure 6). All these data were combined to create the final DEMs at 90 m resolution and mean high water (MHW) depth.

Topographic maps (e.g., from USGS) generally have elevations referenced to orthometric datums, either the NAVD 88 or the older National Geodetic Vertical Datum 1929 (NGVD 29). All GPS positioning data are referenced to one of many 3-D/ellipsoid datums. NOAA's nautical charts have depths referenced to mean lower low water (MLLW), and bridge clearances are referenced to MHW. The legal shoreline in the U.S., which is the shoreline represented on NOAA's nautical charts, is the MHW shoreline; that is, the land-water interface when the water level is at an elevation equal to the MHW datum. For the construction of DEMs a tidal datum—representing a reference plane—is necessary. A tidal datum refers to an average height of the water level at different phases of the tidal cycle ([19]). The MLLW line is also depicted on NOAA's charts[8] [19]. Mean Sea Level (MSL) is usually the average between MLLW and MHHW[9] as provided on the charts. MHW is more challenging, but we assumed the value of 1.52 m (4.98 ft) above MLLW as estimated by NOAA for San Diego for the years between 1993 and 2003 [20].

We also assumed that the highest elevations on Coronado Island have not changed considerably over time (a 6 m (20 ft) contour is the highest elevation contour that appears in early charts; Figure 1, [9]). The plane of reference used in the charts is the average of MLLW (Figures 1 and 2).

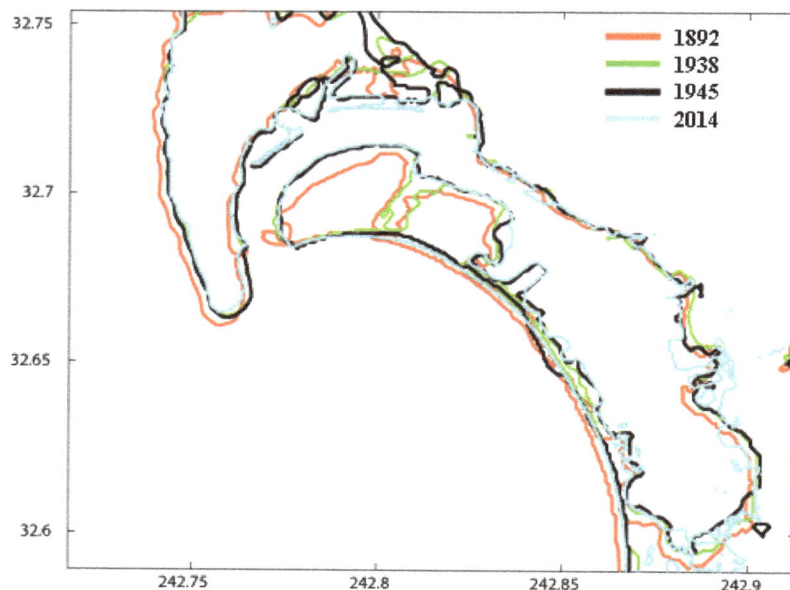

Figure 6. Coastlines of the four digital elevation models (DEMs) used in the numerical simulations.

---

[7] Although contours were matching fairly well with old charts an additional small shift ensured a better fit to the contours in the nautical charts and the coastline.

[8] MLLW is the lower low water height average of each tidal day calculated over a tidal datum epoch (19 year period adopted by the National Ocean Service).

[9] MHHW is the higher high water height average of each tidal day calculated over a tidal datum epoch (19 year period adopted by the National Ocean Service).

5. Numerical Modeling

For the numerical modeling of tsunamis, we use the well-known and tested MOST (Method of Splitting Tsunami) code [21], which solves the depth-averaged, nonlinear shallow-water wave equations that allow wave evolution over variable bathymetry and topography [21,22]. This is a suite of numerical codes that model all phases of the tsunami simulation (generation, propagation, and runup). The initial displacement on the surface of the water is obtained directly from the terminal coseismic deformation of the ocean floor using Okada's formulation for the surface expression of shear and tensile faults in a homogeneous, elastic half space [23]. The equations of motion are then solved numerically to propagate the water surface disturbance across the computational domain [21,22,24]. A 0.0009 Manning's roughness coefficient was used in all simulations.

MOST has been tested extensively, complies with the standards and procedures outlined in [25], and has shown reliability in forecasting inundation. The Tohoku 2011 source used in this article was also selected due to a really good fit to DART® buoy data from the Pacific Ocean [26]. More recently it was also shown to compare fairly well against tsunami currents measured by recently installed acoustic Doppler current profilers (ADCP; [27]).

A set of four scenarios were considered on all numerical grids. A common A and B grid was used for all four DEMs; those are the outermost and intermediate grids used in the simulations for computational efficiency, respectively. We applied two distant scenarios and two local/regional scenarios to investigate the role of modified bathymetry and nearshore topography on our results. Distant scenarios were run for approximately 24 h while local scenarios for 12 h each.

6. Tsunami Sources

San Diego is exposed to tsunamis from both local and distant sources. A small but balanced set of sources was selected (two distant, two local) to allow for the arrival of waves from different directions in order to investigate, among other factors, the contribution of source location to tsunami impact in San Diego. Specifically, two distant tsunamis generated in the south and in the northwest Pacific were considered. With respect to near-field sources, one local tectonic source and one landslide source were included for locations to the southwest, where the waves' approach is directly into the bay.

The distant sources are major historical subduction zone earthquakes (in opposite hemispheres): the 22 May 1960 Chile earthquake (M_w 9.5) and the 11 March 2011 Tohoku, Japan (M_w 9.0) earthquake. The local sources are theoretical and include a large transpressional earthquake (M_w 7.4) on the San Clemente fault (SCF), located about 70 km southwest of San Diego Bay, and a large submarine landslide located in Coronado Canyon about 25 km southwest of San Diego. For a comprehensive tsunami hazard assessment for San Diego harbor, considering all types of local, regional, and distant sources capable of generating tsunamis that may be damaging, see [28]. A more detailed discussion of the sources with impact to San Diego, as well as the most significant tsunami records in San Diego, can also be found in the same article.

6.1. Chile 1960

A simple rectangular source (800 km × 200 km) was used to represent the fault surface that produced the 22 May 1960 M_w 9.5 earthquake and transpacific tsunami [29]. The collision zone of Nazca and South American plates (NazSA[10]) is an important source of tsunamis and hazard studies for the U.S. West Coast, and California in particular. The 22 May 1960 Chile earthquake is known as the largest recorded earthquake and was preceded and followed by large earthquakes within a few minutes (~Mw 8.0 or greater; see [30,31]). Rupture of the NazSA zone is likely more complex and possibly longer, but far-field studies are not affected greatly by such variation. We consider a Chile 1960 type of source to represent a source similar to the 22 May 1960 event, represented by a single fault plane [29] with a slip of 24 m. More details are provided in Table 1.

Table 1. Table of sources.

Source Name (Far-field)	Fault Length (km)	Fault Width (km)	Slip (m)	Dip (degrees)	Rake (degrees)	Strike (degrees)	Depth[11] (km)
Chile 1960[12]	800	200	24	10	90	10	87.73
Tohoku 2011[13]							
Segment 1	100	50	4.66	19	90	185	5
Segment 2	100	50	12.23	19	90	185	5
Segment 3	100	50	26.31	21	90	188	21.28
Segment 4	100	50	21.27	19	90	188	5
Segment 5	100	50	22.75	21	90	198	21.28
Segment 6	100	50	4.98	19	90	198	5
Source Name (Near-field)	**Fault Length (km)**	**Fault Width (km)**	**Slip (m)**	**Dip (degrees)**	**Rake (degrees)**	**Strike (degrees)**	**Depth (km)**
San Clemente Fault Bend Region							
Segment 1	18.52	10	1.6	89	−161.6	330.4	0.5
Segment 2	8.86	10	2.7	89	158.2	310.2	0.5
Segment 3	6.96	10	7.0	80	135.0	307.8	0.5
Segment 4	23.65	10	4.2	89	166.0	304.9	0.5
Segment 5	12.92	10	5.8	85	149.0	314.5	0.5
Segment 6	6.9	10	3.2	89	161.6	312.4	0.5
Segment 7	19.5	10	1.6	70	−161.6	311.8	0.5

Landslide Case	Lat	Lon	Depth (m)	Uplift (m)	Subsidence (m)	Direction (deg)
Coronado Canyon	32.519	−117.338	370	5	5	260

---

[10] We abbreviated the source simply for easy reference in this article.

[11] Depth is measured from ground surface to top of the fault.

[12] [29].

[13] Source from NOAA Center for Tsunami Research.

6.2. Tohoku 2011

The 11 March 2011 M$_w$ 9.0 Tohoku Japan tsunami caused wide devastation on the Japanese coast, damaged facilities and vessels on the U.S. West Coast [4] and was recorded by multiple sea level stations across the Pacific. As a result of the tsunami, at least 20,000 perished [24] and the nearshore topography was completely redefined [32,33]. The tsunami source used in this study is based on real-time forecasting done by NOAA Center for Tsunami Research. Tsunami wave data from two DART® stations near Japan (21418 and 21401; [34]) were used by an inversion algorithm [26] that selected six unit sources from a pre-computed propagation database and its corresponding coefficients [35]. Parameters of each unit source and their corresponding coefficients are listed in Table 1.

6.3. San Clemente Fault (SCF)—Bend Region

We modeled a major left bend in the SCF offshore northern Baja California, Mexico (Table 1). This is a zone of oblique shortening, where the fault impedes the general northwest movement of the Pacific plate. Seafloor uplift resulting from this restraining bend covers a region of about 875 km^2 and represents as much as 720 m of tectonic uplift [36–38]. Recent submersible investigations of the SCF revealed vertical seafloor scarps, representative of seafloor uplift during the most recent large earthquakes on the SCF [39]. The largest scarps are 1–3 m high and would be associated with earthquakes of magnitude 6.5 or greater. Three moderate earthquakes (M6) along the SCF have been recorded by modern seismographs in the past 83 years.

6.4. Coronado Canyon Landslide

Sediment and sedimentary rock slope failures may occur along the Coronado Escarpment within Coronado and La Jolla submarine canyons. A large failure in Coronado Canyon was modeled to represent a submarine landslide close to San Diego Bay. Our landslide case was represented as a dipole (uplift pattern; see Table 1; [40,41]).

7. Results

7.1. Chile 1960

The inundation extent and corresponding inundation pattern varies greatly from year to year in this scenario for San Diego (Figure 7). Inundation for the present-day (2014) configuration of San Diego is primarily constrained to the south. For earlier years we see large inundation on the inner side of the harbor and NNE in the area of what used to be large tidal lands. Large inundation is evident for the 1945 DEM, especially north of the San Diego region and close to Lindbergh Field (San Diego airport), but that was reversed with the expansion of the Dutch Flat. The early bay, mostly unmodified by human activity, suffered relatively minor wave amplitudes beyond the harbor entrance. This is probably a result of shallow water at the entrance, which reduces the surge into the harbor. Inundation was increased locally at narrow embayments north and west of San Diego where dredging created deeper channels for boats, and fill areas further restricted the channel width. In contrast, the low-lying

Silver Strand was overtopped in earlier configurations, but widening and increased elevation reduced the area of potential inundation in more recent years.

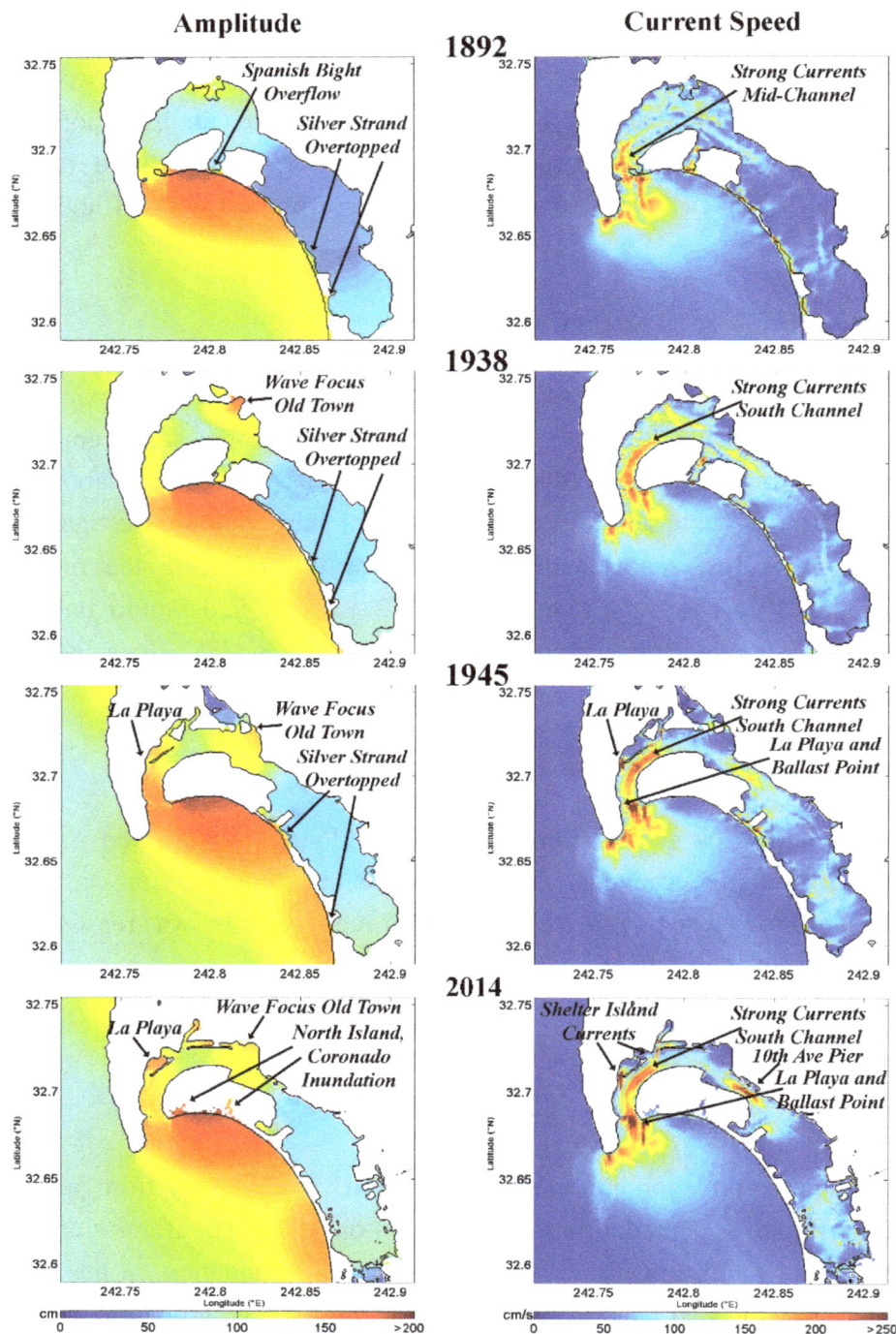

Figure 7. Maximum wave amplitude and maximum current distribution in San Diego harbor for four different configurations (1892, 1938, 1945, and 2014). A Chile 1960 type of scenario was used to investigate changes in wave dynamics as a result of anthropogenic changes. Years indicate time tagging of large changes in the harbor. **Left column** shows wave heights in cm; **right column** tsunami currents in cm/s.

Tohoku 2011 (M_W=9.0)

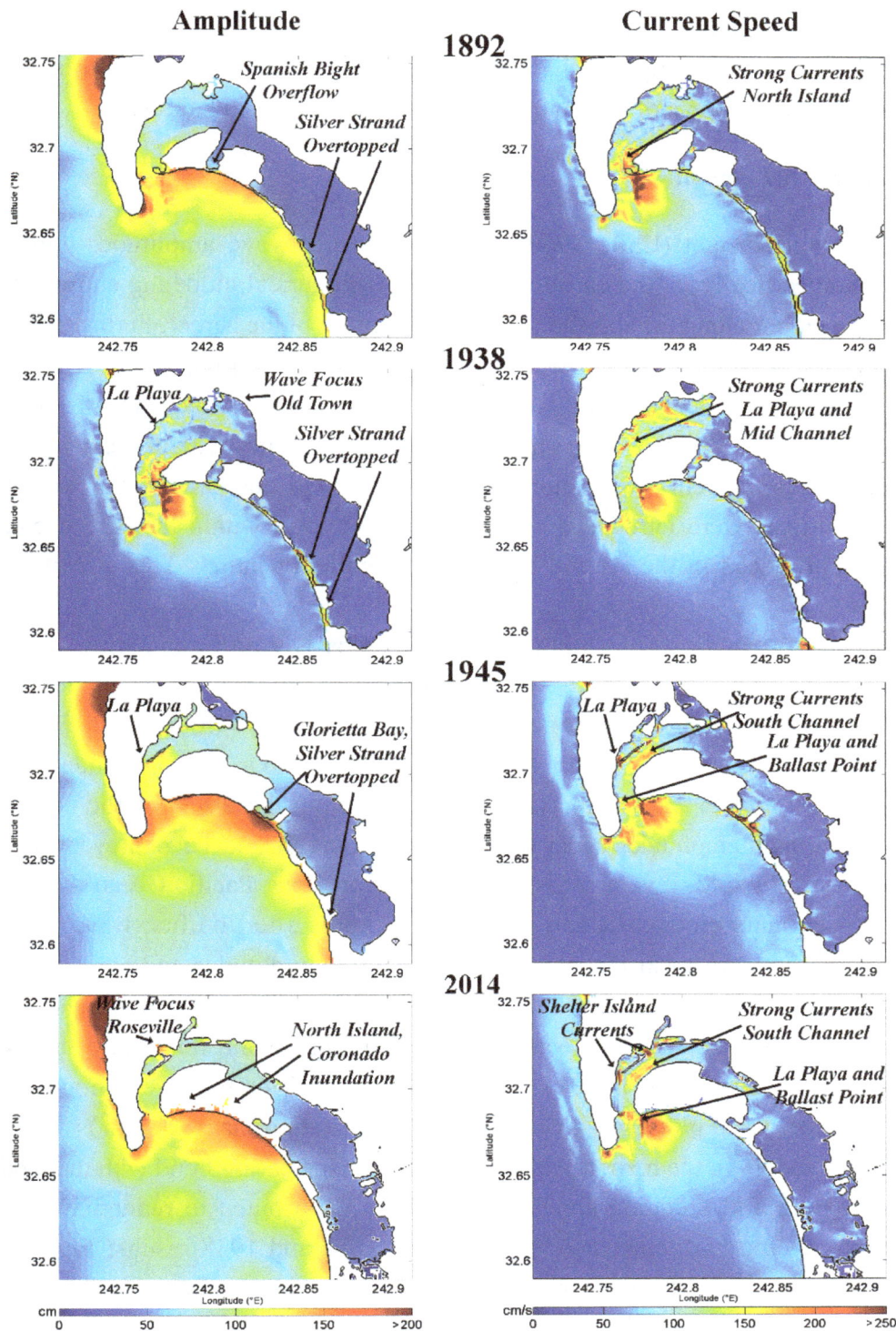

Figure 8. Maximum wave amplitude and maximum current distribution in San Diego harbor for four different configurations (1892, 1938, 1945, and 2014). A tsunami scenario resembling the 11 March 2011 Tohoku tsunami was used to investigate changes in wave dynamics as a result of anthropogenic changes. Years indicate time tagging of large changes in the harbor. **Left column** shows wave heights in cm; **right column** tsunami currents in cm/s.

Generally, the pattern of wave amplitude distribution away from (or outside) the harbor has remained the same over the years but wave amplitudes inside the harbor show a gradual increase as a result of modifications. Currents also show a gradual increase as a function of further modifications for this event. A high current curve following the shape of the harbor is evident for all DEMs in this South American scenario. Additional spots of high current activity also appear in later years.

7.2. 11 March 2011 Japan Tsunami

Similar to the previous distant (Chile 1960) scenario, smaller wave amplitudes appear in older configurations. More specifically, at the earliest year (1892), numerical modeling shows energy spots constrained outside the harbor along the coastline and at the entrance of the harbor. Further development in the bay introduced larger waves inside the harbor around La Playa (Figure 8, 1938). In later years, the creation of the jetty near La Playa changed this pattern and redistributed waves away from the original coastline and along the jetty. Currents also show an increase, but the overall trend is not so clear. Generally, in earlier years, currents were more widely distributed compared to more recent years, while spots of high currents were introduced. The high current spots within the harbor correspond to a new characteristic introduced in the second half of the twentieth century. This is common in both distant scenarios.

The tsunami current distribution from both distant scenarios is worth comparing. In the case of Chile 1960, high currents appear to be more widely distributed along the harbor, while for the Japan tsunami, high currents appear more concentrated near the entrance. This would likely be attributed to the location of the tsunami source and directivity of waves. The maximum amplitudes occur on the south side of Coronado/North Island for the Chile event, showing the effect of the southern source (directivity). In contrast, high amplitudes occur west of Point Loma for the Tohoku event, which arrives from the north and west. Multiple wave arrivals due to refraction along Pacific Ocean waveguides, e.g., Emperor–Hawaiian Seamounts produce strong energy many hours after the first wave arrivals ([42]). Later wave arrivals may involve local resonant effects, so large amplitudes and strong currents were observed during both events. The long period character of these distant subduction megathrust events is very important for emergency planning. Major surges arriving several hours after the first waves caused serious and unexpected damage in 2011 (the surges also persisted for several hours in 1960).

We notice high currents in all plots, but it is clear that the creation of narrow islands and boat basins changes the distribution pattern in the harbor. Although there is an increasing trend with the modifications, enlargement of land space and dredging appear also to have contributed to reduction of currents in some cases while introducing high currents elsewhere (see 1892 and 1938 results; Figures 5 and 6). Currents are high in narrow openings as one might expect, and more such spots of high currents appear in later years as San Diego land expansion led to a narrower opening in the port. The development of a narrow opening combined with channel deepening to the harbor appears also to be responsible for the intensification of the currents that follow a curved wider line along the harbor in 1938 as opposed to 1892. This implies increased danger to vessels not only at the entrances of narrow openings, but also along the "throat[14]" of the San Diego harbor. Strong currents were indeed observed in Harbor Island and Shelter Island during the Tohoku tsunami and in other harbors in California where narrow channels exist (e.g., Catalina harbor, King Harbor in Redondo beach *etc.*).

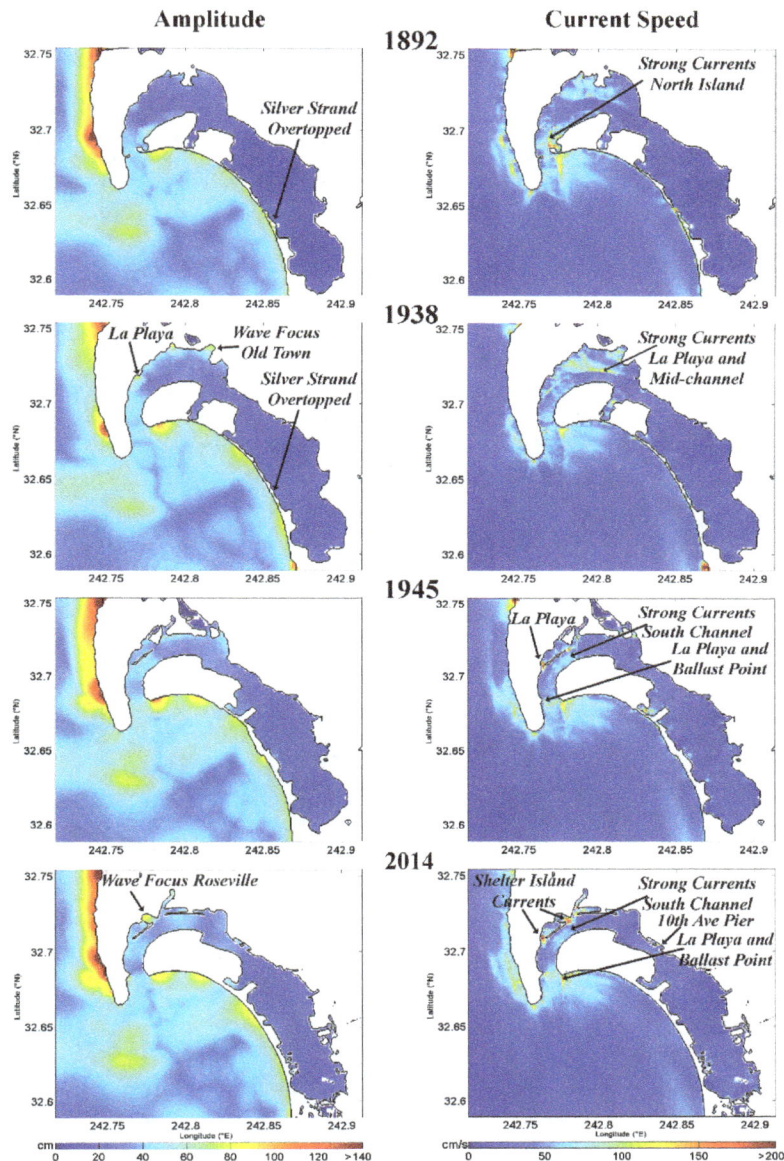

Figure 9. Maximum wave amplitude and maximum current distribution in San Diego harbor for four different configurations (1892, 1938, 1945, and 2014). A San Clemente fault rupture type of tsunami scenario was used to investigate changes in wave dynamics as a result of anthropogenic changes. Years indicate time tagging of large changes in the harbor. **Left column** shows wave heights in cm; **right column** tsunami currents in cm/s.

---

[14] Numerous piers and anchorages exist in the area between San Diego and Coronado. Strong currents in these areas were damaging to Navy ships and other boats, while the ferry service was disrupted during the 1960 event. Docks were damaged and some boats dragged from their moorings [3].

Coronado Canyon - Landslide

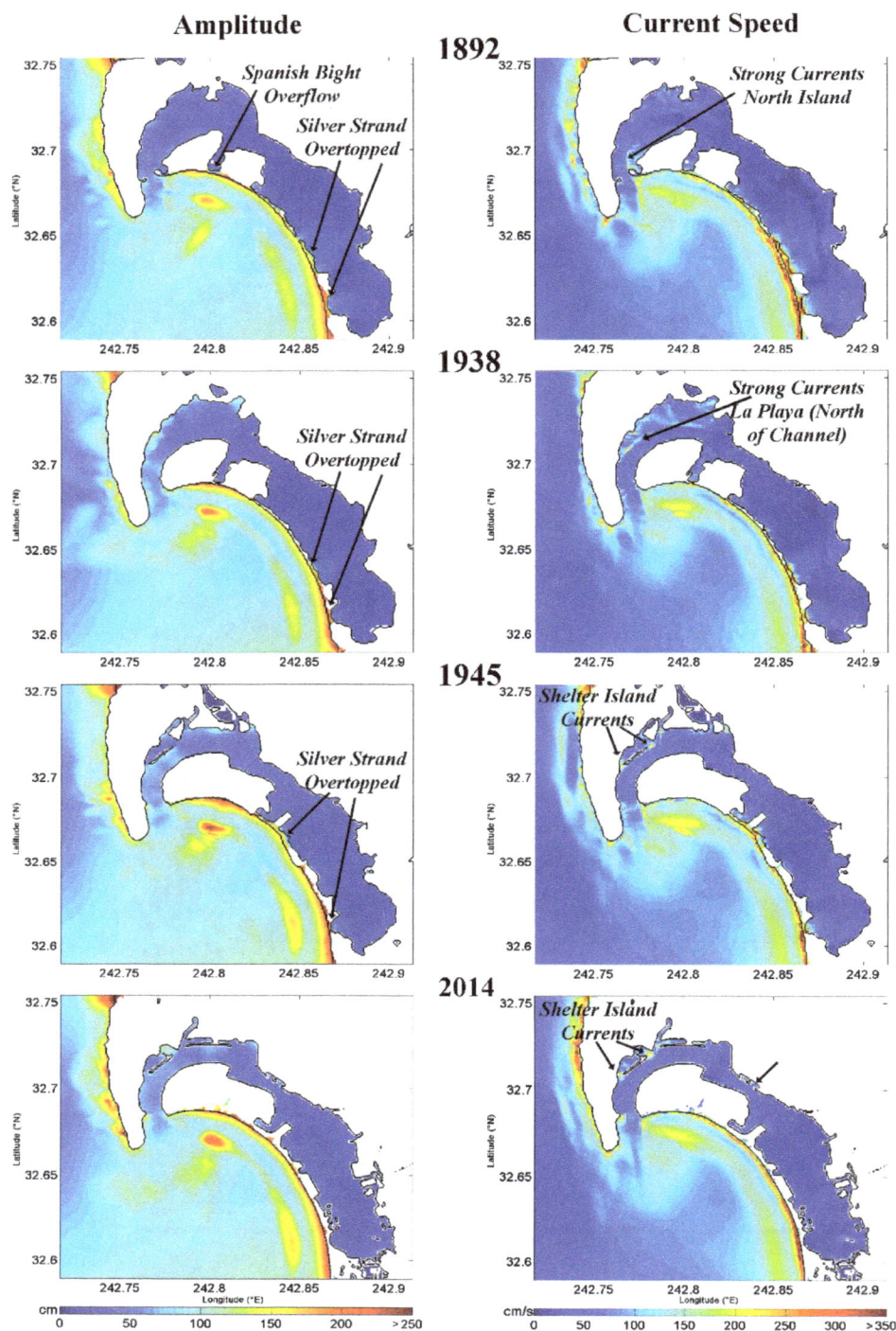

Figure 10. Maximum wave amplitude and maximum current distribution in San Diego harbor for four different configurations (1892, 1938, 1945, and 2014). A hypothetical tsunami scenario generated by a landslide on Coronado Canyon was used to investigate changes in wave dynamics as a result of anthropogenic changes. Years indicate time tagging of large changes in the harbor. **Left column** shows wave heights in cm; **right column** tsunami currents in cm/s.

7.3. San Clemente Fault—Bend Region

Wave amplitudes for this scenario appear relatively low (predominantly 1–2 m). The pattern does not change with modifications. Tsunami currents also appear low for all years, but isolated spots of high energy are introduced after the extended modifications at the entrance of boat basins (e.g., La Playa; see Figure 9). The small changes for this event likely are attributed to waves of shorter wavelength.

7.4. Coronado Canyon Landslide

Wave amplitude pattern does not appear vastly changed during the years for this scenario. Currents also generally maintain their distribution pattern while some spots of high activity get introduced. High currents appear constrained mainly along Coronado Island and Silver Strand outside the harbor (Figure 10). Specifically, we notice similar patterns to the other scenarios with increased currents where channels narrow. Most importantly, for these scenarios, amplitudes are similar to distant events or greater (2 m or higher) due to proximity of the source and the landslide mechanism (larger displacement). The Coronado Canyon landslide source has a nearly direct path into San Diego Bay coming from the south.

8. Discussion/Conclusions

San Diego harbor has undergone extensive modifications and development. Four scenarios were run on four different DEMs to investigate the effects of the major harbor changes to tsunamis. The set of scenarios was selected to ensure a balanced set of scenarios that included both distant and local sources with varying geographical locations. The numerical code MOST was used to model all phases of the tsunami (generation to inundation).

Results show varying inundation patterns for the scenarios with some common characteristics for the distant events when compared to the local events. Inundation within the bay generally was reduced and constrained to locations outside the bay mainly due to expansion of land (channel narrowing) and further filling to elevate tidal flats. Introduction of jetties also caused redistribution of wave heights.

With respect to tsunami currents, it appears that spots of high energy increased in later years. This appears to be associated with the narrowing and deepening of the channel combined with the construction of jetties. In some occasions, changes had a positive effect (reduction of currents), but these positive effects appear to have been counteracted by creation of new high energy spots, due to the creation of narrow entrances to boat basins, in order to accommodate more vessels.

Altering the location of entrances to the boat basins or changing the shape of the boat basins might lead to alternative designs that attenuate currents. While the harbor modifications improve navigation, anchorage and mooring areas, the narrowed channels, especially where elongate islands were created near the harbor entrance, create intense currents during long period wave activity from major distant tsunamis. These currents have created havoc, damaging boats and docking facilities during large events in 1960 and 1988 (Chile), 1964 (Alaska), and 2011 (Japan) [4]. The short period character and shorter duration of tsunamis from local sources reduce these effects. Some narrow low-lying areas along the Silver Strand may be susceptible to overtopping and inundation during large amplitude events, depending also on tidal conditions during the tsunami arrival. In addition, strong currents along beaches may cause severe erosion and scour, and pose serious threat to swimmers and small boats.

The effects of harbor modifications on tsunami impact is a subject that merits further investigation. However, the work presented here leads to the following more general conclusions:

- Enhancement of port space increases exposure to assets within the harbor compared to previous years as more vessels are now accommodated within the port facilities.
- Numerical results also show that the introduction of high currents could threaten navigation, vessels and facilities at narrow openings and also along the harbor "throat".
- Harbor modifications can have both positive and negative effects on tsunami impact. For example, currents can be constrained to areas of minimal impact (e.g., outside of the harbor), thereby exposing less assets to dangerous wave action.
- The long duration of distant subduction megathrust tsunamis (Mw >8 source magnitude) spans several tidal cycles so that serious currents occur for several hours, or even days, after the first wave arrivals.

Acknowledgments

We would like to thank Finn Scheele for help with some of the maps in this manuscript. We would like to thank Cynthia Holder, Deputy District clerk/records manager of the Unified Port of San Diego for giving us access to the historical photographs of San Diego Bay.

This publication is partially funded by the Joint Institute for the Study of Atmosphere and Ocean (JISAO) at the University of Washington under NOAA Cooperative Agreement No. NA17RJ1232, Contribution No. 2445. This is Contribution No. 4364 from the NOAA/Pacific Marine Environmental Laboratory.

Author Contributions

A.B. wrote the majority of the paper, digitized the nautical charts for the simulations and made about half of the figures appearing in the manuscript. M.R.L. had the original idea for this paper, provided information on the local sources and wrote sections of the paper relative to the sources and San Diego Bay history. E.G. ran the simulations, helped with the decision-making of the distant sources and contributed to the editing of this paper.

Conflicts of Interest

The authors declare no conflict of interest.

References

1. Unified Port of San Diego. Available online: http://www.portofsandiego.org (accessed on 7 July 2015).
2. Kemp, K. The Mississippi Levee System and the Old River Control Structure. Available online: http://www.tulane.edu/~bfleury/envirobio/enviroweb/FloodControl.htm (accessed on 7 July 2015).

3. Lander, J.F.; Lockridge, P.A. *United States Tsunami 1890–1988*; National Geophysical Data Center, National Environment Satellite, Data: Boulder, CO, USA, 1989.

4. Wilson, R.I.; Admire, A.R.; Borrero, J.C.; Dengler, L.A.; Legg, M.R.; Lynett, P.; McCrink, T.P.; Miller, K.M.; Ritchie, A.; Sterling, K.; *et al.* Observations and impacts from the 2010 Chilean and 2011 Japanese tsunamis in California (USA). *Pure Appl. Geophys.* **2012**, *170*, 22.

5. Shuto, N. Damages to coastal structures by tsunami-induced currents in the past. *J. Disaster Res.* **2009**, *4*, 462–468.

6. Military.com, Naval Air Station North Island. Available online: http://www.military.com/base-guide/naval-air-station-north-island (accessed on 13 July 2015).

7. 2010 Census—Census.gov, 2010. Available online: http://www.census.gov/2010census/ (accessed on 14 September 2015).

8. California Department of Parks and Recreation. About the Beach. Available online: http://www.parks.ca.gov/?page_id=984 (accessed on 13 July 2015).

9. Office of Coast Survey, NOAA. Historical Charts. Available online: http://historicalcharts.noaa.gov/ (accessed on 8 July 2015).

10. Fischer, P.J.; Mills, G.I. *The Offshore Newport-Inglewood-Rose Canyon Fault Zone, California: Structure, Segmentation and Tectonics*; Environmental Perils of the San Diego Region: San Diego, CA, USA, 1991; pp. 17–36.

11. Harper & Row. *Shepard, Submarine Geology*, 3rd ed.; Harper & Row: New York, NY, USA, 1973.

12. Gohres, S.D.H.C.-D. Tidal Marigrams. Available online: https://www.sandiegohistory.org/journal/64october/marigrams.htm#Gohres (accessed on 10 July 2015).

13. Gohres, H. Tidal Marigrams. *J. San Diego Hist.* **1964**, *10*, 4.

14. Military.com. CNIC|Naval Base Coronado Complex. Available online: http://www.cnic.navy.mil/regions/cnrsw/installations/navbase_coronado.html (accessed on 13 July 2015).

15. NOAA, National Centers for Environmental Information. NOAA Tsunami Inundation Digital Elevation Models (DEMs). Available online: http://www.ngdc.noaa.gov/mgg/inundation/ (accessed on 8 July 2015).

16. NOAA, National Centers for Environmental Information, "ETOPO1 Global Relief Model". Available online: https://www.ngdc.noaa.gov/mgg/global/global.html (accessed on 8 July 2015).

17. Carignan, K.S.; Taylor, L.A.; Eakins, B.W.; Friday, D.Z.; Grothe, P.R.; Love, M. *Digital Elevation Models of San Diego, California: Procedures, Data Sources and Analysis*; National Geophysical Data Center, National Oceanic Atmospheric Administration: Silver Spring, MD, USA, 2012.

18. National Geodetic Survey, NOAA. Vertical Datums. Available online: http://www.ngs.noaa.gov/datums/vertical/ (accessed on 8 July 2015).

19. NOAA, Center for Operational Oceanographic Products and Services. *Tidal Datums and Their Applications*; US Department of Commerce: Silver Spring, MD, USA, 2000.

20. NOAA, Tides and Currents-Superseded Bench Mark Sheet for 9410170, San Diego CA. Available online: http://beta.tidesandcurrents.noaa.gov/benchmarks.html?id=9410170&type=superseded (accessed on 10 July 2015).

21. Titov, V.V.; Synolakis, C.E. Numerical Modeling of tidal waver runup. *J. Waterw. Port Coast. Ocean Eng.* **1998**, *124*, 15.

22. Titov, V.V.; Gonzalez, F.I. *Implementation and Testing of the Method of Splitting Tsunami (MOST) Model*; Technical Memorandum ERL PMEL-112; NOAA: Seattle, WA, USA, 1997.

23. Okada, Y. Surface deformation due to shear and tensile faults in a half-space. *Bull. Seismol. Soc. Am.* **1985**, *75*, 20.

24. International Tsunami Information Center (ITIC), m.yamamoto@unesco.org (accessed on 9 October 2012).

25. Synolakis, C.E.; Bernard, E.N.; Titov, V.V.; Kanoglu, U.; Gonzalez, F.I. Validation and Verification of Tsunami Numerical Models. *Pure Appl. Geophys.* **2008**, *165*, 38.

26. Percival, D.B.; Denbo, D.W.; Eble, M.C.; Gica, E.; Mofjeld, H.O.; Spillane, M.C.; Tang, L.; Titov, V.V. Extraction of tsunami source coefficients via inversion of DART® buoy data. *J. Int. Soc. Prev. Mitig. Nat. Hazards* **2011**, *58*, 567–590.

27. Admire, A.R.; Dengler, L.A.; Crawford, G.B.; Uslu, B.U.; Borrero, J.C.; Greer, S.D.; Wilson, R.I. Observed and modeled currents from the Tohoku-oki Japan and other recent tsunamis in Northern California. *Pure Appl. Geophys.* **2014**, *171*, 18.

28. Barberopoulou, A.; Legg, M.R.; Uslu, B.; Synolakis, C.E. Reassessing the tsunami risk in major ports and harbors of California I: San Diego. *Nat. Hazards* **2011**, *58*, 18.

29. Kanamori, H.; Cipar, J.J. Focal Process of the great Chilean earthquake 22 May 1960. *Phys. Earth Planet. Inter.* **1974**, *9*, 9.

30. Kanamori, H. The energy release in great earthquakes. *J. Geophys. Res.* **1977**, *82*, 7.

31. Barrientos, S.E.; Ward, S.N. The 1960 Chile earthquake: Inversion for slip distribution from surface deformation. *Geophys. J. Int.* **1990**, *103*, 9.

32. Tanaka, H.; Tinh, N.X.; Umeda, M.; Hirao, R.; Pradjoko, E.; Mano, A.; Udo, K. Coastal and estuarine morphology changes induced by the 2011 Great East Japan Earthquake Tsunami. *Coast. Eng. J.* **2012**, *54*, 25.

33. Taylor, A. Japan Earthquake, 2 Years Later: Before and After. Available online: http://www.theatlantic.com/photo/2013/03/japan-earthquake-2-years-later-before-and-after/100469/#img18 (accessed on 13 July 2015).

34. NOAA, National Data Buoy Center. Available online: http://www.ndbc.noaa.gov/dart.shtml (accessed on 10 July 2015).

35. Gica, E.; Spillane, M.C.; Titov, V.V.; Chamberlin, C.D.; Newman, J.C. *Development of the forecast propagation database for NOAA's Short-Term Inundation Forecast for Tsunamis (SIFT)*; NOAA Technical Memorandum, OAR PMEL-139; NOAA: Seattle, WA, USA, 2008.

36. Legg, M.R. Geologic Structure and Tectonics of the Inner Continental Borderland Offshore Northern Baja California, Mexico, Santa Barbara, CA. Ph.D. Dissertation, University of California, Santa Barbara, CA, USA, 1985.

37. Legg, M.R.; Borrero, J.C. Tsunami potential of major restraining bends along submarine strike-slip faults. In Proceedings of the International Tsunami Symposium 2001, NOAA/PMEL, Seattle, WA, USA, 7–10 August 2001.

38. Legg, M.R.; Goldfinger, C.; Kamerling, M.J.; Chaytor, J.D.; Einstein, D.E. Morphology, structure and evolution of California Continental Borderland restraining bends. *Tecton. Strike-Slip Restraining Releas. Bends Cont. Oceanic Settings* **2007**, *290*, 143–168.

39. Goldfinger, C.; Legg, M.R.; Torres, M.E. *New Mapping and Submersible Observations of Recent Activity on the San Clemente Fault*; EOS, Transactions of the American Geophysical Union: Washington, DC, USA, 2000.

40. Synolakis, C.E.; Borrero, J.C.; Eisner, R. Developing inundation maps for Southern California. In Proceedings of the 2002 Coastal Disasters Conference, San Diego, CA, USA, 24–27 February 2002.

41. Synolakis, C.E.; Bardet, J.P.; Borrero, J.; Davies, H.; Okal, E.; Silver, E.; Sweet, J.; Tappin, D. Slump origin of the 1998 Papua New Guinea tsunami. *Proc. R. Soc. Lond. A* **2002**, *458*, 763–789.

42. Barberopoulou, A.; Legg, M.R.; Gica, E. *Multiple Wave Arrivals Contribute to Damage and Tsunami Duration on the US West Coast*; Springer: Berlin, Germany; Heidelberg, Germany, 2014.

10

Reducing Reliability Uncertainties for Marine Renewable Energy

Sam D. Weller *, Philipp R. Thies, Tessa Gordelier and Lars Johanning

University of Exeter, Penryn Campus, Cornwall, TR10 9FE, UK;
E-Mails: P.R.Thies@exeter.ac.uk (P.R.T.); tjg206@exeter.ac.uk (T.G.);
L.Johanning@exeter.ac.uk (L.J.)

* Author to whom correspondence should be addressed; E-Mail: S.Weller@exeter.ac.uk

Academic Editor: Bjoern Elsaesser

Abstract: Technology Readiness Levels (TRLs) are a widely used metric of technology maturity and risk for marine renewable energy (MRE) devices. To-date, a large number of device concepts have been proposed which have reached the early validation stages of development (TRLs 1–3). Only a handful of mature designs have attained pre-commercial development status following prototype sea trials (TRLs 7–8). In order to navigate through the aptly named "valley of death" (TRLs 4–6) towards commercial realisation, it is necessary for new technologies to be de-risked in terms of component durability and reliability. In this paper the scope of the reliability assessment module of the DTOcean Design Tool is outlined including aspects of Tool integration, data provision and how prediction uncertainties are accounted for. In addition, two case studies are reported of mooring component fatigue testing providing insight into long-term component use and system design for MRE devices. The case studies are used to highlight how test data could be utilised to improve the prediction capabilities of statistical reliability assessment approaches, such as the bottom–up statistical method.

Keywords: component reliability testing; "valley of death"; reliability uncertainties

1. Introduction

It is widely acknowledged that marine renewable energy (MRE) has a significant role to play in the transition towards a global green economy. The 20% target for electricity generation set within the European Commission's *Europe 2020* strategy [1] includes 200–300 MW of installed MRE capacity in the United Kingdom [2] which could equate to the creation of 10,000 jobs by 2020 and be worth £6.1 billion by 2035 [3]. With the present level of installed capacity around 9 MW, such projections are ambitious considering the current nascent state of the industry in which only a few notable projects have reached the field demonstration stage (TRLs 7–8) prior to economic validation (TRL 9, as defined by [4]). As yet no device operators have deployed large scale array projects, although the current installed capacity will be bolstered by two tidal array projects that have recently been approved (MeyGen's 10 MW Phase 1 and Scottish Power Renewables 10 MW Sound of Islay Tidal Array) [5].

Despite the forecasted growth of the industry over the next two decades and support of funding incentives in the UK (e.g., the Marine Energy Array Demonstrator scheme, Marine Renewables Commercialisation Fund, Saltire Prize and various Technology Strategy Board and EU initiatives) a number of barriers have been identified which must be overcome before large scale deployments are realised. The *Wave and Tidal Energy in the UK. Conquering Challenges, Generating Growth* report produced by RenewableUK [3] identified four key risk areas which could hamper progress: finance, technology development, grid and consenting. Confidence in the ability of the MRE industry to deliver a localised cost of energy (LCOE) which is competitive with other forms of power generation in an acceptable time-frame is essential for continued investment in the sector. An operational availability threshold of 75% has been identified [5] to achieve this. In order for the sector to progress towards higher TRLs the reliability of components and sub-systems must be demonstrated as this plays a key role in the overall availability of the device [6] as well as shaping efficient maintenance intervals [7].

The DTOcean (Optimal Design Tools for Ocean Energy Arrays: www.dtocean.eu) project aims to accelerate the deployment of the first generation of wave and tidal energy arrays through the development of a "Design Tool" which will be able to assess the: (i) economics; (ii) reliability; and (iii) environmental impact of wave or tidal energy arrays. This paper will therefore focus on reliability assessment methods currently employed within the sector and the approach selected for the DTOcean Tool. In addition, several case studies are provided to highlight the role of component reliability testing in understanding component performance and reducing reliability uncertainties.

2. Reliability Assessment within the DTOcean Tool

2.1. DTOcean Design Tool Overview

The aim of the FP-7 funded DTOcean project is to develop an open-source Design Tool which will provide a number of solutions for MRE array design. The Tool will comprise a suite of design modules which will be used to analyse and optimise several key aspects; array layout, electrical system architecture, mooring and foundation systems in addition to lifecycle logistics. The role of the electrical architecture and mooring and foundation modules will be to identify optimum electrical layouts (including radial, single- and double-sided strings) as well as suitable mooring and foundation systems based on user-provided device and site parameters. For the purpose of reliability assessment the components which make up each designed sub-system will be displayed to the user via a Reliability Block Diagram contained within the Tool's graphical user interface (GUI). These diagrams are a powerful method to calculate system reliability when sub-systems exist which have inter-dependencies or where provision has been made for redundancy. In order to provide a full assessment of system reliability, the other parts of the system (*i.e.*, power take-off, structure and condition-monitoring) will be represented by a generic block named "device" in the Reliability Block Diagram and assigned a failure rate by the user, perhaps based on sea-trials of a single device or component testing. The user will be able to provide more detailed information about these sub-systems if available.

2.2. Assessment Method

A widely used metric for assessing the expected operating life of a component, sub-system or global system is mean time to failure (MTTF). The basis of MTTF calculations are reliability functions $R(t)$ which are based on statistical probability density functions (PDFs) generated from numerical modelling or physical test/field data:

$$MTTF(T) = \int_0^\infty R(t)\, dt \tag{1}$$

An exponential PDF is typically used to represent the constant rate of random failures which tend to occur during the "useful" or operational life of the component or sub-system. This interval is the focus of reliability assessment within the first release version of the DTOcean Tool. However, the option to include more sophisticated failure distributions and adjustment mechanisms representative of the "burn-in" and "wear-out" periods of component operation may be included in a later release. Within the Tool system reliability is calculated based on the hierarchical links between each component and sub-system that make up devices within the array, including any unique relationships such as the m of n mooring system shown in Figure 1. The time to failure (TTF) of each component or assembly is also calculated and used by the System Control and Operation module in order to plan maintenance actions based on time-domain stochastic (Poisson process) simulations. If the calculated MTTF is lower than the limit set by the user, the design modules will be re-invoked to identify a different solution with a higher reliability level. The design modules will also be re-run if the associated LCOE or environmental impact of a solution is deemed to be unacceptable when compared to user-defined criteria.

2.3. Data Provision

As yet a common failure database has not been established within the MRE industry, despite the development of initial reliability models [8]. A similar endeavour to the offshore wind SPARTA (System performance, Availability and Reliability Trend Analysis [9]) project has been recently initiated for tidal turbine drivetrains (TiPTORS; Tidal Turbine Power Take-Off Reliability Simulation programme [10]). The current absence of a common database is due to a lack of design convergence within the sector (particularly for wave energy devices), the use of custom-made components and also the commercial confidentiality of designs. At present, developers must therefore rely on the adaptation of existing reliability assessment methods, as well as knowledge gained during earlier TRL stages to predict the operational performance of devices.

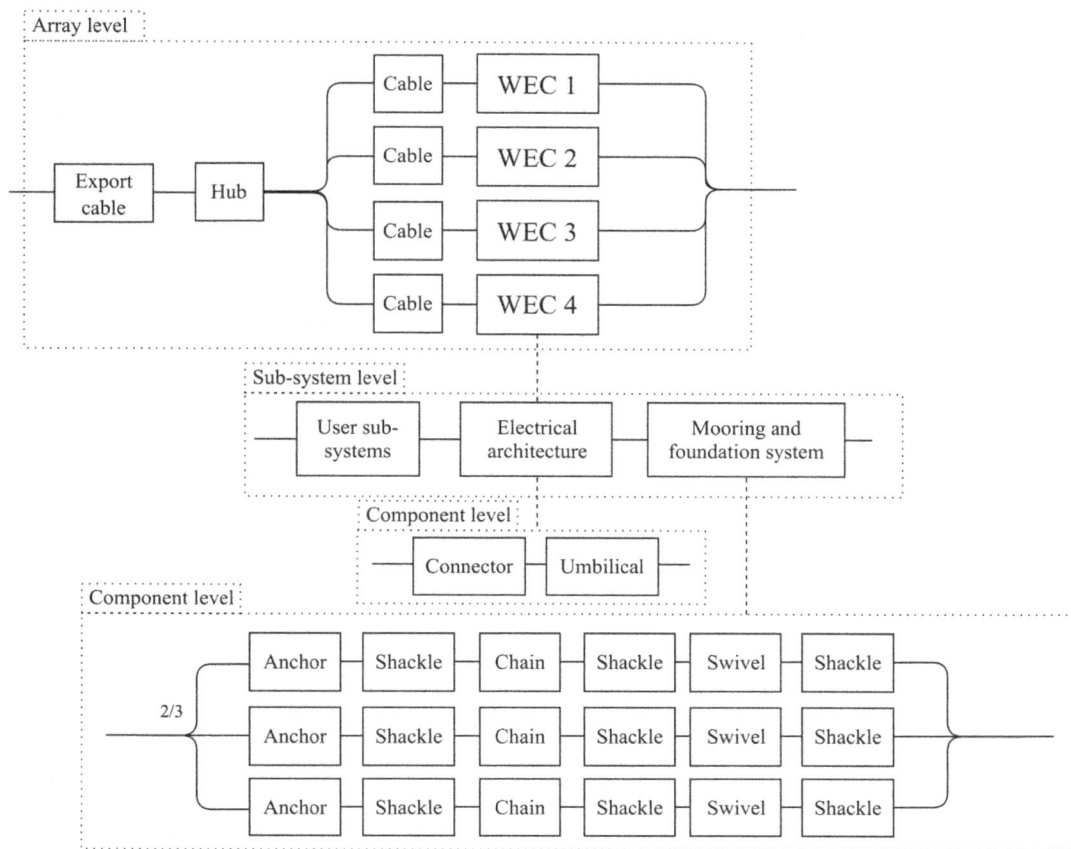

Figure 1. Reliability Block Diagram for an array of four generic wave energy converters, each possessing three mooring lines.

The challenge for any reliability assessment is to accurately quantify or estimate the underlying failure rates. There are two distinct scenarios. First, established industries with large-scale production tend to have considerable operational experience (e.g., the offshore wind industry can now rely on as much as 4000 years of cumulative turbine operation [11]). Although no reliability information is publicly available for offshore wind operations, turbine manufacturers and asset operators can rely on this experience for their statistical analysis to determine failure rates and governing failure modes.

The second situation exists where sufficient operating experience does not exist and reliability assessments rely on existing statistical information which is then adjusted to account for the sensitivity of component failure rates in different applications. These *bottom–up statistical methods* apply influence factors to existing base failure rates to account for variations in quality, environment and stress (*i.e.*, temperature, use rate or applied load to name but a few). Using this approach, reliability calculations can be carried out with a relatively low amount of information such as the type and number of components and operating and environmental conditions. Whilst this method is straightforward to implement it is highly reliant on the underlying data and although influence factors are available in databases such as Mil-Hdbk-217F [12] for electronic components or OREDA® database [13] produced by the oil and gas industry, the use of factors designed for other applications introduces uncertainty to the predicted failure rate. To illustrate this point Table 1 lists environmental influence factors for a commercial-off-the-shelf (COTS) piece of equipment; a power transformer. Clearly the use of the *naval, sheltered* factor will result in a conservative failure rate which is representative of equipment exposed to a non-benign environment and indeed it would be prudent to use field data even if it is from a different industry (*i.e.*, offshore wind [14]). In the absence of actual performance data a failure estimation based on this approach is highly simplified and resulting predictions are heavily influenced [15] by the following assumptions: (i) it focuses only on the central portion of the "bathtub" curve; (ii) it does not take into account developments in manufacturing or design to improve reliability; and (iii) it treats failures as independent events in a system (*i.e.*, it ignores cascade failures).

Table 1. Environmental influence factors for power transformers adapted from Mil-Hdbk-217F [12].

Environment	Factor	Description
Ground, benign	1.0	Non-mobile, temperature and humidity controlled environment, readily maintainable
Naval, sheltered	5.0	Sheltered or below deck conditions
Naval, unsheltered	16.0	Unprotected, surface equipment exposed to weather conditions and salt water immersion

Although it is acknowledged that the widely used bottom–up statistical method [16] does have some shortcomings, given the current lack of detailed operational data in the MRE sector (particularly shared data) this method is suited to the present state of the sector. Its inclusion within the Tool will take the form of reliability assessment algorithms which will draw upon failure rates for each component in addition to relevant influence factors from a centralised database. These values will be sourced from appropriate failure rate databases, including those identified in this paper. If the user has conducted their own analysis (such as from field trials), the Tool will have the functionality to accept other failure rates, overwriting the default values held in the database.

2.4. Uncertainty

Reliability calculation uncertainties will be addressed by the application of uncertainty ranges which will be dependent on the status of the technology and application area (e.g., Table 2). This approach will be based on technology classification assessment procedures outlined in [17]. For example a proven technology in a known application will be assigned a much smaller uncertainty range than a new or unproven technology used in a new application.

Table 2. Technology assessment classifications according to DNV-OSS-213 [17].

Application Area	Technology Status		
	1 (Proven)	2 (Limited Field History)	3 (New or Unproven)
1 (known)	1	2	3
2 (new)	2	3	4

To assist the user in identifying components which have a high consequence of failure and/or failure frequency it has been proposed that colour-coded risk priority number [18] charts will also be included in the GUI. Whilst the focus of this paper has been on mooring components, the generic approach used in the DTOcean Tool is applicable to other MRE sub-systems.

3. Improving Reliability Predictions through Component Reliability Testing

Technology developers at TRLs 4–6 are subject to competing demands; the need to prove that their technology is reliable to strengthen investor confidence whilst at the same time reducing costs to make their technology commercially viable. Necessary savings of 50%–75% by 2025 have been proposed by the Low Carbon Innovation Coordination Group to ensure that the sector is commercially competitive with other forms of generation by 2025 [19]. Striking the difficult balance between robust yet affordable designs is a key engineering challenge for developers at this stage. Component reliability testing either in the field or laboratory is a cost effective means of establishing component or sub-system performance in a controlled, low-risk environment before heading offshore [20].

3.1. Testing Facilities: The South West Mooring Test Facility (SWMTF) and Dynamic Marine Component Test Facility (DMAC)

Whilst there are costs associated with running laboratory equipment and employing trained personnel, funding programmes such as MaRINET (Marine Renewables Infrastructure Network: http://www.fp7-marinet.eu/) have provided technology developers with transnational access to a wide range of facilities. Two such MaRINET facilities owned and operated by the University of Exeter are the South West Mooring Test Facility (SWMTF) [21] and the Dynamic Marine Component test rig (DMaC) [21].

Subsea components such as those used in mooring systems (e.g., [22]), risers and cables (e.g., [23]) are prime examples of components which require testing prior to use. These components have to be highly reliable to be fit for purpose (*i.e.*, to ensure that the device is kept on station in the case of mooring components), whilst being cost-effective. Novel solutions may offer lower lifecycle costs or functionality which is not present in conventional components but these require thorough testing to ensure that target performance and reliability levels are adequately met. The following sub-sections will summarise two examples of component testing carried out by the University of Exeter with collaborating partners, both focused on TRLs 4; technology validation in laboratory conditions.

3.2. Reliability Assessment: A Case Study to Review Safety Factors in Mooring Design

In this case study a combined approach for reviewing safety factors and reliability was developed using three key techniques:

- Numerical modelling using finite element software
- Accelerated testing using DMaC
- Field trials at the SWMTF.

Data collected from previous field trials at the SWMTF provided realistic load data to inform the case study. A review of how these techniques can be used to speed up the reliability verification process was conducted.

Component and assembly numerical models of the shackle bow and pin were developed and a range of load cases were reviewed, including the maximum load measured at the SWMTF (53 kN) and the supplier specified minimum breaking load of the shackle (MBL = 122.6 kN).

Controlled break load and accelerated fatigue performance of the shackles was investigated using DMaC. The break tests established an average break load of 210 kN; a safety factor of 8.6 on the shackle working load limit (WLL) and a safety factor of 1.7 on the MBL. Both failures occurred on the thread of the pin. The break tests also allowed identification of the yield point of the shackles; just over 100 kN. This was used to specify the fatigue trials, ensuring they were conducted within the elastic range of the shackles. Force driven cyclical loading of 10–90 kN was specified for the fatigue trials at a frequency of 2 Hz. A total of 11 shackles were fatigue tested resulting in failures ranging from 19,380 cycles to 109,470 cycles and a variety of failure locations including on both the pin and bow (Figure 2 and Table 3). The measured number of cycles to failure is approximately equal to 11% expected by the S-N curve for shackles in air specified in DNV-RP-C203 [24]. This large difference may be due to manufacturing defects in addition to the fact that the shackles were tested with an applied pre-tension, as opposed to the fully reversed load cycles used to construct the S-N curves.

Table 3. Tabulated values of the number of cycles to failure for each shackle.

Shackle	Exposure Level	Failure Location and Type	Log_{10} Load Cycles
10	Low	None	3.70
11	Low	None	3.70
4	Medium	None	4.29
5	Medium	None	4.29
6	Medium	Break (bow)	4.29
7	High	Fatigue crack (pin)	4.39
8	High	Fatigue crack (pin)	4.39
9	High	Break (centre of pin)	4.39

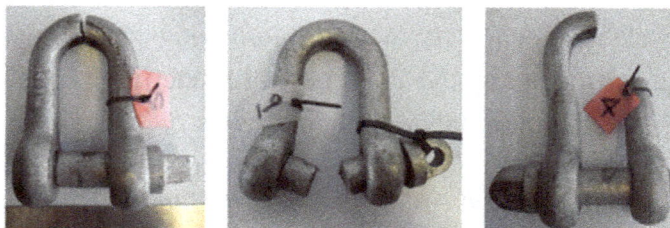

Figure 2. Examples of failed shackles.

Several new and pre-aged shackles were deployed at the SWMTF for a period approaching 6 months, with maximum loads reaching just over 10 kN. Failures were not anticipated at this load range and none were observed. Dye penetrant testing was used to investigate damage; no damage was observed. Following the sea trials, the shackles were subject to further fatigue testing at DMaC.

The numerical modelling correctly identified areas of weakness in the shackle, but significantly underestimated the strength of both the pin and the bow. The physical testing showed that large safety factors are present in static loading situations with the shackle being substantially stronger than the supplier specification or that predicted by the numerical models. Safety factors were significantly reduced in fatigue loading with failures occurring from 20,000 cycles when the 90 kN load was applied cyclically; this loading level is below the MBL specified by the supplier. Further analysis is required regarding the sea trial data.

In this case study the physical testing allowed accurate figures to be established for the failure modes predicted by the numerical modelling. The ability to perform accelerated testing at 2 Hz allowed a large number of cycles to be applied to the shackles for a detailed assessment of fatigue performance. The mean stress applied during these trials was found to have a significant effect on the rate of failure when comparing data to DNV recommended guidance [25]. Further details of the study can be found in [26].

3.3. Synthetic Rope Yarn Durability Assessment

With a proven track record in the offshore industry, synthetic ropes have the potential to be an enabling technology for the MRE sector in terms of the specification of economic and durable mooring components [27]. The response of synthetic ropes is complex, because they display viscoelastic and viscoplastic behaviour which is dependent on time and prior load history [28]. Significant effort over the past two decades has been made into characterising this behaviour through testing and the development of numerical models (e.g., [29]). However, the loading regimes used during testing have reflected the main application of synthetic ropes to-date (large equipment, such as oil and gas exploration drilling platforms or support vessels), for example tests involving low frequency sinusoidal loading. The loading regimes experienced by dynamically responsive MRE devices such as WECs are clearly different and may indeed be sensitive to mooring characteristics such as damping [30]. Therefore a new approach to performance testing and analysis is required for MRE applications.

As part of the MERiFIC (Marine Energy in Far Peripheral and Island Communities: http://www.merific.eu/) project, tests were conducted at the University of Exeter and L'Institut français de recherche pour l'exploitation de la mer (IFREMER) to ascertain the performance of nylon ropes subjected to loading conditions relevant to MRE devices. This built upon earlier characterization work by Ridge *et al.* in [31]. In the first part of the study [28] several new rope samples (of the same construction to that shown in Figure 3a) were subjected to harmonic and irregular loading regimes using the DMaC test rig based on tension measurements recorded by the SWMTF. The focus of the study was to determine the influence of load history on response, characterised through three performance metrics which are important to MRE mooring system design: rope strain, axial stiffness and axial damping.

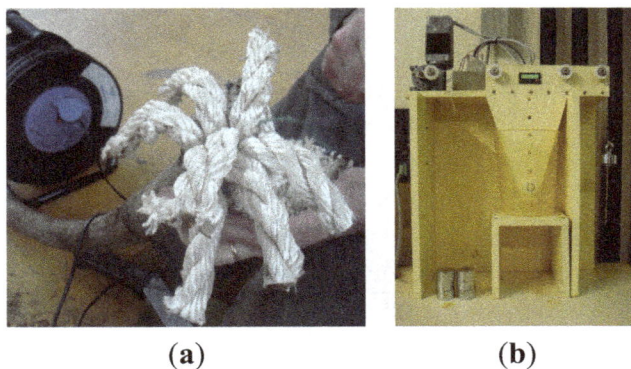

(a)   (b)

Figure 3. (a) Rope sample with outer jacket removed showing construction (parallel-stranded); (b) Yarn-on-yarn abrasion test machine at IFREMER [32].

The second part of the study looked into the performance of aged samples [32] after 18 months use as part of the SWMTF mooring system. Whilst it is important for a device developer to know the short-term performance of mooring components, it is also crucial that the long-term durability is well understood. After subjecting the aged sample to the same loading conditions as the new samples in the first part of the study, the investigation revealed small changes to the properties of the rope. The analysis was extended to include yarn-on-yarn fatigue testing at IFREMER (Figure 3b) in addition to tension testing of yarns and scanning electron micron imaging of fibres [32]. From the dataset of yarn results presented in Figure 4 and Table 4 it is possible to define a linear trend between the number of cycles to failure and mean load (T). Calculated trend line parameters are listed in Table 5 including Cordage Institute recommended number of cycles to failure (based on [33]).

Table 4. Tabulated values of the number of cycles to failure for aged and new yarns.

Yarn Condition	Mean Load (g/dTex)	Log_{10} Cycles to Failure		
		Minimum	Maximum	Average
New	0.12	4.69	5.09	4.87
	0.21	3.62	4.26	4.06
	0.31	3.24	3.71	3.45
	0.40	2.61	3.04	2.81
Aged	0.12	3.4	4.95	4.07
	0.21	1.48	3.22	2.17
	0.31	1.30	2.79	1.79
	0.41	1.34	2.15	1.74

Table 5. Calculated linear trend parameters for number of cycles to failure.

Yarn Condition	Log_{10} Cycles to Failure	R^2
New	$-7.2396T + 5.6824$	0.9345
Aged	$-7.7221T + 4.479$	0.5542
Cordage Institute [33]	$\log_{10}\left(20220e^{-6.332T}\right)$	-

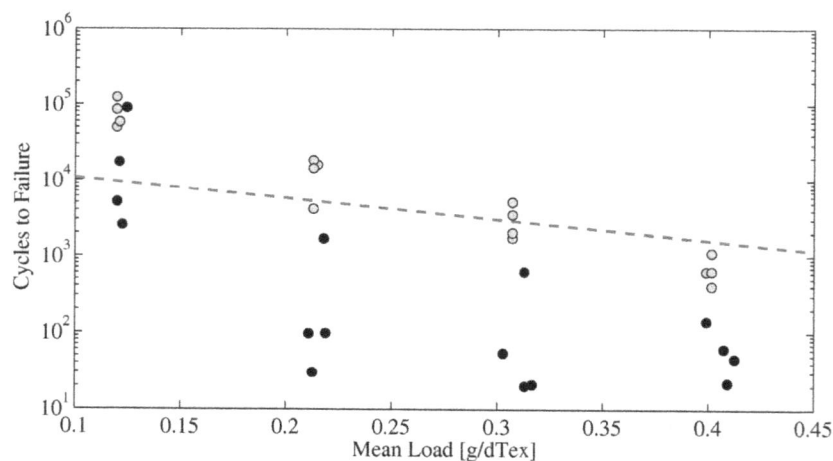

Figure 4. Yarn-on-yarn abrasion tests of aged and new yarns (black and grey markers respectively, from [32]). Mean loads have been normalised by dTex, a measure of mass per unit length (units: g/10 km). Minimum cycles to failure as defined in Cordage Institute guidelines [33] are shown as a dashed red line.

The scatter in the aged yarn results is considerably greater than for the new yarns as demonstrated through the calculated coefficients of variation (R^2) listed in Table 5. Such variability was also observed in break and cyclic tension tests and this is likely to be symptomatic of mild fatigue wear and structural rearrangement sustained during deployment.

3.4. The Value of Physical Component Test Data to Reliability Prediction

Physical component test data provides a means of reducing the lifetime estimate uncertainties associated with using generic failure rates (which may be unverified and potentially unsuitable for the application) prior to gaining operational experience gained in the field. In particular testing allows failure modes to be established prior to installation, providing insight into component wear and reliability and thus allowing confidence levels of long-term installation durability to increase [15]. In the early design stages (TRLs <5) the expected operating conditions may not be fully defined and component performance may be unknown. In this case full fatigue limit state analysis using damage accumulation methods (e.g., the Pålmgren-Miner rule) is typically not possible and instead a statistical approach to reliability assessment combined with experimental fatigue data could be used to estimate the MTTF of the component. The case studies reported in the previous two sections present characteristic failure modes of shackles and synthetic rope yarns. In order to fully define a usable set of fatigue curves more data points are required; a minimum number of 15 samples using at least three stress ranges are required for steel components according to DNV-RP-C203 [24]. Having achieved this, the number of cycles to failure can be estimated if a dominant stress or tension range (or mean load) can be identified from numerical analysis. For example from the yarn-on-yarn test data, the independent variable; mean load is analogous with the influence factor described in Section 2.3., enabling predictions to be made on synthetic rope yarn abrasion. Loading cycles are inherently stochastic in nature, varying in mean load, load amplitude and load rate. It is therefore acknowledged that the estimate of component life provided by this approach would be conservative compared to full damage accumulation analysis.

4. Conclusions

Marine Renewable Energy has the potential to make a significant contribution to the supply of electricity for countries with sufficient wave resource. In order for it to be financially competitive with other forms of electricity generation, significant, but not insurmountable barriers must be overcome before large scale array deployments are realised. Whilst the MRE sector is currently seen as high-risk investment option, the rewards are potentially large. In order to encourage continued investment in the sector confidence in the long-term operating availability and durability of designs is required for a range of stakeholders including certification agencies, insurers and investors. Indeed given the choice of investing in one of several technologies, the ability to operate almost continuously, *i.e.*, having very high reliability target levels to achieve high availability, may be more important in the early stages of development than device performance rating.

For devices which have reached TRL 4, efforts within the sector are currently focused on de-risking technologies through incremental development work with the aim of achieving designs which are both cost effective and reliable. This involves laboratory testing, numerical modelling and prototype testing at benign sites in order to "iron-out" issues before full-scale prototypes are deployed. Once installed, the consequence of failure could range from inconvenient and costly to catastrophic. De-risking not only includes scrutinising novel component designs but also COTS equipment used in different applications in order to determine suitability. Component testing has a key role to play in this process. By subjecting components to representative operating conditions that are likely to be seen in service (*i.e.*, loading regimes, usage rates, electrical loads *etc.*), greater confidence can be gained about failure rates, marking a departure from using generic (and perhaps unsuitable) database values. Testing also enables the causes and effects of failure to be investigated in greater detail further contributing to risk mitigation.

Given the nascent state of the MRE sector and shortage of available operational data it is unsurprising that reliability assessments are currently being conducted using simple statistical methods. The bottom-up statistical method has been adopted by the DTOcean project for inclusion in a Tool for MRE array design. The Tool will be packaged with a database of failure rates which can be overridden by the user should more appropriate data be available. Although simple, this approach will provide a fast estimate of array reliability, allowing weaknesses in the system to be identified through several graphical interfaces. At present device deployment intervals are usually short with the main aim to gain proof of concept. As more operating hours are accumulated, higher accuracy reliability estimates will be possible. In the meantime, a method to improve statistical reliability predictions using physical component test data has been suggested in this paper.

Acknowledgments

The research leading to the results presented in this paper has received funding from the European Community's Seventh Framework Programme. Several projects have been funded by this programme including DTOcean, MERiFIC and MaRINET. In addition, the research mentioned in Section 3.2 received funding from SuperGEN UKCMER through the EPSRC (EP/I027912/1, EP/M014738/1).

Author Contributions

S.D.W. compiled the paper, conducted the synthetic rope study (Section 3.3) and is leading the development of the reliability assessment tool used in the DTOcean project. T.G. carried out the shackle study reported in Section 3.2. P.R.T. and L.J. provided scientific advice and content for the manuscript, in particular the sections on reliability theory.

Conflicts of Interest

The authors declare no conflict of interest.

References

1. European Commission. *Smarter, Greener, More Inclusive? Indicators to Support the Europe 2020 Strategy*; Publications Office of the European Union: Luxembourg, Luxembourg, 2015.
2. European Ocean Energy. *Industry Vision Paper*; European Ocean Energy: Brussels, Belgium, 2013.
3. RenewableUK. *Wave and Tidal Energy in the UK. Conquering Challenges, Generating Growth*; RenewableUK: London, UK, 2013.

4. European Commission. *HORIZON 2020—Work Programme 2014–2015. Annex G. Technology Readiness Levels (TRL)*; Extract from Part 19-Commission Decision C (2014)4995, European Commission, Brussels, Belgium, 2014.

5. Magagna, D.; MacGillivray, A.; Jeffrey, H.; Hanmer, C.; Raventos, A.; Badcock-Broe, A.; Tzimas, E. *Wave and Tidal Energy Strategic Technology Agenda*, 2nd ed.; SIOcean, Joint Research Centre, Brussels, Belgium, 2014.

6. Wolfram, J. On assessing the reliability and availability of marine energy converters: The problems of a new technology. *Proc. Inst. Mech. Eng. O J. Risk Reliabil.* **2006**, *220*, 55–68.

7. Abdulla, K.; Skelton, J.; Doherty, K.; O'Kane, P.; Doherty, R.; Bryans, G. Statistical Availability Analysis of Wave Energy Converters. In Proceedings of the 21st International Offshore and Polar Engineering Conference, Maui, HI, USA, 19–24 June 2011.

8. University of Edinburgh. *SuperGEN Marine Energy Research. Final Report*; University of Edinburgh: Edinburgh, UK, 2011.

9. Catapult Offshore Renewable Energy and the Crown Estate. SPARTA: The Performance Data Exchange Platform for Offshore Wind. Available online: https://ore.catapult.org.uk/sparta (accessed on 5 August 2014).

10. Offshore Renewable Energy (ORE) Catapult. Tidal turbine powertrain reliability project. Available online: https://ore.catapult.org.uk/our-projects/-/asset_publisher/fXyYgbhgACxk/content/tidal-turbine-powertrain-reliability-project (accessed on 5 August 2015).

11. Crabtree, C.J.; Zappalá, D.; Hogg, S.I. Wind energy: UK experiences and offshore operational challenges. *Proc. Inst. Mech. Eng. A J. Power Energy* **2015**, doi:10.1177/0957650915597560.

12. U.S. Department of Defense. *Military Handbook: Reliability Prediction of Electronic Equipment*; MIL-HDBK-217F; U.S. Department of Defense: Washington, DC, USA, 1991.

13. OREDA. Offshore REliability Data Home Page. Available online: http://www.oreda.com (accessed on 10 July 2015).

14. Bala, S.; Pan, S.; Das, D.; Apeldoorn, O.; Ebner, S. Lowering Failure Rates and Improving Serviceability in Offshore Wind Conversion-Collection Systems. In Proceedings of the 2012 IEEE Power Electronics and Machines in Wind Applications, Denver, CO, USA, 16–18 July 2012.

15. Thies, P.R.; Smith, G.H.; Johanning, L. Addressing failure rate uncertainties of marine energy converters. *Renew. Energy* **2012**, *44*, 359–367.

16. McAuliffe, F.D.; Macadré, L.; Donovan, M.H.; Murphy, J.; Lynch, K. Economic and Reliability Assessment of a Combined Marine Renewable Energy Platform. In Proceedings of the 11th European Wave and Tidal Energy Conference, Nantes, France, 6–11 September 2015.

17. Det Norske Veritas. *Offshore Service Specification DNV-OSS-312 Certification of Tidal and Wave Energy Converters*; Det Norske Veritas: Oslo, Norway, 2008.

18. Snowberg, D.; Weber, J.; *MHK Technology Development Risk Management Framework*; NREL/TP-5000-63258; National Renewable Energy Laboratory, Denver, USA, 2015.

19. Low Carbon Innovation Coordination Group. Technology Innovation Needs Assessment (TINA) Marine Energy Summary Report. Available online: http://www.carbontrust.com/media/168547/tina-marine-energy-summary-report.pdf (accessed on 12 September 2014).

20. Thies, P.R.; Johanning, L.; Smith, G.H. Towards component reliability testing for marine energy converters. *Ocean Eng.* **2011**, *34*, 1918–1934.

21. Johanning, L.; Thies, P.R.; Smith, G.H. Component test facilities for marine renewable energy converters. In Proceedings of the Marine Renewable and Offshore Wind Energy Conference, London, UK, 21 April 2010.

22. Gordelier, T.; Parish, D.; Johanning, L. A novel mooring tether for highly dynamic offshore applications; mitigating peak and fatigue loads via reduced and selectable axial stiffness. *J. Mar. Sci. Eng.* **2015**, *3*, 1287–1310.

23. Thies, P.R.; Johanning, L.; Karikari-Boateng, K.A.; Ng, C.; McKeever, P. Component reliability test approaches for marine renewable energy. *Proc. Inst. Mech. Eng. O J. Risk Reliabil.* **2015**, *229*, 403–416.

24. Det Norske Veritas. *Fatigue Design of Offshore Steel Structures*; Det Norske Veritas: Oslo, Norway, 2011.

25. Det Norske Veritas. *Offshore Standard DNV-OS-E301 Position Mooring*; Det Norske Veritas: Oslo, Norway, 2013.

26. Gordelier, T.; Johanning, L.; Thies, P.R. Reliability verification of mooring components for floating marine energy converters. In Proceedings of the SHF Marine Renwable Energy—MRE 2013, Brest, France, 9–10 October 2013.

27. Weller, S.D.; Johanning, L.; Davies, P.; Banfield, S.J. Synthetic mooring ropes for marine renewable energy applications. *Renew. Energy* **2015**, *83*, 1268–1278.

28. Weller, S.D.; Davies, P.; Vickers, A.W.; Johanning, L. Synthetic rope responses in the context of load history: Operational performance. *Ocean Eng.* **2014**, *83*, 111–124.

29. Chailleux, E.; Davies, P. Modelling the Non-Linear Viscoelastic and Viscoplastic Behaviour of Aramid Fibre Yarns. *Mech. Time Depend. Mater.* **2003**, *7*, 291–303.

30. Johanning, L.; Smith, G.H.; Wolfram, J. Measurements of static and dynamic mooring line damping and their importance for floating WEC devices. *Ocean Eng.* **2007**, *34*, 1918–1934.

31. Ridge, I.M.L.; Banfield, S.J.; Mackay, J. Nylon fibre rope moorings for wave energy converters. In Proceedings of the OCEANS 2010 Conference, Seattle, WA, USA, 20–23 September 2010.

32. Weller, S.D.; Davies, P.; Vickers, A.W.; Johanning, L. Synthetic rope responses in the context of load history: The Influence of Aging. *Ocean Eng.* **2015**, *96*, 192–204.

33. Flory, J.F. Cordage Institute guidelines for marine grade nylon and polyester rope-making yarns. In Proceedings of the OCEANS 2013 Conference, Bergen, Norway, 6–8 May 2013.

11

Longer-Term Mental and Behavioral Health Effects of the Deepwater Horizon Gulf Oil Spill

Tonya Cross Hansel *, Howard J. Osofsky, Joy D. Osofsky and Anthony Speier

Department of Psychiatry, Louisiana State University Health Sciences Center, 1542 Tulane Avenue, New Orleans, LA 70433, USA; E-Mails: HOsofs@lsuhsc.edu (H.J.O.); JOsofs@lsuhsc.edu (J.D.O.); Aspei1@lsuhsc.edu (A.S.)

* Author to whom correspondence should be addressed; E-Mail: tcros1@lsuhsc.edu

Academic Editor: Merv Fingas

Abstract: Mental health issues are a significant concern after technological disasters such as the 2010 Gulf Oil Spill; however, there is limited knowledge about the long-term effects of oil spills. The study was part of a larger research effort to improve understanding of the mental and behavioral health effects of the Deepwater Horizon Gulf Oil Spill. Data were collected immediately following the spill and the same individuals were resampled again after the second anniversary ($n = 314$). The results show that mental health symptoms of depression, serious mental illness and posttraumatic stress have not statistically decreased, and anxiety symptoms were statistically equivalent to immediate symptoms. Results also showed that the greatest effect on anxiety is related to the extent of disruption to participants' lives, work, family, and social engagement. This study supports lessons learned following the Exxon Valdez spill suggesting that mental health effects are long term and recovery is slow. Elevated symptoms indicate the continued need for mental health services, especially for individuals with high levels of disruption resulting in increased anxiety. Findings also suggest that the longer-term recovery trajectories following the Deepwater Horizon Gulf Oil Spill do not fall within traditional disaster recovery timelines.

Keywords: behavioral modifications; oil spill; anxiety

---

1. Introduction

Existing research suggests a number of negative mental health consequences for communities directly affected by oil spills [1]. In a community survey carried out in 1989, one year after the Exxon Valdez oil spill, Palinkas, Petterson, Russell and Downs [2] found a significant increase in rates of anxiety, posttraumatic stress disorder, and depression in residents with a high level of exposure to the spill and subsequent cleanup efforts. They also found a relationship between exposure to the oil spill and increased alcohol and substance use, domestic violence, chronic physical conditions, and a decline in social relationships. Those most vulnerable were groups with significant exposure and dependence on fishing and oil work for subsistence [3]. In an earlier study of the Sea Empress Oil Spill in Wales [4], the social and economic consequences following the spill resulted in increased concerns about health, finances, and perceived environmental risk; all of these factors resulted in increases in mental health symptoms [5]. Greater exposure resulting in increased behavioral health symptoms was also evident in the research done after the 2002 Prestige Oil Spill in Spain [6,7]. While several earlier studies of behavioral health following oil spills suggest an immediate negative impact, few studies explore the longer-term effects following oil spills.

1.1. Longer Term Effects of Oil Spills

Because most of the research concerning mental health effects following oil spills has been conducted within one year of the spill, there is a significant gap in the literature on how communities respond to the continued stress and changing environment following oil spills. Studies following the Exxon Valdez oil spill provide limited understanding of long-term mental and behavioral health effects indicating that the impact of oil spills persists for extended periods of time [8,9]. Eight years after the spill, Picou and Arata [10] found elevated levels of depression, intrusive stress, avoidance, and family conflict. Lessons learned from the Exxon Valdez spill show that individual and community effects lasted for decades, with at least part of the fishing industry unable to completely recover. In addition, destruction of the ecosystem occurs with oil spills that impacts on individuals and communities dependent on natural resources for their social and economic livelihood [11], thus, disrupting the usual networks of support that communities depend on to cope with adversities. With loss of jobs and livelihood, families may have few choices; they either have to move or live apart [3,9,12].

1.2. Longer-Term Disaster Recovery

There is limited research on longer-term mental health outcomes following oil spills, however findings increased mental health concerns for almost a decade following Exxon Valdez [10], suggest reevaluation of national disaster recovery timelines [13]. The Substance Abuse and Mental Health Services Administration's (SAMHSA) Disaster Kit suggests that the initial expected response and recovery trajectory focuses on the phases of heroism, honeymoon, and disillusionment, with reconstruction and the new beginning coinciding with the first year anniversary [14]. The surge in initial recovery efforts is often remarkable within the first year providing the boost needed for individuals, families, and communities to begin to move forward with the more prolonged recovery tasks. In most instances, the vast majority of those impacted have dealt with their recovery requirements within 12–18 months after the incident [15]. The 18-month timeline for disaster recovery is also evident in the Federal Emergency Management Agency (FEMA) recovery work timeline, where the final date for permanent work ends at 18 months, marking the end of recovery [16]. However, FEMA also states that time extensions may be granted for complicated disasters [16]. Depending on disaster experiences, personal history and recovery environment, behavioral health effects can linger far beyond the physical recovery and cleanup.

In addition, the Centers for Disease Control have noted that the phases and timelines of disaster recovery have been observed and developed based on natural disasters [15]. For natural disasters, studies attempting to understand the longer-term mental and behavioral health consequences are varied. In a recent review article, MacFarlane and Williams [15] noted anxiety disorder rates ranging from 2% to 29% in longitudinal studies. While many disaster studies report a natural remission [15], population studies have shown that diagnosis of PTSD can be chronic and take upwards of 72 months to remit [17]. Specific to the Gulf South, rates of probable PTSD remained elevated two years following Hurricane Katrina with over 40% endorsing symptoms [18]. The variance in rates and length is due to many factors including sampling, longevity of disaster (*i.e.*, whether it had a clear beginning or ending), magnitude of disaster, preparedness, co-morbidity, and subsequent traumas. Clearly there is a need for more research understanding the longer-term recovery trajectories following all disasters and specifically for technological disasters. Given the historical presence of disasters along the Gulf Coast some individuals may remain in the stage of disillusionment as recovery becomes increasingly more elusive. This outcome seems to have occurred following the Deepwater Horizon Incident otherwise known as the Gulf Oil Spill, where environmental, ecological, and economic effects of the oil spill are still largely unknown.

2. Deepwater Horizon Gulf Oil Spill

The Deepwater Horizon (DWH) Gulf Oil Spill, caused by an offshore oil platform explosion about 50 miles southeast of the Mississippi River delta, occurred on 20 April 2010. Deepwater Horizon spewed an estimated five million barrels of oil for three consecutive months, and is the largest marine oil spill in history [11]. Given the uniqueness of the spill, especially its size and occurrence less than five years following the worst natural disaster in United States' history, Hurricane Katrina, it is difficult to make assumptions about the impact on areas affected.

The Louisiana State University Health Sciences Center Department of Psychiatry conducted a study designed to assess the immediate mental health impact on residents in Southeastern Louisiana heavily impacted by the Gulf Oil Spill using telephone and face-to-face interviews. The results showed that the factors having the greatest effect on mental health were the extent of disruption on participants' lives, work, family, and social engagement resulting in increased symptoms of anxiety, depression, and posttraumatic stress. Given that the location of the oil spill affected individuals and communities with prior devastation from Hurricane Katrina, results also revealed that losses from Hurricane Katrina were highly associated with negative mental health outcomes, however the oil spill distress had unique variance in the analyses supporting that the DWH Gulf Oil Spill represents a complex recovery [19]. Additional studies conducted across the Gulf States have concurred with these findings and support the need for continued mental health treatment of children and adults, due to increased mental health concerns and symptoms [20–24]. In contrast, findings from a federal studies found a lack of increase in mental health symptoms following the oil spill; however, the authors note that a limitation with their study is that the broad population based surveillance methods may underestimate prevalence due to individuals directly affected living in smaller sub-communities [25].

The DWH Gulf Oil Spill studies demonstrate the immediate mental and behavioral health impact and subsequent needs following the disaster. Based on both clinical experience and supportive work done in communities along the Louisiana Gulf Coast, the current study hypothesized that negative mental health symptoms would remain elevated longer than the traditional one-year disaster recovery timeline. Consistent with disaster literature, it was hypothesized further that continued symptomatology would be associated with greater perceived disruption from the DWH Gulf Oil Spill. This study aims to explore recovery of a sample of Gulf Coast residents assessed in the first year following the spill and again just after the second Anniversary. The overall goal is to improve understanding of the longer-term impact of oil spills.

3. Experimental Design

This study was part of a larger research effort designed to improve understanding of the mental and behavioral health effects on individuals following the DWH Gulf Oil Spill. The first set of data was collected 1 year following the spill (Time 1) and the second set was gathered one year later after the second anniversary (Time 2). Time 1 began in August 2010 and with funding provided by the Louisiana Department of Social Services and ended in October 2011 with funding from the Louisiana State Department of Health and Human Services, Office of Behavioral Health. Coinciding with changes in funding and to increase the comprehensiveness of symptoms assessed, additional measures assessing depression and anxiety were added mid surveillance in Time 1. A total of 2093 participants were surveyed in Time 1 using both random telephone and purposive sampling. Participants from Time 1 were resampled following the second anniversary of the spill beginning in April 2012 and ending in August 2012. Interviews were conducted over the telephone using valid numbers provided in Time 1. Three attempts were made to contact each person by telephone and a total of 769 successful contacts were made. Of those contacted, a total sample of 314 agreed to participate, were matched based on last name and birthdate, and provided valid responses. The minimum time between surveys was 5 months and the maximum was 22 months ($M = 13.89$, $SD = 4.76$). The research protocol was approved by the Louisiana State University Health Sciences Center Institutional Review board.

4. Measures

The Deepwater Horizon Psychosocial Assessment was developed with consultation from stakeholders, local leaders, and state and national consultants. The assessment was comprised of the following sections measuring: socio-demographics, Hurricane Katrina losses, oil spill concerns and disruption, and mental health.

Hurricane Katrina experiences: Respondents were asked to endorse if they had experienced the following as a result of Hurricane Katrina in 2005: house destroyed, house damaged, injured, loss of business, loss of income, family members injured, family members killed, loss of personal property other than house, became seriously ill, victimized, friends/family members house destroyed/damaged, friends injured, and friends killed. A Hurricane Katrina experience index was created where 1 point was given for endorsement of each variable. The minimum score was 0 and the maximum was 11 (M = 4.04, SD = 2.38).

Oil spill concerns and disruption: Respondents were asked to endorse if they had concerns or problems with the following as a result of the DWH Gulf Oil Spill: damage to wildlife and environment; health and food concerns; loss of usual way of life; loss of job opportunities; loss of tourism; personal health effects; loss of personal or family business; and needing to relocate. An oil spill concern index was created where 1 point was given for endorsement of each variable. The minimum score was 0 and the maximum was 8 (Time 1, M = 4.64, SD = 2.26; Time 2, M = 4.69, SD = 2.42). A modified version of the Sheehan Disability Scale (SDS) was used to assess overall disruption of life from the oil spill [26]. Participants were asked to rate the extent to which the oil spill disrupted their work, school work, social life and leisure activities, and family life and home responsibilities on a five-point Likert scale ranging from 1 *(not at all)* to 5 *(extremely)*. The minimum score was 3 and the maximum was 15 (Time 1, M = 7.92, SD = 4.21; Time 2, M = 7.22, SD = 4.10).

Mental health: Mental health was assessed using the K6 [27], Posttraumatic Symptom Checklist for Civilians (PCL-C) [28], Center for Epidemiologic Studies Depression Scale (CESD) [29], and General Anxiety Disorder (GAD-7) [30].

The K6 was used to assess overall well-being and, specifically, symptoms related to anxiety and depression. Respondents were asked to rate on a five-point Likert scale ranging from 0 *(none of the time)* to 4 *(all of the time)* if they felt: nervous, hopeless, restless or fidgety, so depressed that nothing could cheer them up, that everything was an effort, and if they felt worthless. Scores range from 0 to 24 and the minimum score for the current sample was 0 and the maximum was 24 (Time 1, M = 6.80, SD = 6.61, α = 0.94; Time 2, M = 6.19, SD = 6.56, α = 0.92). A cut-off score of 13+ was used to determine significant symptoms of serious mental illness; 63 (21%) met the cut off at Time 1 and 62 (20%) at Time 2.

Posttraumatic stress symptoms were assessed using the PCL-C. The 17 item scores range from 1 *(not at all)* to 5 *(extremely)* and total scores can range from 17 to 85. The minimum score for the current sample was 17 and the maximum was 85 (Time 1, M = 34.93, SD = 16.80, α = 0.97; Time 2, M = 34.45, SD = 18.42, α = 0.96). A cut-off score of 50 was used to determine significant symptoms of posttraumatic stress; 59 (20%) met the cut off at Time 1 and 66 (21%) at Time 2.

Depression was measured using the CESD. The 10 item scores are assigned values from 0 *(none of the time)* to 3 *all of the time* and total score ranges from 0 to 30. The minimum score for the current sample was 0 and the maximum was 30 (Time 1, M = 9.60, SD = 9.32, α = 0.94; Time 2, M = 9.06, SD = 9.13, α = 0.95). A cut-off score of 10+ was used. At Time 1 (n = 172), 73 (42%) met the cut off and at Time 2 (n = 313), 53 (17%) met the cut off for depression.

Anxiety was measured using the GAD-7. The 7 item scores are assigned values from 0 *(not at all)* to 3 *(nearly every day)*; total score for the 7 items ranges from 0 to 21. Scores of 5, 10, and 15 are taken as the cut off points for mild, moderate, and severe anxiety, respectively. The minimum score for the current sample was 0 and the maximum was 21 (Time 1, M = 7.97, SD = 7.21, α = 0.96; Time 2, M = 7.40, SD = 7.06, α = 0.94). At Time 1 (n = 172), 44 (26%) met the cut off for mild anxiety, 21 (12%) for moderate anxiety and 37 (22%) for severe anxiety. At Time 2 (n = 314), 53 (17%) met the cut off for mild anxiety, 42 (13%) for moderate anxiety and 68 (22%) for severe anxiety.

5. Participants

Two hundred ten (67%) participants were female and 104 (33%) were male; the minimum age was 18 and the maximum age was 80 (M = 49.15, SD = 14.39). The majority of the participants were: married/cohabitating (n = 188, 60%); white (n = 214, 68%); and reported a 2009 annual income of less than \$40,000 (n = 185, 64%). Forty participants (13%) reported occupations affected by the oil spill, including hospitality and tourism; seafood related industries; fishing; and oil/drilling support. The majority of participants were from parishes (counties) legally defined as most exposed to the DWH Gulf Oil Spill (n = 270, 86%), which include Lafourche, St. Bernard, Plaquemines, Terrebonne, Jefferson, and Orleans [20]. Seventy-nine or 25% applied for financial assistance following the oil spill. Participants were asked if they were a litigant in the BP lawsuit; 34 (11%) replied yes and 275 (89%) said no.

6. Statistical Analysis

To answer the primary hypothesis—sample equivalence on somatic complaints, posttraumatic stress, serious mental illness, anxiety and depression—two one-sided test (TOST) procedures were used with confidence intervals based on the Cohen's d t-test effect sizes to determine the margin of equivalence [31,32]. TOST procedures utilize traditional hypothesis difference testing (paired sample t-test), but extend the application to equivalence testing by asking whether the non-significant difference is small enough to determine that the samples are indeed similar [31]. United States Food and Drug equivalence determination was used and based on whether the mean difference lies within the confidence interval of equivalence [32]. To answer the secondary hypothesis—continued symptomatology would be associated with greater perceived disruption from the DWH Gulf Oil Spill—ordinary least squares regression was used. Regression was used to explore additional factors that may also contribute to continued levels of anxiety, such as demographics, Hurricane Katrina losses, and additional oil spill variables.

7. Results

The first step in assessing the hypotheses—sample equivalence on posttraumatic stress, serious mental illness, anxiety and depression—was to conduct five paired sample t-tests. Results are presented in Table 1, where results failed to reveal a significant difference on posttraumatic stress, serious mental illness, anxiety and depression. Next confidence intervals of equivalence were calculated based on Cohen's d to assess if the non-significant difference is small enough (see Table 1). Results revealed that the mean difference for anxiety lay within the margin of equivalence. While there was no statistical difference among posttraumatic stress, serious mental illness, and depression, the margin of equivalence did not include the mean difference between Time 1 and Time 2.

Given partial support of the primary hypothesis with no change in anxiety symptoms, regression analyses were used to explore which factors (being married or cohabitating, pre-oil spill income, oil/Gulf dependent occupation, litigation status, oil spill concerns, oil spill disruption, post spill funding requests, Hurricane Katrina losses) predict continued levels of anxiety. Preliminary analyses revealed that gender, race (white *vs.* other), parish (most impacted *vs.* other) r^{pb}-values (314) −0.01 to 0.03, p-values 0.63 to 0.86, and age, r (314) −0.01, p = 0.90, were not associated with anxiety thus these were not included in the regression. The enter method was used and with all variables accounted for 37% (adjusted $R^2 = 0.355$) of the variance in anxiety, $F(8, 305) = 22.51$, $p < 0.001$. Beta coefficients are presented in Table 2, where marital status, applied for financial assistance following spill, Hurricane Katrina losses, and oil spill disruption individually predicted anxiety. Results suggest that as individuals tend to be married or cohabitate, anxiety scores decrease by 0.11. For individuals that applied for financial assistance following the oil spill, anxiety scores decrease by 0.12. For individuals reporting a pre-oil spill income under $40,000, anxiety scores increase by 0.16. As Hurricane Katrina losses increase by 1, anxiety scores increase by 0.19 and as oil spill disruption increases by 1, anxiety scores increase by 0.42.

Table 1. Paired Sample Statistics and TOST Procedures for Mental Health Symptoms.

Mental Health	Time 1		Time 2		95% CI Difference						90% CI Equivalence	
	M	SD	M	SD	M^Δ (change)	L	U	t	df	p	d	L	U
Anxiety	8.0	7.2	8.0	7.1	−0.05	−1.01	0.90	−0.11	171	0.91	0.01	−0.14	0.16
Serious Mental Illness	6.8	6.6	6.2	6.6	0.65	−0.04	1.34	1.84	301	0.07	0.11	−0.01	0.22
Depression	9.7	9.3	10.3	9.5	−0.63	−1.98	0.73	−0.91	170	0.36	0.07	−0.08	0.22
Posttraumatic Stress	35.0	16.6	34.0	18.2	0.94	−0.95	2.83	0.98	292	0.33	0.06	−0.06	0.17

Table 2. Beta Coefficients Predicting Anxiety.

	B	SE	β	T	p	95% CI Lower	95% CI Upper
Married or cohabitating	−1.53	0.69	−0.11	−2.21	0.028	−2.89	−0.17
Oil/Gulf dependent occupation	1.36	1.13	0.06	1.20	0.231	−0.87	3.58
Litigant	1.35	1.14	0.06	1.18	0.237	−0.89	3.59
Oil spill concerns	0.10	0.16	0.03	0.59	0.559	−0.23	0.42
Oil spill disruption	0.73	0.10	0.42	7.16	0.000	0.53	0.93
Hurricane Katrina Losses	0.59	0.15	0.19	3.89	0.000	0.29	0.90
Income above 40,000	−2.41	0.72	−0.16	−3.36	0.001	−3.82	−1.00
Post spill financial assistance	−2.02	0.89	−0.12	−2.26	0.025	−3.78	−0.26

8. Discussion

During the first 18 months following the Deepwater Horizon (DWH) Gulf Oil Spill, residents of Southeastern Louisiana reported increased symptoms of anxiety, depression, and posttraumatic stress [19,33]. The current study resampled individuals from the initial responders and results failed to reveal significant changes in anxiety, depression, serious mental illness, and posttraumatic stress two years post spill. Analyses further revealed that immediate anxiety symptoms were statistically equivalent to the elevated anxiety symptoms over two years following the disaster. While posttraumatic stress, serious mental illness, and depression did not statistically decrease, they were not statistically equivalent either. An explanation for anxiety statistically remaining at the same rates over two years post disaster may be contributed to the nature of the disaster. The role of uncertainty and unknown outcomes in a human caused disaster leads to anxiety on how, when, and if recovery will happen [1,22]. These findings suggest that the longer-term recovery trajectories for the DWH Gulf Oil Spill do not fall within the more traditional 18-month disaster recovery timeline [15,16,34].

Variables associated with continued symptoms of anxiety included marital status, application for financial assistance following the spill, Hurricane Katrina losses, and oil spill disruption. As with the initial study of immediate mental health symptoms following the spill [19], oil spill disruption was the most significant contributor to increased symptomotology, and accounted for the largest proportion of variance in anxiety symptoms. Interestingly an indirect association was revealed, where individuals that applied for financial assistance following the oil spill reported fewer symptoms of anxiety. This finding may support reports that the application process was overly complicated and was unattainable for the business practices of self-employed individuals in the fishing industries [23,24]. Contrary to the Exxon Valdez findings of Picou, Marshall and Gill [35], the low association between anxiety and litigation was no longer significant when accounting for the other variables. However, similar to their study, socioeconomic status predicted anxiety [35]; for individuals with incomes below $40,000 reporting more symptoms.

With rates of serious mental illness at 20%, depression at 35%, posttraumatic stress at 21% and moderate to severe anxiety at 35%, the rates of longer-term mental health symptoms continue to be elevated well above national norms of 6% for serious mental illness [36], 10% depression [37], 3% posttraumatic stress [38], and 18% for anxiety [39]. Mental health services are currently provided on a limited basis through the Gulf Region Health Outreach Program as part of the Deepwater Horizon Medical Benefits Class Action Settlement, which was approved by the U.S. District Court in New Orleans on 11 January 2013 and became effective on 12 February 2014. Four institutions from each of the four most impacted states collaborate to carry out the Mental and Behavioral Health Capacity Project (MBHCP), including the University of South Alabama, University of West Florida, Louisiana State University Health Sciences Center, and the University of Sothern Mississippi. A recent report on the Louisiana component of the project, supports the findings from the current study, and indicates a continued need for mental and behavioral health treatment [40].

The primary limitations with this study, consistent with disaster research [41], are the lack of pre-disaster data and reliance on self-report measures. While purposive sampling allowed for better representation of those directly affected by the spill, it does limit the generalizability to the larger populations. Other limitations include the relatively low response rate and the range of 17 months between Time 1 and Time 2. Analyses comparing respondents (33%) *versus* non-respondents (28%) on anxiety cut-off scores failed to reveal a significant group difference $\chi^2 = 2.3$, $p = 0.12$ or an association among time and anxiety ($r = -0.06$). Nonetheless, the low response rate and time between studies may have impacted findings in unknown ways. Another reason for lack of response may have been the ongoing litigation and fear that participation may be used against them in the settlement procedures. This limitation may have contributed to the lack of association among litigation and anxiety. Finally, the lack of litigation association may also suggest a limitation with timing due to ongoing legal action possibly influencing respondents to be hesitant to acknowledge their involvement. Continued longitudinal community surveys would help to better understand the overall recovery trajectory for individuals affected by the DWH Gulf Oil Spill. Further in-depth investigation of individuals that were most disrupted would provide more information to inform methods of how to address elevated symptoms.

9. Conclusions

This study supports many of the lessons learned from the Exxon Valdez spill, [3,8–10] suggesting that the indirect effects of the DWH Gulf Oil Spill are long term and recovery is slow. With mental health symptoms of anxiety, depression, PTSD and serious mental illness elevated above national rates, the need for continued mental health services is evident. Based on the above research mental health services should be targeted toward individuals with high levels of disruption and anxiety. In addition this study highlights the need for policy discussions around disaster recovery timelines and established norms [13].

Author Contributions

H.J.O., J.D.O. and A.S. conceived of the study. H.J.O., J.D.O. and T.C.H. designed the sampling and analysis plan. T.C.H compiled, summarized, interpreted the results and drafted the manuscript. All authors read, contributed to, and approved the manuscript.

Conflicts of Interest

The authors declare no conflict of interest.

References

1. Goldstein, B.D.; Osofsky, H.J.; Lichtveld, M.Y. The Gulf oil spill. *N. Engl. J. Med.* **2011**, *364*, 1334–1348.
2. Palinkas, L.A.; Petterson, J.S.; Russell, J.C.; Downs, M.A. Community patterns of psychiatric disorders after the Exxon Valdez oil spill. *Am. J. Psychiatry* **1993**, *150*, 1517–1524.
3. Palinkas, L.A.; Russell, J.C.; Downs, M.A.; Petterson, J.S. Ethnic difference in stress: Coping and depressive symptoms after the Exxon Valdez oil spill. *J. Nerv. Mental Disorder* **1992**, *180*, 287–295.
4. Lyons, R.A.; Temple, J.M.; Evans, D.; Fone, D.L.; Palmer, S.R. Acute health effects of the Sea Empress oil spill. *J. Epidemiol. Commun. Health* **1999**, *53*, 306–310.
5. Gallacher, J.; Brostering, K.; Palmer, S.; Fone, D.; Lyon, R.S. Symptomatology attributable to psychological exposure to a chemical incident: A natural experiment. *J. Epidemiol. Commun. Health* **2007**, *61*, 506–512.
6. Carrasco, J.M.; Perez-Gomez, B.; Garcia-Mendizabal, M.J. Health-related quality of life and mental health in the medium-term aftermath of the Prestige oil spill in Galiza (Spain): A cross-sectional study. *BMC Public Health* **2007**, *7*, 245–252.
7. Sabucedo, J.M.; Arce, C.; Senra, C.; Seoane, G.; Vazquez, I. Symptomatic profile and health-related quality of life of persons affected by the Prestige catastrophe. *Disasters* **2009**, *34*, 809–820.
8. Palinkas, L.A.; Petterson, J.S.; Russell, J.C.; Downs, M.A. Ethnic differences in symptoms of posttraumatic stress after the Exxon Valdez oil spill. *Prehospital Disaster Med.* **2004**, *19*, 102–112.
9. Picou, S.; Formichella, C.; Marshall, B.; Arata, C. Community Impacts of the Exxon Valdez Oil Spill: A synthesis and elaboration of social science research. In *Synthesis: Three Decades of Research on Socioeconomic Effects Related to Offshore Petroleum Development in Coastal Alaska*; Stephen R. Braund & Associates: Anchorage, AK, USA, 2009; pp. 279–310.

10. Picou, S.; Arata, C. Chronic Impacts of the Exxon Valdez Oil Spill: Resource Loss and Commercial Fishers. In *Coping with Technological Disasters*; Prince William Sound Regional Citizens' Advisory Council: Anchorage, AK, USA, 1997; pp. J2–J43.

11. National Commission on the BP Deepwater Horizon Oil Spill and Offshore Drilling. *Deep Water: The Gulf Oil Disaster and the Future of Offshore Drilling*; Report to the President of the USA, New Orleans, LA, USA, 2011.

12. Gill, D.A.; Picou, J.S. Technological disaster and chronic community stress. *Soc. Nat. Resour.* **1998**, *11*, 795–815.

13. McFarlane, A.C.; Williams, R. Mental Health Services Required after Disasters: Learning from the Lasting Effects of Disasters. *Depress Res. Treat.* **2012**, *2012*, doi:10.1155/2012/970194.

14. Danya Institute. Disaster Mental Health Responder Certification Training at the DC Department of Behavioral Health. Available online: http://www.danyainstitute.org/2014/02/disaster-mental-health-responder-certification-training-at-the-dc-department-of-behavioral-health/ (accessed on 19 October 2015).

15. Substance Abuse and Mental Health Services Administration (SAMHSA). *Field Manual for Mental Health and Human Service Workers in Major Disasters*; ERIC: Washington, DC, USA, 2000.

16. Federal Emergency Management Agency. Deadlines & Timelines. Available online: http://www.fema.gov/public-assistance-local-state-tribal-and-non-profit/deadlines-timelines (accessed on 19 October 2015).

17. Kessler, R.C.; Sonnega, A.; Bromet, E.; Hughes, M.; Nelson C.B. Posttraumatic stress disorder in the National Comorbidity Survey. *Arch. General Psychiatry* **1995**, *52*, 1048–1060.

18. Kessler, R.C.; Galea, S.; Gruber, M.J.; Sampson, N.A.; Ursano, R.J.; Wessely, S. Trends in mental illness and suicidality after Hurricane Katrina. *Mol. Psychiatry* **2008**, *13*, 374–384.

19. Osofsky, H.J.; Osofsky, J.D.; Hansel, T.C. Deepwater Horizon Oil Spill: Mental health effects on residents in heavily affected areas. *Disaster Med. Public Health Prep.* **2011**, *5*, 280–286.

20. Abramson, D.M.; Redlener, I.E.; Stehling-Ariza, T.; Sury, J.; Banister, A.N.; Park, Y.S. *Impact on Children and Families of the Deepwater Horizon Oil Spill: Preliminary Findings of the Costal Population Impact Study*; Report for National Center for Disaster Preparedness: New York, NY, USA, 2010.

21. Witters, D. *Gulf Coast Residents Worse of Emotionally after BP Oil Spill*; Gallup: Washington, DC, USA, 2010.

22. Morris, J.G.; Grattan, L.M.; Mayer, B.M.; Blackburn, J.K. Psychological responses and resilience of people and communities impacted by the Deepwater Horizon oil spill. *Trans. Am. Clin. Climatol.* **2013**, *124*, 191–201.

23. Buttke, D.; Vagi, S.; Bayleyegn, T.; Sircar, K.; Strine, T.; Morrison, M.; Allen, M.; Wolkin, A. Mental health needs assessment after the gulf coast oil spill—Alabama and Mississippi, 2010. *Prehosp. Disaster Med.* **2012**, *27*, 401–408.

24. Grattan, L.M.; Roberts, S.; Mahan, W.T.; McLaughlin, P.K.; Otwell, W.S.; Morris, J.G. Early psychological impacts of the Deepwater Horizon oil spill on Florida and Alabama communities. *Environ. Health Perspect.* **2011**, *119*, 838–843.

25. Substance Abuse and Mental Health Services Administration and Centers for Disease Control and Prevention. *Behavioral Health in the Gulf Coast Region Following the Deepwater Horizon Oil Spill*; HHS Publication No. (SMA) 13-4737; Rockville, M.D., Ed.; Substance Abuse and Mental Health Services Administration and Centers for Disease Control and Prevention: Atlanta, GA, USA, 2013.

26. Sheehan, D.V.; Harnett-Sheehan, K.; Raj, B.A. The measurement of disability. *Int. Clin. Psychopharmacol.* **1996**, *11*, 89–95.

27. Kessler, R.C.; Andrews, G.; Colpe, L.J.; Hiripi, E.; Mroczek, D.K.; Normand, S.-L.T.; Walters, E.E.; Zaslavsky, A. Short screening scales to monitor population prevalence and trends in nonspecific psychological distress. *Psychol. Med.* **2002**, *32*, 959–976.

28. Weathers, F.W.; Litz, B.T.; Herman, D.S.; Huska, J.A.; Keane, T.M. The PTSD Checklist (PCL): Reliability; Validity; and Diagnostic Utility. In *Annual Convention of the International Society for Traumatic Stress Studies*; International Society for Traumatic Stress Studies: San Antonio, TX, USA, 1993.

29. Radloff, L.S.; Locke, B.Z. The community mental health assessment survey and the CES-D Scale. In *Community Surveys of Psychiatric Disorders*; Weissman, M.M., Myers, J.K., Ross, C.E., Eds.; Rutgers University Press: New Brunswick, NJ, USA, 1986; pp. 177–189.

30. Spitzer, R.L.; Kroenke, K.; Williams, J.B.W.; Lowe, B. A brief measure for assessing generalized anxiety disorder. *Arch. Inern. Med.* **2006**, *166*, 1092–1097.

31. Wuensch, K.L. *Confidence Intervals; Pooled and Separate Variances T*; Department of Psychology, East Carolina University: Greenville, NC, USA, 2010.

32. Lachenbruch, P.A. *Equvalence Testing*; United States Food and Drug Administration: Silver Spring, MD, USA, 2001.

33. Osofsky, H.J.; Hansel, T.C.; Osofsky, J.D.; Speier, A. Factors Contributing to Mental and Physical Health Care in a Disaster-Prone Environment. *Behav. Med.* **2015**, *31*, 131–137.

34. Center for Disease Control and Prevention. Disaster Mental Health Primer: Key Principles, Issues and Questions. 2012. Available online: http://emergency.cdc.gov/mentalhealth/primer.asp (accessed on 19 October 2015).

35. Picou, J.S.; Marshall, B.K.; Gill, D.A. Disaster; Litigation; and the Corrosive Community. *Soc. Forces* **2004**, *82*, 1493–1522.

36. Kessler, R.C.; Chiu, W.T.; Demler, O.; Walters, E.E. Prevalence; severity; and comorbidity of twelve-month DSM-IV disorders in the National Comorbidity Survey Replication (NCS-R). *Arch. General Psychiatry* **2005**, *62*, 617–627.

37. Center for Disease Control and Prevention. An Estimated 1 in 10 U.S. Adults Report Depression. Available online: http://www.cdc.gov/features/dsdepression/ (accessed on 19 October 2015).

38. National Institute of Mental Health. The Numbers Count: Mental Disorders in America 2010. Available online: http://www.nimh.nih.gov/health/publications/the-numbers-count-mental-disorders -in-america/index.shtml (accessed on 19 October 2015).

39. Anxiety and Depression Association of America. Facts & Statistics. Available online: http://www.adaa.org/about-adaa/press-room/facts-statistics (accessed on 19 October 2015).

40. Osofsky, H.J.; Osofsky, J.D.; Wells, J.H.; Weems, C. Integrated care: Meeting mental health needs after the Gulf oil spill. *Psychiatr. Serv.* **2014**, *65*, 280–283.

41. Masten, A.S.; Osofsky, J.D. Disasters and their impact on child development: Introduction to the special section. *Child Dev.* **2010**, *81*, 1029–1039.

Marine Microphytobenthic Assemblage Shift along a Natural Shallow-Water CO₂ Gradient Subjected to Multiple Environmental Stressors

Vivienne R. Johnson [1], Colin Brownlee [2], Marco Milazzo [3] and Jason M. Hall-Spencer [1,*]

[1] Marine Biology and Ecology Research Centre, School of Marine Science and Engineering, Plymouth University, Plymouth PL4 8AA, UK; E-Mail: vivjo27@googlemail.com

[2] The Marine Biological Association of the United Kingdom, Citadel Hill, Plymouth PL1 2PB, UK; E-Mail: cbr@mba.ac.uk

[3] Department of Earth and Marine Sciences, University of Palermo, I-90123 Palermo, Italy; E-Mail: marco.milazzo@unipa.it

* Author to whom correspondence should be addressed; E-Mail: jhall-spencer@plymouth.ac.uk

Academic Editor: Magnus Wahlberg

Abstract: Predicting the effects of anthropogenic CO_2 emissions on coastal ecosystems requires an understanding of the responses of algae, since these are a vital functional component of shallow-water habitats. We investigated microphytobenthic assemblages on rock and sandy habitats along a shallow subtidal pCO_2 gradient near volcanic seeps in the Mediterranean Sea. Field studies of natural pCO_2 gradients help us understand the likely effects of ocean acidification because entire communities are subjected to a realistic suite of environmental stressors such as over-fishing and coastal pollution. Temperature, total alkalinity, salinity, light levels and sediment properties were similar at our study sites. On sand and on rock, benthic diatom abundance and the photosynthetic standing crop of biofilms increased significantly with increasing pCO_2. There were also marked shifts in diatom community composition as pCO_2 levels increased. Cyanobacterial abundance was only elevated at extremely high levels of pCO_2 (>1400 µatm). This is the first demonstration of the tolerance of natural marine benthic microalgae assemblages to elevated CO_2 in an ecosystem subjected to multiple environmental stressors. Our observations indicate that

Mediterranean coastal systems will alter as pCO_2 levels continue to rise, with increased photosynthetic standing crop and taxonomic shifts in microalgal assemblages.

Keywords: cyanobacteria; diatoms; Mediterranean; microphytobenthos; ocean acidification; multiple stressors

---

1. Introduction

The current rate of CO_2 release into the atmosphere is driving geochemical changes in the ocean that are thought to be unparalleled in the last 300 million years [1]. Researchers around the world are now striving to find out how ongoing increases in surface seawater pCO_2 and bicarbonate (HCO_3^-) will affect photoautotrophs [2]. Most microalgae have evolved a carbon concentration mechanism (CCM) which increases the CO_2 concentration at the active site of ribulose-1,5-bisphosphate carboxylase/oxygenase (RUBSICO) [3]. Down-regulation of CCM activity has been observed when microalgae have been grown in high CO_2 conditions [4,5]. Due to the high metabolic costs of CCMs, it is thought that microalgae could benefit from CO_2 enrichment since down regulation of the CCM may allow for optimized energy and resource allocation [4–7]. The responses of microalgae to ocean acidification caused by rising pCO_2 levels could have wide-ranging ramifications for ocean health globally since they drive biogeochemical cycles and contribute significantly to global primary production [8]. However, ocean acidification is not proceeding in isolation; interactive effects of elevated CO_2 with other changing environmental conditions such as temperature [9,10], light [11], nutrients [12], metal toxicity [13] and pollution (e.g., sewage, petroleum wastes, pesticides, agricultural run-off) [14] may occur, complicating the prediction of ecosystem responses to rising CO_2 levels.

"Microphytobenthos" are assemblages of microalgae and photosynthetic bacteria that colonise benthic substrata. We decided to investigate assemblages living on rock and on sandy sediments at increasing levels of pCO_2 since microphytobenthic assemblages contribute significantly to primary production in shallow coastal ecosystems, providing a key source of food and playing important roles in biogeochemical processes, as well as in invertebrate and macroalgal settlement [15,16]. Benthic microalgal biomass may be much higher in sediments than in the water column (particularly in muddy sediments) and provide a primary source of fixed carbon to marine food webs; at the sediment-water interface these assemblages also stabilise sediment and modify nutrient and oxygen flux [17].

Laboratory and field studies have indicated that macroalgal assemblages are likely to show profound shifts in taxonomic composition as CO_2 levels rise this century due to alterations in the competitive outcomes between calcified and non-calcified species, and between those with and without-CCMs [18,19]. However no previous studies have addressed the effects of high CO_2 on natural microphytobenthic assemblages; trials using artificial settlement substrata revealed dramatic alterations in the composition of bacterial and microalgal biofilm communities [20–22]. Despite the advantage of being highly standardised, the use of artificial substrata does not accurately represent natural communities [23]. Sampling communities that have colonised natural surfaces along CO_2 gradients may provide more realistic insights into the effects of ocean acidification than studies that used artificial substrata [24]. A recent investigation of natural marine sediment bacterial communities documented increased bacterial diversity and a shift in community composition with increasing levels of pCO_2 [25].

Here, sub-tidal epilithic biofilms and the microphytobenthos associated with sandy sediments (both epipelic and epipsammic components) were investigated along a coastal CO_2 gradient in the Mediterranean Sea. Certain carbon dioxide seeps provide useful proxies for ocean acidification, revealing the structure of marine communities that are resilient to the long-term effects of elevated pCO_2 [26,27]. *In situ* assessments of these communities provide an opportunity for investigating the effects of multiple environmental stressors on ecosystems because, in addition to CO_2-enrichment, the ecological communities found here are subjected to the typical stressors of a coastal habitat in the Anthropocene, such as coastal pollution [28] and overfishing [29]. Elevated levels of metals mercury, cadmium and zinc have been located at many seep systems and although such areas are normally avoided in field studies of the effects of ocean acidification, they could be used to investigate the interactive effects of acidification and metal toxicity [30].

Since biofilm formation on hard substrata is a ubiquitous process in the marine environment, and much of the surface layer of sediments on continental shelves receive sufficient light to sustain primary production, these microbial habitats are important components of coastal and global carbon cycles. Benthic net community production has been estimated to occur over 33% of the global shelf area (based on the compensation irradiance for microphytobenthic community metabolism) and theupper limits of the global net and gross primary production of microphytobenthos have been reported at 2.7×1013 and 3.5×1014 moL C year^{-1}, respectively [31]. The aim of this study was to investigate changes in microphytobenthic biomass and composition along a natural pCO_2 gradient, subject to multiple coastal stressors (e.g., fishing, coastal pollution, heat waves), to better inform predictions of coastal ecosystem resilience to ocean acidification. We hypothesise that microphytobenthic photosynthetic standing crop will be enhanced in acidified areas due to an increase in inorganic carbon availability for photosynthesis.

2. Material and Methods

2.1. Study Site and Sampling Stations

Microphytobenthic assemblages were sampled along a rocky coast off Vulcano (38°25′ N, 14°57′ E, part of the Aeolian Island chain, NE Sicily, Figure 1). This is a microtidal region where volcanic CO_2 seeps acidify the seawater producing a gradient ranging from ~pH 8.2 to ~pH 6.8, running parallel to the coast.

The carbonate chemistry of the gradient (at a depth of ~0.5 m) was monitored over a two year period (September–October 2009, April 2010, September–October 2010, May 2011 and September 2011–October, $n = 22$–27) using methods previously published [21,32]. In brief, a calibrated YSI (556 MPS) pH (NBS scale) meter was used to measure temperature, pH and salinity. Total Alkalinity was measured at each station, from a water sample that had been passed through a 0.2-μm pore size filter (stored in the dark at 4 °C), using an AS-Alk 2 Total Alkalinity Titrator (Apollo SciTech Inc., Newark, GA, USA). The remaining parameters of the carbonate system were then calculated using CO2 SYS software [33].

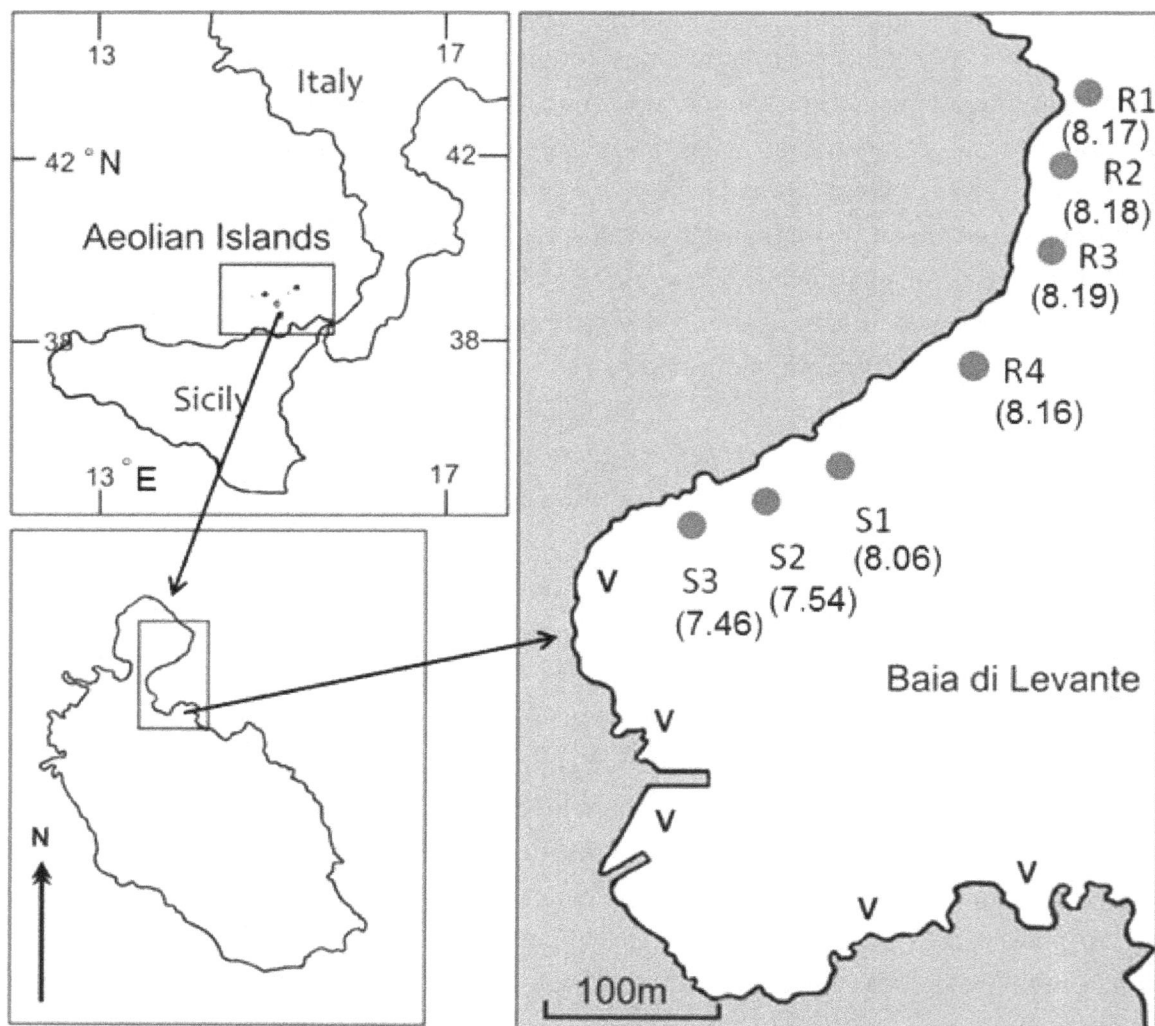

Figure 1. Location of Baia di Levante (Vulcano Island, NE Sicily), showing sampling stations S1–S3 and R1–R4, V = CO_2 seeps. Data in brackets show mean pH of each station ($n = 22–27$).

Three reference stations were located outside the CO_2 seep area with normal, relatively stable pH (R1 mean pH 8.17, 95% confidence intervals (CI) as percentage of the mean pH = 0.42%; R2 pH 8.18, CI = 0.32%; R3 pH 8.19, CI = 0.28%, $n = 22–24$) representative of present-day pCO_2 conditions. Three stations with increasing proximity to the seeps were characterised by intermediate to low mean pH (S1 mean pH 8.06, CI = 0.59%; S2 pH 7.54, CI = 1.59%; S3 pH 7.46, CI = 2.03%, $n = 24–27$). An additional reference station was used to investigate sediment assemblages; this station also had normal, stable pH (R4 mean pH 8.16, CI = 0.32%, $n = 22$) with the same black volcanic sand as found throughout the bay, deposited by the last volcanic explosion on the island in 1888. The seven stations provided a gradient of increasing pCO_2 (Table 1). Due to dynamic seep activity and advection of high CO_2 water, the stations close to the CO_2 seeps were characterised by high pH variability as similarly reported in other seep studies [34–36].

Table 1. Seawater pH, Total Alkalinity (±SE, $n = 3$) measurements and calculated pCO_2 off Vulcano. Temperature (range 18.6–27.7 °C; April–October), pH (NBS scale) and salinity ($n = 38$) were measured on several occasions between September 2009 and September 2011.

Station		pH (NBS Scale)	TA mmol kg^{-1}	pCO_2 (µatm)	HCO$_3^-$ mmol kg^{-1}
	max	8.35		241	2.206
R1	med	8.17	2.682	388	2.341
	min	8.06	(±0.12)	513	2.405
	max	8.29		274	2.177
R2	med	8.18	2.591	365	2.251
	min	8.08	(±0.03)	471	2.311
	max	8.29		272	2.165
R3	med	8.18	2.579	364	2.247
	min	8.10	(±0.04)	446	2.288
	max	8.26		295	2.189
R4	med	8.12	2.582	424	2.28
	min	8.08	(±0.05)	470	2.307
	max	8.22		355	2.401
S1	med	8.08	2.79	510	2.499
	min	7.76	(±0.08)	1119	2.627
	max	8.10		474	2.436
S2	med	7.71	2.742	1244	2.601
	min	7.07	(±0.07)	5628	2.697
	max	8.24		337	2.392
S3	med	7.66	2.796	1428	2.662
	min	6.80	(±0.12)	10,730	2.762

Temperature, total alkalinity, and salinity were relatively constant in the shallow subtidal region along this gradient [21] and geochemical monitoring of the bay has confirmed the suitability of the chosen stretch of coastline for ocean acidification studies, since H_2S and other toxic volcanic compounds are undetectable in the water column [36]. There was no significant difference in midday light intensities between stations R2 (mean lux = 36,935 ± 3641, $n = 13$) and S3 (mean lux = 38,895 ± 4234, $n = 13$) [21]. Water samples for dissolved nutrient analysis (nitrite, nitrate, silicate and phosphate) were collected from stations S1–S3 and R1 (as a representative of all reference stations) at <1 m depth. Samples were collected in 60 mL Nalgene bottles which had been pre-rinsed three times with the sample whilst wearing nitrile-free gloves. Between 5 and 6 replicate samples were taken between May 25–26 2011 and frozen prior to colorimetric analysis on a multi-channel AutoAnalyser (Bran + Luebbe GmbH SPX Process Equipment, Norderstedt, Germany). At stations S1–S3 and R4, three replicate 100 g sediment samples were collected at a depth of ~1 m and air dried in May 2011. The sediment was analysed to determine grain size distribution, sediment sorting and skewness, sediment type and texture as described by Johnson [32].

2.2. In Situ Microphytobenthic Sampling

See Figure S1 for an overview of the experimental approach. Epilithic biofilms were sampled from large boulders at 0.5 m depth within 5 m × 3 m plots at stations R1–R3 and S1–S3 between September and October 2010. Samples were separated by a gap of least 30 cm and to reduce potentially confounding biological and physical effects, the samples were removed from central surfaces of relatively flat-topped rocks where macroflora and fauna were absent at the time of sampling.

For the majority of the analyses (except for chl*a* and SEM analysis which were sampled from rock chippings), biofilm was sampled using a brush mounted in an acrylic cylinder. Once an adequate seal was created between the open end of the cylinder and a flat rock surface, the brush was twisted continuously for 2 min to dislodge biofilm material from a 7 cm^2 area. The suspended material was then collected through an attached 60 mL syringe. To avoid cross sample contamination, the device was thoroughly rinsed between each use and a new brush head attached at each station.

Sediment cores (area = 56.7 cm^2, depth = 3 cm) were taken at 0.5–1.5 m depth from 5 × 3 m plots at R4, S1 S2 and S3 in May 2011 (it was not possible to sample epipsammic and epipelic communities at sites R1–R3 due to the lack of sandy sediment at shallow depth (<1.5 m) comparable to S1–S3). For each set of analyses, ten cores were taken haphazardly with a gap of at least 30 cm between samples. To account for patchy distributions of the microphytobenthos, and to ensure sufficient material for each analysis, five sub-samples (area = 0.64 cm^2, depth = 5 mm) from the surface of each core were pooled and treated as one sample.

2.3. Chlorophyll-a Extraction

The photosynthetic standing stock of epilithic biofilms was assessed by removing (by hammer and chisel) 30 rock chips (~2 cm × 2 cm) from the sampling plots. The photosynthetic standing stock of sand microphytobenthos was assessed from sediment cores (n = 10 per station). Rock chips and sediment samples were immediately frozen (−20 °C) on collection then stored at −80 °C until analysis (<2 weeks) to prevent chlorophyll degradation.

The surface area of the rock chips were measured from photographs using Image J software (v 1.43, National Institutes of Health, Bethesda, MD, USA) and the biofilm material was removed from the surface. Sediment samples were first lyophilised to remove water content (which can affect absorbance readings) and to improve the extraction efficiency [37]. Chlorophyll was extracted from the rock chip biofilm material and sediment samples using 100% hot ethanol, and the absorbance of each sample at 632, 649, 665, 969 and 750 nm was measured in a spectrophotometer (Cecil CE2011, Cecil Instruments Ltd., Cambridge, UK). Three replicate readings were taken at each absorbance to calculate an average value of chl*a* for each sample. The total amount of chl*a* was calculated using the quadrichroic equation of Ritchie [38] and expressed per unit surface area ($\mu g\ cm^{-2}$) for the rock chips and per unit dry weight for the sediment ($\mu g\ g^{-1}$).

2.4. Diatom Abundance

To determine diatom abundance within the biofilm samples (dislodged biofilm material suspended in 60 mL seawater) and sediment (pooled sediment samples in 20 mL seawater), the samples (n = 10) were first preserved with Lugol's Iodine solution and stored in a cool, dark place until counting. Bottles containing the biofilm were shaken vigorously to break up clumps and re-suspend diatoms. The epipelic and epipsammic diatom components of the sediment samples were separated following the methods of Hickman and Round [39]. The non-attached epipelon were isolated by shaking the sample in filtered seawater and removing the supernatant, a process repeated five times. The epipsammic diatoms were removed from the sand grains by sonication, an optimum sonication period of 10 min was set to ensure sufficient diatom removal (~90%) with minimum cell damage [40]. After sonication the sample was gently shaken to re-suspend the diatoms which were then removed in the supernatant.

Diatom abundances were determined using a Sedgewick-Rafter plankton counting chamber [41]. A 1 mL aliquot from each sample was put into the counting chamber and were observed under 400× magnification using an inverted light microscope (Diaphot, Nikon, Japan). Diatoms were counted in grids across randomly selected columns in the chamber until approximately 100–200 diatoms had been counted. The number of diatoms per unit surface area (cm^{-2}) was then calculated as follows:

$$\text{Diatom cells/mL} = \frac{\text{diatoms counted} \times 1000}{\text{no. of observed grids}}$$

$$\text{Diatom cell density (diatoms/cm}^{-2}) = \frac{\text{diatoms/mL} \times \text{storage volume}}{\text{surface area of benthos sampled}}$$

Three replicate counts were conducted to give an average diatom density for each sample.

2.5. Epi-Fluorescence Microscopy

A Biorad Radiance 2000 confocal laser scanning microscope (CLSM) was used to determine the abundance of cyanobacteria within epilithic biofilms. Cyanobacteria disperse their pigments throughout their cytoplasm (as opposed to diatoms where they are enclosed in plastids) thereby making epi-fluorescence a suitable technique for quantifying coverage of these microorganisms within a biofilm. In was not possible to use CLSM on the rock chips because the rugose microtopography made focusing difficult, limiting the accuracy of the technique. Instead, 60 mL samples of biofilm suspension ($n = 10$ per station) were filtered on to 0.2 μm cyclopore polycarbonate membranes (Whatman, GE Healthcare Life Sciences, Buckinghamshire, UK).

A CLSM was also used to determine the abundance of cyanobacteria within the sediment samples. Dead cells and empty diatom valves in the sediment (often the remains of previous epipelic and epipsammic assemblages and those of surrounding epiphytic and planktonic populations which settle on the bottom sediment after death) can lead to a significant source of error when estimating microphytobenthic populations under bright field light microscopy [42]. Therefore, as epifluorescence measurements discriminate against dead cells, diatom epifluorescence was also measured as a proxy for diatom abundance. This technique was not useful for examining the epipsammon, as the uneven microtopography of the sand grains made focusing difficult; therefore, only the epipelic component was analysed with this technique.

The filtered biofilm samples and pooled sediment samples were fixed in 2.5% glutaraldehyde (diluted with filtered seawater) for approximately 1 h in the dark. The filter papers containing the biofilm were then rinsed in distilled water and mounted on microscope slides. The epipelon was separated from the sediment by the process described above. The supernatant was filtered through 0.2 µm filter papers which were then mounted on microscope slides. All slides were stored at −20 °C (in the field <48 h) to ensure the chloroplasts retained their autofluorescence for examination at a later date [43]. When longer storage periods (<3 weeks) were required i.e., between processing of samples in the UK, storage took place at −80 °C [44]. The slides were viewed using a CLSM; excitation 488 nm; emission 570–590/70 and 660–700 nm. A total of 30 images were taken (×10 magnification) at random locations across the filter papers and the percentage cover of cyanobacteria fluorescence (and epipelic diatoms) was digitally quantified per image using Image J software. The mean percentage cover was then calculated for each sample from the 30 images.

2.6. SEM Analysis

The composition of the epilithic and epipelic diatom community was analysed by a scanning electron microscope (SEM). Five replicate rock chips (~2 cm × ~2 cm) were removed from stations R2, S1–S3 and six sediment samples were taken from R4, S1–S3. Only one reference station was sampled in this component of the study; however, since all reference stations exhibited similar physical and chemical seawater properties, selecting one representative reference station was sufficient for comparative purposes. All samples (the epipelic component was isolated as above, filtered onto filter papers and stored in a phosphate buffer solution) were fixed (2.5% glutaraldehyde, one hour), air dried and coated with gold prior to SEM observation. Between 200 and 400 cells were counted and identified to genus from randomly positioned images on the colonised areas of rock chips/filter papers. The relative composition of diatom genera was then averaged for each station. Diatoms that could not be accurately identified were assigned to numbered groups (e.g., unidentified pennate 1, 2, etc.).

2.7. Statistical Analysis

Differences in microphytobenthic assemblages between stations were tested using one-way ANOVA and multiple pairwise comparison post hoc tests (Tukey's Test) were performed where differences were significant. Data that failed tests for normality (Shapiro-Wilk) and homogeneity (Levene Median test) were transformed (arc sin and ln) until assumptions were met. When transformations were unsuccessful, data were analysed by Kruskal-Wallis one way analysis of variance on ranks and multiple pairwise comparison post hoc tests (Dunn's method). These statistical analyses were performed using SigmaPlot 11.0 (Systat Software, Inc., San Jose, CA, USA).

The abundances of diatom genera were used to calculate Shannon diversity (H'), Pielou's evenness (E) and Simpsons index of dominance (D) for each rock chip/sediment sample from each station. The similarity of community assemblages across the epilithic biofilms (total $n = 20$) and sediment samples (total $n = 24$) was examined by hierarchical cluster analysis using IBM SPSS Statistics 18 (IBM Corp, NY, USA). Only genera representing over 1% abundance were included in this analysis including any of the numbered unidentified diatoms groups. Assemblages were clustered using a dissimilarity coefficient (squared Euclidian distance) and Ward's method (minimum variance clustering).

3. Results

3.1. Dissolved Nutrient Concentrations and Sediment Properties

Phosphate concentrations remained at low, undetectable levels (<10 nmol L^{-1}) and there were no significant differences in nitrate concentrations (ANOVA: $F_{(3,20)} = 1.827$, $p = 0.175$). There were significant differences in nitrite (Kruskal-Wallis: $H_{(3)} = 8.327$, $p < 0.05$) and silicate (Kruskal-Wallis: $H_{(3)} = 12.780$, $p < 0.05$) concentrations along the CO_2 gradient (Table 2); S3 had significantly higher nitrite and silicate concentrations than R1 (Dunn, $p < 0.05$), but there were no significant differences in these nutrients between the remaining stations (Dunn, $p > 0.05$). The sedimentary stations all comprised well-sorted coarse black volcanic sand.

Table 2. Mean ($\pm$ SE; $n = 5–6$) dissolved nutrient concentrations along a CO_2 gradient off Vulcano. Phosphate was below the detection limit of the AutoAnalyser (~10 nmol/L) at all stations.

Station	Nitrite (µM)	Nitrate (µM)	Silicate (µM)
R1	0.01 ± 0.001	0.24 ± 0.05	3.43 ± 0.05
S1	0.01 ± 0.005	0.13 ± 0.03	8.34 ± 2.73
S2	0.02 ± 0.003	0.16 ± 0.06	15.12 ± 2.45
S3	0.02 ± 0.001	0.33 ± 0.1	19.39 ± 1.24

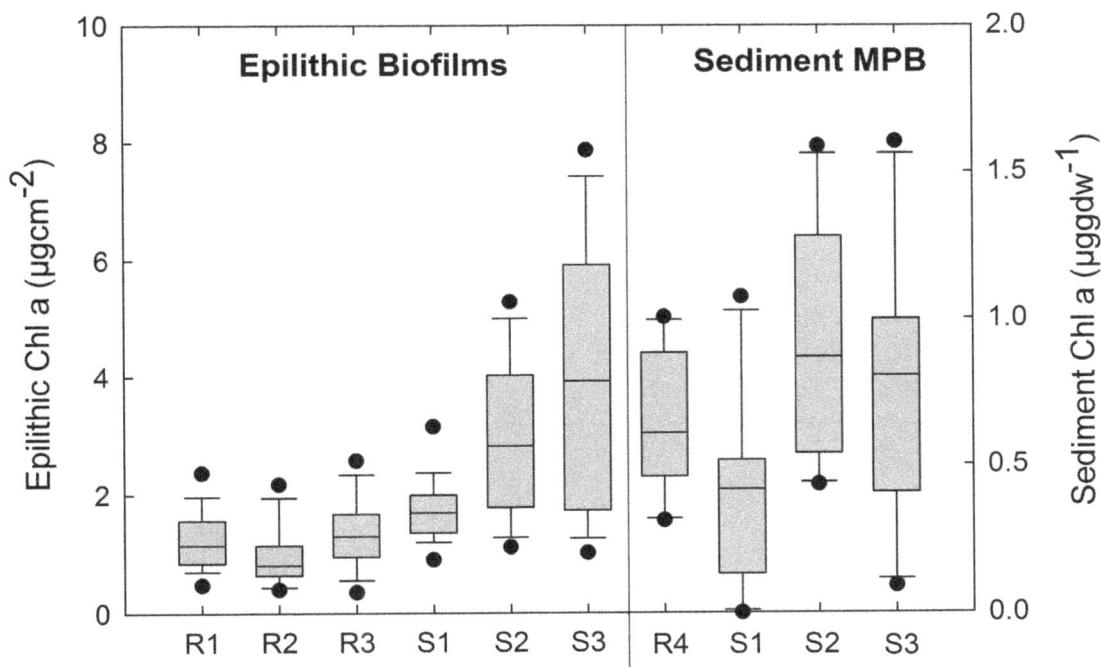

Figure 2. Chl*a* concentration of epilithic biofilms and chl*a* content of the microphytobenthos (MPB) in surface sediment sampled along a CO_2 gradient off Vulcano, Italy (median = horizontal line, 25th and 75th percentiles = vertical boxes, 10th and 90th percentiles = whiskers and dots = min/max values, epilithic; $n = 30$ per station, sediment; $n = 10$ per station).

3.2. Photosynthetic Standing Crop (Chla)

Chla concentration altered significantly in both rocky (Kruskal-Wallis, $H_5 = 84.219$, $p < 0.001$) and sedimentary habitats (ANOVA: $F_{(3,36)} = 3.908$, $p < 0.05$) along the gradient of increasing pCO_2 (Figure 2). The Chla concentration in epilithic biofilms was significantly greater (Dunn, $p < 0.05$) in the CO_2 enriched stations S2 (mean Chla = 2.9 $\mu gcm^{-2} \pm$ SE 0.2, $n = 10$) and S3 (4.0 $\mu g\ cm^{-2} \pm$ SE 0.4, $n = 10$) compared to the reference stations R1 (1.3 $\mu gcm^{-2} \pm$ SE 0.1 $n = 10$), R2 (1.0 $\mu gcm^{-2} \pm$ SE 0.09 $n = 10$) and R3 (1.4 $\mu gcm^{-2} \pm$ SE 0.10, $n = 10$) which did not differ significantly from one another (Dunn, $p > 0.05$). On sand, the highest mean values of Chla were recorded in the CO_2 enriched stations S2 and S3 (0.92 $\mu g\ gdw^{-1} \pm 0.13$ and 0.76 $\mu g\ gdw^{-1} \pm 0.14$, respectively). *Post hoc* pairwise comparisons however, revealed that Chla content only differed significantly between stations S1 and S2 (Tukey, $p < 0.05$).

3.3 Benthic Diatom Abundance

Diatoms increased significantly in abundance along the gradient of increasing CO_2 (Figure 3) in epilithic (ANOVA, $F_{(5,54)} = 34.554$, $p < 0.001$), epipsammic (ANOVA: $F_{(3,36)} = 36.187$, $p < 0.001$) and epipelic assemblages (ANOVA: $F_{(3,36)} = 24.653$, $p < 0.001$). There were three times as many epilithic diatoms at S3 (189,448 cells $cm^{-2} \pm$ SE 11,494) than at the reference stations where diatom abundances did not significantly differ from each other (Tukey $p > 0.05$; $R1 = 65,437$ cells $cm^{-2} \pm$ SE 5527 $n = 10$, $R2 = 52,135 \pm$ SE 4114 $n = 10$, $R3 = 73,639 \pm$ SE 7904 $n = 10$). The abundances of epipsammic and epipelic diatoms were also significantly greater at S3 (11,384 cells $cm^{-2} \pm$ SE 702 $n = 10$ and 84,247 cells $cm^{-2} \pm$ SE 12,233 $n = 10$, respectively) compared to the other stations (Tukey, $p < 0.05$) where diatom abundances were not statistically different from each other (Tukey, $p > 0.05$).

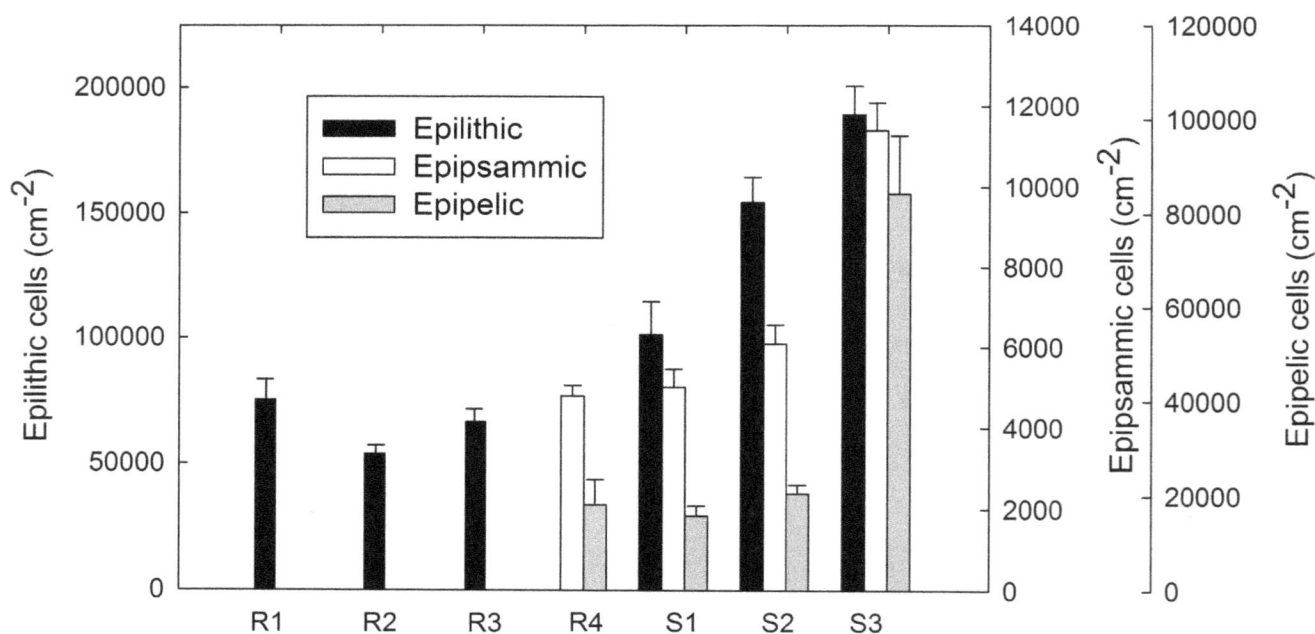

Figure 3. Mean diatom abundance (+SE) of epilithic, epipsammic and epipelic assemblages along a natural CO_2 gradient ($n = 10$ per station).

3.4. Epi-Fluorescence Analysis

Filamentous rhodophytes were ubiquitous in the biofilms sampled (Figure S2). Their photosynthetic pigments emit fluorescence at the same wavelength as cyanobacteria (~660 nm) so it was not possible to use this technique to accurately differentiate between the groups as both can occur in filamentous, branching forms. Therefore, areal coverage of phycobilin fluorescence was used to indicate changes in photosynthetic biomass along the CO_2 gradient. There was a significant difference detected in the percentage cover of phycobilin related fluorescence within biofilms sampled along the gradient (ANOVA: $F_{(5,54)} = 9.09$, $p < 0.001$), but there were no significant differences detected between R1, R2, R3, S1 and S2 (Tukey, $p > 0.05$). Only biofilm at S3 had a significantly higher cover of algae containing phycobilins (15.4% ± SE 1.8%, $n = 10$) than the remaining stations (Tukey, $p < 0.001$), the cover of which ranged between 6% and 9%.

Epipelic diatom epi-fluorescence also varied significantly between stations (Figures S3 and S4; ANOVA: $F_{(3,36)} = 25.065$, $p < 0.001$). The percentage cover of diatom epi-fluorescence was significantly greater at stations S2 and S3 (2.7% ± SE 0.39% and 3.3% ± SE 0.23% respectively) than R4 (Tukey, $p < 0.05$). The cover of cyanobacteria epifluorescence (in this case there was an absence of the filamentous epilithic rhodophytes observed in biofilm samples) was significantly different along the CO_2 gradient (Figure 4; ANOVA: $F_{(3,36)} = 52.936$, $p < 0.001$). The cover at S3 (2.3% ± SE 0.08%) was significantly greater than at S1, S2 and R4 (Tukey, $p < 0.05$) which were all found to be statistically similar (Tukey, $p > 0.05$).

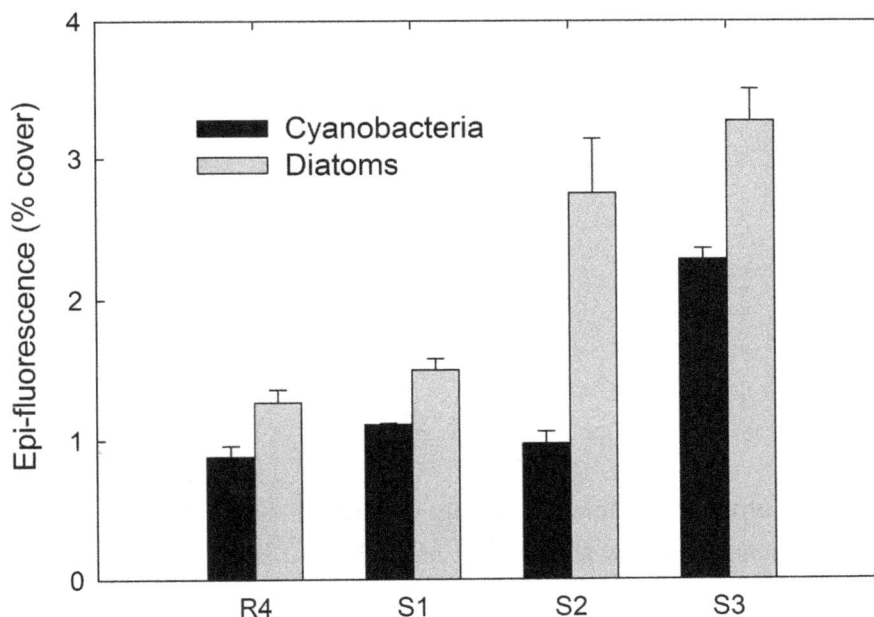

Figure 4. Mean (+SE) epi-fluorescence of epipelic diatoms and cyanobacteria sampled from the surface sediment at each station ($n = 10$).

3.5. Benthic Diatom Assemblage Composition

Both rocky shore and sedimentary diatom assemblages altered significantly along the CO_2 gradient (Figure 5). Epilithic assemblages were similar at the reference stations and S1, where there was a small increase in pCO_2, but markedly different from the stations S2 and S3 which had high levels of pCO_2 (Figure 5). Biofilm SEM images (Figure S4) and cell counts of the most numerous taxa (Figure S5) reveal which epilithic genera proliferated at elevated CO_2 levels (*Rhabdonema*, unidentified Bacillariaceae, *Cocconeis*, *Amphora*, *Striatella*), which became more scarce (*Licmophora*, *Tabularia*) and those that appear to be unaffected by the changes in seawater carbonate chemistry (*Synedra*, *Mastogloia*, *Navicula*, *Nitzschia*). The CO_2 enriched stations had a larger proportion of chain-forming genera (*Rhabdonema*, *Striatella*, unidentified Bacillariaceae) than the reference station. The diversity of epilithic diatom genera (Figure S6) was higher at the reference station H' $2.12 \pm$ SE 0.07 $n = 5$) than at the CO_2 enriched stations (S1 $= 1.83 \pm$ SE 0.06 $n = 5$, S2 $= 1.55 \pm$ SE 0.24 $n = 5$, S3 $= 1.77 \pm$ SE 0.07 $n = 5$) although differences in diversity indices between these stations were not statistically significant (H' ANOVA: $F_{(3,16)} = 2.927$, $p = 0.066$; evenness J ANOVA: $F_{(3,16)} = 3.113$, $p = 0.056$; dominance D ANOVA: $F_{(3,16)} = 0.718$, $p = 0.556$).

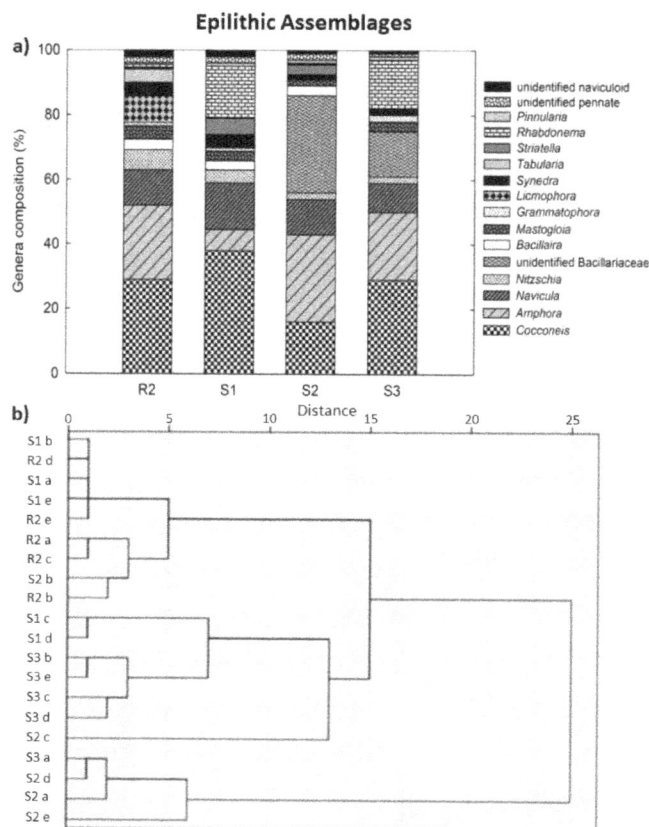

Figure 5. Epilithic diatom assemblages along a natural CO_2 gradient. (**a**) Relative composition of epilithic assemblages at different pCO_2 levels. Bar charts include all genera present over 1% and all unidentified diatoms grouped as unidentified pennate or naviculoid (*n* = 5 per station,); (**b**) Hierarchical cluster analyses of the similarity of epilithic diatom assemblage composition based on Ward's method with squared Euclidian distance for all the biofilms sampled along a natural CO_2 gradient (*n* = 20). Analysis consists of all genera present over 1%, including any of the unidentified groups.

In epipelic assemblages, *Cocconeis* and *Striatella* were more prevalent at the high CO_2 stations (S2 and S3) but several genera were less abundant than in reference conditions (*Mastogloia*,

Grammatophora, *Synedra*, *Nitzschia* and *Amphora*, Figure 6a). The assemblages differed markedly between the reference station and the CO_2 enriched stations (Figure 6b). At elevated pCO_2 there was a proliferation of some sand-dwelling genera (*Cocconeis*, *Striatella*), reduced abundance of several others (*Navicula*, *Nitzschia*, *Grammatophora*, *Mastogloia*, *Synedra*, *Amphora*), while the abundance of other genera showed relatively little change along the CO_2 gradient (*Licmophora*, *Bacillaria*). The diversity of epipelic diatom genera decreased significantly at stations with high CO_2 (ANOVA, $F_{(3,20)} = 48.120$, $p < 0.001$), the dominance index was significantly higher at S2 and S3 than the other two stations (ANOVA: $F_{(3,20)} = 47.516$, $p < 0.001$; Tukey, $p < 0.05$) and the evenness of the assemblage decreased significantly at elevated CO_2 (ANOVA: $F_{(3,20)} = 19.877$, $p < 0.001$), with the reference station having significantly greater evenness than S2 and S3 (Tukey, $p < 0.05$).

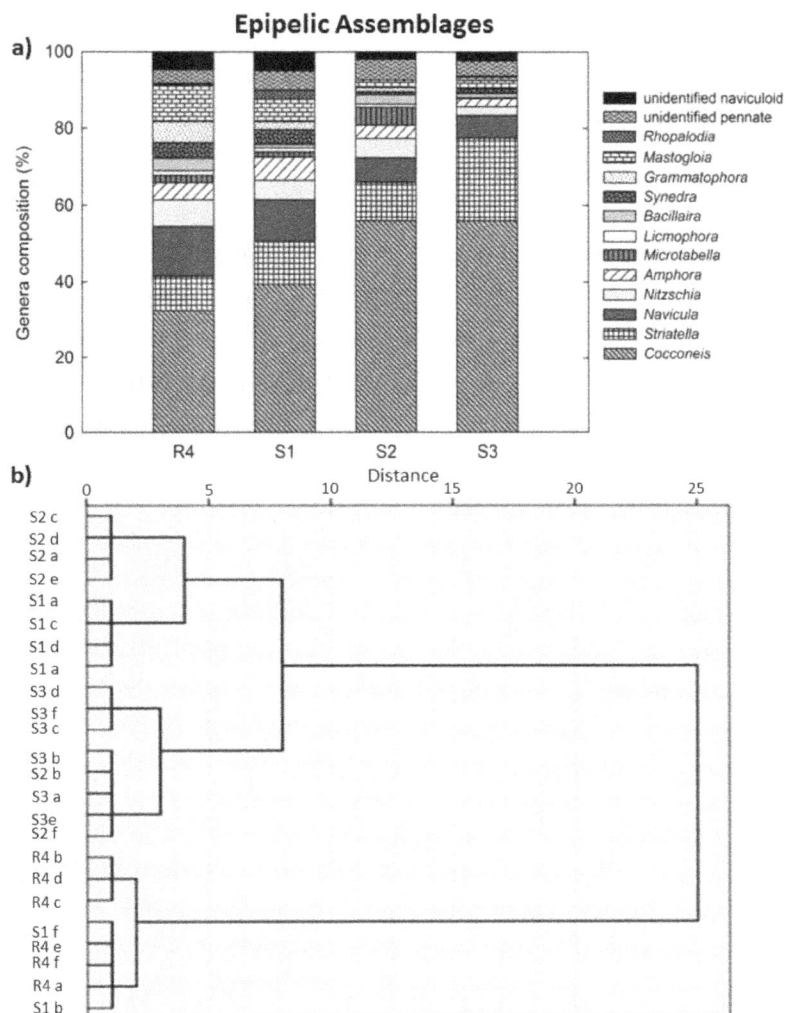

Figure 6. Epipelic diatom assemblages along a natural CO_2 gradient. (**a**) Relative composition of epipelic assemblages at different pCO_2 levels. Bar charts include all genera present over 1% and all unidentified diatoms grouped as unidentified pennate or naviculoid (*n* = 6 per station,); (**b**) Hierarchical cluster analyses of the similarity of epipelic diatom assemblage composition based on Ward's method with squared Euclidian distance for all the sediment samples collected along a natural CO_2 gradient (*n* = 24). Analysis consists of all genera present over 1%, including any of the unidentified groups.

4. Discussion

Researchers have recently begun to examine the potential effects of ocean acidification on microphytobenthic assemblages since they are known to play a crucial role in coastal ecosystems [21,22]. Since ocean acidification is occurring alongside a variety of other anthropogenic changes, studies of marine photoautotrophs have also started to address the interactive effects of multiple stressors [45]. To the best of our knowledge, the present study provides the first field observations of natural populations along a gradient of pCO_2 subjected to multiple coastal stressors (e.g., fishing, coastal pollutionand heat waves) and revealed considerable differences between epilithic, epipelic and epipsammic microalgal assemblages in acidified *vs.* ambient conditions. The standing crop of microalgae on rock and sediment was significantly higher in areas with elevated pCO_2 compared to reference conditions, with a proliferation of diatoms at moderate increases in pCO_2 and of cyanobacteria in areas with extremely large increases in pCO_2. This study validates the findings of previous work on the using artificial substrata [21]; by evaluating natural epilithic, epipelic and epipsammic communities this study provides a more realistic indication of the future changes expected in shallow sub-tidal rocky habitats as oceans acidify.

It is clear from our study, and others [4,46], that many coastal diatoms are tolerant of wide fluctuations in pH. However, it is important to consider the potential internal buffering capacity within these assemblages that occurs as a result of the high pH generated in dense photoautotrophic biofilms, since this may reduce the potential negative effects of low pH on the growth of microalgal cells [47]. In addition, the pH may vary in different layers of a biofilm; future studies should incorporate the use of pH microelectrodes to measure the potential gradient through the microalgal layers. Our observations contradict studies that report small or negative responses to CO_2 in diatoms [48–50] but support those that found that elevated CO_2 stimulated diatom growth [51–55]. Diatoms may benefit from seawater carbonation through down regulation of CCM capacity and the associated reductions in carbon fixation energy costs [4,6,7]. Hopkinson *et al.* [56] predicted that a doubling of ambient CO_2 would save around 20% of the CCM expenditure, reducing the amount of energy expended on carbon fixation by 3% to 6% and increasing primary production. In contrast to our findings, mesocosm experiments investigating benthic microalgae in more cohesive sediments have failed to detect any influence of pCO_2 levels on biomass [57,58]. Sediment type, however, has a major influence on microphytobenthic biomass, depth distribution, species diversity and assemblage composition [59,60]. The microphytobenthos at our coarse sediment stations will experience greater water exchange with the overlying water column than occurs in mud [61] and the microphytobenthos in muddy sediments may have access to other dissolved carbon sources; therefore, the responses observed in our study may not apply to all types of coastal sediments.

The use of natural analogues to predict effects of ocean acidification requires careful consideration of factors that may mask or enhance effects of elevated pCO_2 levels [30,62]. Geochemical monitoring was carried out to choose rocky and sandy habitats that were very similar along a gradient in seawater carbonate chemistry but were away from the influence of heated water, H_2S or anomalies in alkalinity and salinity, in order to provide a consistent basis for comparison of the effects of seawater CO_2 enrichment [36]. Sediment analysis revealed similarities in the physical properties of the volcanic sand at each sampling station. Dissolved nutrient levels and light are potentially important confounding variables to consider in these *in situ* experiments as it is difficult to isolate these effects from CO_2

elevations on algal growth. However, light intensity was constant between CO_2 enriched and reference sites, and phosphate and nitrate levels were similar along the gradient, while the increase in nitrite at stations S2 and S3 was only very small (0.01 μM). This suggests that pCO_2 was the main factor responsible for the observed differences in microphytobenthic communities between the CO_2 enriched and reference stations. Silicate levels were the only parameter that we found co-varied significantly with increasing pCO_2 levels; Dando et al. [63] also found that silicate is often elevated at submarine CO_2 seeps. Silicate limits diatom growth below ~2 μM, above which diatom growth and abundance remain relatively constant as silicate concentrations are further elevated [64–66]. Since the background silicate concentration at Vulcano was well above 2 μM we may infer that increases in pCO_2 (as opposed to silicate) were responsible for the significant increases in diatoms observed. However, while silicate uptake in diatoms has been typically characterised by saturation kinetics, non-saturable uptake kinetics have been observed in a few pelagic diatom species [67]. Consequently, not all diatom species in the microphytobenthos along the CO_2 gradients may have been silicate-saturated; this clearly warrants further in-depth investigation. Furthermore, a limitation of this study was that nutrient analyses were only conducted in one season and are therefore not representative of the whole year. In addition, ammonium levels, that can constitute a significant source of inorganic nitrogen, were not assessed.

The high variability in the carbonate system of volcanic seep sites may be considered a drawback to in situ studies because accurate dose-response relationships become difficult to determine. Furthermore, surface waters are not thought to be characterised by such rapid variability as the oceans acidify [68] and this may complicate the use of the information derived from vent studies in projecting future high CO_2 scenarios [69]. However, CO_2 vent systems provide an opportunity to examine the ecological effects of pH variability. This is essential for forecasting organism responses to acidification in habitats exposed to large natural diel, semi-diurnal and stochastic fluctuations in the carbonate system. The pH of the oceans, particularly coastal regions, is not constant [70,71], over diurnal scales pH shows strong systematic variation as a result of CO_2 uptake during photosynthesis and CO_2 release during respiration [72]. A compilation of high resolution time series of upper ocean pH collected over a variety of ecosystems has highlighted the natural temporal fluctuations (over a period of one month) and environmental heterogeneity associated with coastal seawater pH [73]. This natural variability was seldom considered in the early stages of ocean acidification research, as perturbation experiments mainly investigated the responses of organisms to constant levels of lowered pH.

Acidification of coastal seawater can enhance iron bioavailability through pH-induced changes in iron chemistry [74] and increase concentrations of one of the most toxic and bioavailable forms of copper [75]. Laboratory experiments replicating near-future ocean acidification scenarios have found that the susceptibility of benthic ecosystems to metal contaminants increases at high pCO_2 [13]. Natural CO_2 vent gradients provide an opportunity to assess the combined effects of changes in ocean pH including increased solubility and bioavailability of trace elements [13,76] although care is needed to avoid areas that have higher levels of trace elements than are predicted due to ocean acidification [30]. The effects of ocean acidification on shallow ecosystems will depend on interactions between algae and their consumers since grazers have a major influence on the microflora of rocks and sediments [77]. Some ecologically important groups of grazers (e.g., gastropods and sea urchins) decrease in abundance along CO_2 seep gradients [26] and whilst some micrograzers (e.g., amphipods) tolerate areas enriched with CO_2, many do not [24,78]. Research into the biogeochemical influence of ocean acidification on grazer-primary producer interactions is in its infancy; Arnold *et al.* [79] found that several seagrass species are less able to chemically defend themselves from grazers when subjected to chronic exposures to elevated CO_2 levels and Rossoll *et al.* [8] found that ocean acidification may degrade the food quality of phytoplankton. Preliminary work on grazer-microphytobenthos interactions at increased CO_2 levels produced contradictory results from short-term and long-term experiments so field observations will be useful to determine likely effects of ocean acidification [22]. As we did not quantify herbivore abundance it is not possible to determine the influence of grazers on microphytobenthic communities in this study. Our findings confirm observations of stimulated microphytobenthic growth in CO_2 enriched areas on artificial substrata where grazers were absent [20,21], increasing our confidence in the hypothesis that microbial primary producers will thrive on rocky and coarse sedimentary coastal habitats as pCO_2 levels continue to rise.

Phycobylin fluorescence (from rhodophytes and cyanobacteria) was only significantly enhanced at >1400 µatm pCO_2. Kübler *et al.* [80] observed positive effects of high CO_2 on the red alga *Lomentaria articulata* and Porzio *et al.* [81] recorded an increase in abundance of fleshy rhodophytes along a natural volcanic CO_2 gradient. There were no measurable effects of moderate increases in pCO_2 levels on phycobilin fluorescence, which is consistent with the findings of Johnson *et al.* [21] and similar to those reported by Tribollet *et al.* [82] who did not detect a significant effect of 750 ppm CO_2 on endolithic assemblages. Cyanobacteria possess a form of RUBSICO that has a very low affinity for CO_2 [83,84] and as a result have developed a highly effective CCM that is thought to be one of the most effective of any photosynthetic organism [85]. This may explain why cyanobacteria were less sensitive to increasing pCO_2 levels along the vent gradient than diatoms. Furthermore, in microbial mats dominated by cyanobacteria, photosynthesis has been shown to be non-sensitive to pH changes in the range of 5.6–9.6 [86]. Our findings contrast with several studies of oceanic species which reported significant, positive interactions of elevated CO_2 on cyanobacterial photosynthesis, nitrogen fixation and growth [87–89]. These experiments appear to have been conducted in nutrient replete conditions; in oligotrophic conditions, such as off Sicily and in mid-ocean gyres, cyanobacteria may not respond positivity to levels of CO_2 enrichment predicted this century [90].

The genus-specific differences in microphytobenthic abundance that we found tie-in with genus-specific diatom responses to changing CO_2 levels [6,50–52]. While some genera (e.g., *Amphora*, *Cocconeis* and *Navicula*) remained relatively constant across the CO_2 gradient, others became more scarce in areas of elevated pCO_2 (*Licmophora*, *Tabularia*, *Synedra*) and some (including chain-forming genera, *Rhabdonema*, *Striatella* and an unidentified Bacillariaceae) increased in abundance. Elevations in CO_2 have been shown to selectively enhance the growth of large planktonic diatom species over smaller species [91]. Furthermore, a greater dominance of large, chain forming periphytic species at elevated CO_2 have also been reported by Tortell *et al.* [51] and by Johnson *et al.* [21]. Collectively, these findings indicate that elevated CO_2 influences the competitive abilities of different size classes/morphologies of diatoms, causing shifts in assemblages. This may be related to differences in CO_2 diffusion limitation between various diatom size groups and morphologies [92]. In addition to pCO_2, changes in pH may also influence diatom community composition. Reductions in pH can affect diatom growth rate, silicon metabolism and intracellular pH homeostasis, which is believed to vary among different species according to species-specific intrinsic buffering capacity and adaptive capabilities [93]. Ocean acidification induced alterations in microphytobenthic assemblages may have wide-reaching ecological consequences as assemblage composition is known to mediate the structure of the overlying macrobenthos [94].

In summary, the microphytobenthic assemblages on natural substrata showed significant changes along a CO_2 gradient that mirror those recorded on artificial substrata [16]. Photosynthetic standing crop of both epilithic biofilms and microphytobenthic assemblages found in surface layers of sandy sediment was greatest in areas with long-term exposures to elevated CO_2 levels. Since these responses may have been modulated by concurrent stressors such as trace metal enrichment, overfishing and coastal pollution, they are valuable for formulating more realistic predictions of ocean acidification. Such alterations in microphytobenthic assemblages may have important consequences for benthic trophic webs and larger-scale biogeochemical processes as anthropogenic CO_2 emissions continue to rise and interact with multiple environmental stressors.

Acknowledgments

We thank C. Totti and T. Romagnoli at Università Politecnica delle Marche, Ancona, Italy for assistance with diatom identification and Lisa Al-Moosawi at Plymouth Marine Laboratory for nutrient analyses. Funding was provided by the Marine Institute (Plymouth University) and EU FP7 "Mediterranean Sea Acidification under a changing climate" (grant 265103).

Author Contributions

Conceived and designed the experiments: V.R.J., J.M.H.-S., C.B., M.M. Performed the experiments: V.R.J.; Analysed the data: V. R. J.; Wrote the paper: V. R. J., J.M.H.-S., C.B., M.M.

Conflicts of Interest

The authors declare no conflict of interest.

References

1. Hönisch, B.; Ridgwell, A.; Schmidt, D.N.; Thomas, E.; Gibbs, S.J.; Sluijs, A.; Zeebe, R.; Kump, L.; Martindale, R.C.; Greene, S.E.; *et al.* The geological record of ocean acidification. *Science* **2012**, *335*, 1058–1063.

2. Connell, S.D.; Kroeker, K.J.; Fabricius, K.E.; Kline, D.I.; Russell, B.D. The other ocean acidification problem: CO_2 as a resource among competitors for ecosystem dominance. *Philos. Trans. R. Soc. B* **2013**, *368*, doi:10.1098/rstb.2012.0442.

3. Giordano, M.; Beardall, J.; Raven, J.A. CO_2 concentrating mechanisms in algae: Mechanisms, environmental modulation, and evolution. *Annu. Rev. Plant Biol.* **2005**, *56*, 99–131.

4. Beardall, J.; Giordano, M. Ecological implications of microalgal and cyanobacterial CO_2 concentrating mechanisms and their regulation. *Funct. Plant Biol.* **2002**, *29*, 335–347.

5. Collins, S.; Bell, G. Phenotypic consequences of 1000 generations of selection at elevated CO_2 in a green alga. *Nature* **2004**, *431*, 566–569.

6. Trimborn, S.; Wolf-Gladrow, D.; Ritcher, K-L.; Rost, B. The effect of pCO_2 on carbon acquisition and intracellular assimilation in four marine diatoms. *J. Exp. Mar. Biol. Ecol.* **2009**, *376*, 26–36.

7.   Rost, B.; Zondervan, I.; Wolf-Gladrow, D. Sensitivity of phytoplankton to future changes in ocean carbonate chemistry: Current knowledge, contradictions and research directions. *Mar. Ecol. Prog. Ser.* **2008**, *373*, 227–237.

8.   Rossoll, D.; Bermúdez, R.; Hauss, H.; Schulz, K.G.; Riebesell, U.; Sommer, U.; Winder, M. Ocean acidification-induced food quality deterioration constrains trophic transfer. *PLoS ONE* **2012**, *7*, doi:10.1371/journal.pone.0034737.

9.   Sinutok, S.; Hill, R.; Doblin, M.A.; Wuhrer, R.; Ralph, P.J. Warmer more acidic conditions cause decreased productivity and calcification in subtropical coral reef sediment-dwelling calcifiers. *Limnol. Oceangr.* **2011**, *56*, 1200–1212.

10. Connell, S.D.; Russell, B.D. The direct effects of increasing CO_2 and temperature on non-calcifying organisms: Increasing the potential for phase shifts in kelp forests. *Proc. R. Soc. Lond. B* **2010**, *277*, 1409–1415.

11. Gao, K.; Xu, J.; Gao, G.; Li, Y.; Hutchins, D.A.; Huang, B.; Wang, L.; Zheng, Y.; Jin, P.; Cai, X.; *et al.* Rising CO_2 and increased light exposure synergistically reduce marine primary productivity. *Nat. Clim. Chang.* **2012**, *2*, 519–523.

12. Lefebvre, S.C.; Benner, E.; Stillman, J.H.; Parker,A.E.; Drake, M.K.; Rossignol, P. Nitrogen source and pCO$_2$ synergisitically affect carbon allocation, growth and morphology of the coccolithophore Emiliania huxleyi: Potential implications of ocean acidification for the carbon cycle. *Glob. Chang. Biol.* **2012**, *18*, 493–503.

13. Roberts, D.A.; Birchenough, S.N.R.; Lewis, C.; Sanders, M.B.; Bolam, T.; Sheahan, D. Ocean acidification increases the toxicity of contaminated sediments. *Glob. Chang. Biol.* **2013**, *19*, 340–351.

14. Zeng, X.; Chen, X.; Zhuang, J. The positive relationship between ocean acidification and pollution. *Mar. Pollut. Bull.* **2015**, *91*, 14–21.

15. MacIntyre, H.L.; Geider, R.J.; Miller, D.C. Microphytobenthos: The ecological role of the "secret garden" of unvegetated, shallow-water marine habitats. 1. Distribution, abundance and primary production. *Estuaries* **1996**, *19*, 186–201.

16. Magalhães, C.M.; Bordalo, A.A.; Wiebe, W.J. Intertidal biofilms on rocky substratum can play a major role in estuarine carbon and nutrient dynamics. *Mar. Ecol. Prog. Ser.* **2003**, *258*, 275–281.

17. Underwood, G.J.C.; Paterson, D.M. The importance of extracellular carbohydrate production by marine epipelic diatoms. *Adv. Bot. Res.* **2003**, *40*, 184–240.

18. Brodie, J.; Williamson, C.J.; Smale, D. The future of the northeast Atlantic benthic flora in a high CO_2 world. *Ecol. Evol.* **2014**, *4*, 2787–2798.

19. Koch, M.; Bowes, G.; Ross, C.; Zhang, X.H. Climate change and ocean acidification effects on seagrassess and marine macroalgae. *Glob. Chang. Biol.* **2013**, *19*, 103–132.

20. Lidbury, I.; Johnson, V.R.; Hall-Spencer. J.M.; Munn. C.B.; Cunliffe, M. Community-level response of coastal microbial biofilms to ocean acidification in a natural carbon dioxide vent system. *Mar. Pollut. Bull.* **2012**, *64*, 1063–1066

21. Johnson, V.R.; Brownlee, C.; Rickaby, R.E.M.; Graziano, M.; Milazzo, M.; Hall-Spencer, J.M. Responses of marine benthic microalgae to elevated CO_2. *Mar. Biol.* **2013**, *160*, 1813–1824.

22. Russell, B.D.; Connell, S.D.; Findlay, H.S.; Tait, K.; Widdicombe, S.; Mieszkowska, N. Ocean acidification and rising temperatures may increase biofilm primary productivity but decrease grazer consumption. *Philos. Trans. R. Soc. B* **2013**, *368*, doi:10.1098/rstb.2012.0438.

23. Snoeijs, P. Monitoring pollution effects by diatom community composition. A comparison of sampling methods. *Arch. Hydrobiol.* **1991**, *121*, 497–510.

24. Kroeker, K.J.; Micheli, F.; Gambi, M.C. Ocean acidification causes ecosystem shifts via altered competitive interactions. *Nat. Clim. Chang.* **2012**, *3*, 156–159.

25. Kerfahi, D.; Hall-Spencer, J.M.; Tripathi, B.M.; Milazzo, M.; Lee, J.; Adams, J. Shallow water marine sediment bacterial community shifts along a natural CO_2 gradient in the Mediterranean Sea off Vulcano, Italy. *Microb. Ecol.* **2014**, *67*, 819–828.

26. Hall-Spencer, J.M.; Rodolfo-Metalpa, R.; Martin, S.; Ransome, E.; Fine, M.; Turner, S.M.; Rowley, S.J.; Tedesco, D.; Buia, M-C. Volcanic carbon dioxide vents show ecosystem effects of ocean acidification. *Nature* **2008**, *454*, 96–99.

27. Fabricius, K.E.; De'ath, G.; Noonan, S.; Uthike, S. Ecological effects of ocean acidification and habitat complexity on reef-associated macroinvertebrate communities. *Proc. R. Soc. B* **2014**, *281*, doi:10.1098/rspb.2013.2479.

28. Vikas, M.; Dwarakish, G.S. Coastal pollution: A review. *Aquat. Procedia* **2015**, *4*, 381–388.

29. Vasilakopoulos, P.; Maravelias, C.D.; Tserpes, G. The alarming decline of Mediterranean fish stocks. *Curr. Biol.* **2014**, *24*, 1643–1648.

30. Vizzini, S.; Leonardo, R.D.; Costa, V.; Tramati, C.D.; Luzzu, F.; Mazzola, A. Trace element bias in the use of CO_2 vents as analogues for low pH environments: Implications for contamination levels in acidified oceans. *Estuar. Coast. Shelf Sci.* **2013**, *134*, 19–30.

31. Gattuso, J.P.; Gentill, B.; Duarte, C.M.; Kleypas, J.A.; Middelburg, J.J.; Antione, D. Light availability in the coastal ocean: Impact on the distribution of benthic photosynthetic organisms and contribution to primary production. *Biogeosciences* **2006**, *3*, 489–513.

32. Johnson, V.R. Using volcanic CO_2 Gradients to Investigate the Responses of Marine Benthic Algae to Ocean Acidification. PhD Thesis, School of Biomedical and Biological Sciences, Marine Biology and Ecology Research Centre, Plymouth University, Plymouth, UK, 2012.

33. Lewis, E.; Wallace, W.R. *Program Developed for CO_2 System Calculations*; Carbon dioxide information analysis center, Oak Ridge National Laboratory, U.S. Department of Energy: Oak Ridge, TN, USA, 1998.

34. Fabricius, K.E.; Langdon, C.; Uthicke, S.; Humphrey, C.; Noonan, S.; De'ath, G.; Okazaki, R.; Muehllehner, N.; Glas, M.S.; Lough, J.M. Losers and winners in coral reefs acclimatized to elevated carbon dioxide concentrations. *Nat. Clim. Chang.* **2011**, *1*, 165–169.

35. Kerrison. P.; Hall-Spencer, J.M.; Suggett, D.; Hepburn, L.J.; Steinke, M. Assessment of pH variability at a coastal CO2 vent for ocean acidification studies. *Estuar. Coast. Shelf Sci.* **2011**, *94*, 129–137.

36. Boatta, F.; D'Alessandro, W.; Gagliano, A.; Liotta, M.; Milazzo, M.; Rodolfo-Metalpa, R.; Hall-Spencer, J.M.; Parello, F. Geochemical survey of Levante Bay, Vulcano Island (Italy) and its suitability as a natural laboratory for the study of ocean acidification. *Mar. Pollult. Bull.* **2013**, *73*, 485–494.

37. Hansson, L.-A. Chlorophyll-*a* determination of periphyton on sediments: Identification of problems and recommendation of method. *Freshw. Biol.* **1998**, *20*, 347–352.

38. Ritchie, R.J. Universal chlorophyll equations for estimating chlorophylls *a*, *b*, *c*, and d and total chlorophylls in natural assemblages of photosynthetic organisms using acetone, methanol, or ethanol solvents. *Photosynthetica* **2008**, *46*, 115–1126.

39. Hickman, M.; Round, F.E. Primary production and standing crops of epipsammic and epipelic algae. *Br. Phycol. J.* **1970**, *5*, 247–255.

40. Hickman, M. Methods for determining the primary productivity of epipelic and epipsammic algal associations. *Limnol. Oceanogr.* **1969**, *14*, 936–941.

41. McAlice, B.J. Phytoplankton sampling with the Sedgewick Rafter Cell. *Limnol. Oceanogr.* **1971**, *16*, 19–28.

42. Eaton, J.W.; Moss, B. The estimation of numbers and pigment content in epipelic algal populations. *Limnol. Oceanogr.* **1996**, *11*, 584–595.

43. Nagarkar, S.; Williams, G.A. Comparative techniques to quantify cyanobacteria dominated epilithic biofilms on tropical rocky shores. *Mar. Ecol. Prog. Ser.* **1997**, *154*, 281–291.

44. Thompson, R.C.; Tobin, M.L.; Hawkins, S.J.; Norton, T.A. Problems in extraction and spectrophotometric determination of chlorophyll from epilithic microbial biofilms: Towards a standard method. *J. Mar. Biol. Assoc. UK* **1999**, *79*, 551–558.

45. Celis-Plá1, P.S.M.; Hall-Spencer, J.M.; Antunes Horta, P. Macroalgal responses to ocean acidification depend on nutrient and light levels. *Front. Mar. Sci.* **2015**, *2*, doi:10.3389/fmars.2015.00026.

46. Hinga, K.R. Effects of pH on coastal marine phytoplankton. *Mar. Ecol. Prog. Ser.* **2002**, *238*, 281–300.

47. Hama, T.; Kawahima, S.; Shimotori, K.; Satoh, Y.; Omori, Y.; Wada, S.; Adachi, T.; Hasegawa, S.; Midorikawa, T.; Ishii, M.; *et al.* Effect of ocean acidification on coastal phytoplankton composition and accompanying organic nitrogen production. *J. Oceanogr.* **2012**, *68*, 183–194.

48. Burkhardt, S.; Zondervan, I.; Riebesell, U. Effect of CO_2 concentration on C:N:P ratio in marine phytoplankton: A species comparison. *Limnol. Oceanogr.* **1999**, *44*, 683–690.

49. Crawfurd, K.J.; Raven, J.A.; Wheeler, G.L.; Baxter. E.J.; Joint, I. The response of *Thalassiosira pseudonana* to long-term exposure to increased CO_2 and decreased pH. *PLoS ONE* **2011**, *6*, e26695.

50. Torstensson, A.; Chierici, M.; Wulff, A. The influence of increased temperature and carbon dioxide levels on the benthic/sea ice diatom *Navicula directa*. *Polar Biol.* **2011**, *35*, 205–214.

51. Tortell, P.D.; DiTullio, G.R.; Sigman, D.M.; Morel, F.M.M. CO_2 effects on taxonomic composition and nutrient utilisation in an Equatorial Pacific phytoplankton assemblage. *Mar. Ecol. Prog. Ser.* **2002**, *236*, 37–43.

52. Tortell, P.D.; Payne, C.D.; Li, Y.; Trimborn, S.; Rost, B.; Smith, W.O.; Riesselman, C.; Dunbar, R.B.; Sedwick, P.; DiTullio, G.R. CO_2 sensitivity of southern ocean phytoplankton. *Geophysi. Res. Lett.* **2008**, *35*, doi:10.1029/2007GL032583.

53. Feng, Y.; Hare, C.E.; Leblanc, K.; Rose, J.M.; Zhang, Y.; DiTullio, G.R.; Lee, P.A.; Wilhelm, S.W.; Rowe, J.M.; Nemcek, N.; *et al.* Effects of increased pCO_2 and temperature on the North Atlantic spring bloom. I. The phytoplankton community and biogeochemical response. *Mar. Ecol. Prog. Ser.* **2009**, *388*, 13–25.

54. Sun, J.; Hutchins, D.A.; Feng, Y.; Seubert, E.L.; Caron, D.A.; Fu, F-X. Effects of changing pCO_2 and phosphate availability on domoic acid production and physiology of the marine harmful bloom diatom *Pseudo-nitzschia multiseries*. *Limnol. Oceanogr.* **2011**, *56*, 829–840.

55. Gao, K.; Xu, J.; Gao, G.; Li, Y.; Hutchins, D.A.; Huang, B.;. Wang, L.; Zheng, Y.; Jin,P.; Cai, X.; *et al.* Rising CO_2 and increased light exposure synergistically reduce marine primary productivity. *Nat. Clim. Chang.* **2012**, *2*, 519–523.

56. Hopkinson, B.M.; Dupont, C.L.; Allen, A.E.; Morel, F.M.M. Efficiency of the CO_2-concentrating mechanism of diatoms. *Proc. Nat. Acad. Sci. USA* **2011**, *108*, 3830–3837.

57. Hicks, N.; Bulling, M.T.; Solan, M.; Raffaelli, D.; White, P.C.L.; Paterson, M.P. Impact of biodiversity-climate futures on primary production and metabolism in a model benthic estuarine system. *BMC Ecol.* **2011**, *11*, 7, doi:10.1186/1472-6785-11-7.

58. Alsterberg, C.; Eklöf, J.S.; Gamfeldt, L.; Havenhand, J.N.; Sundbäck, K. 2013. Consumers mediate the effects of experimental ocean acidification and warming on primary producers. *Proc. Nat. Acad. Sci. USA* **2013**, *110*, 8603–8608.

59. Underwood, G.J.C.; Barnett, M. What Determines Species Composition in Microphytobenthic Biofilms? In *Functioning of Microphytobenthos in Estuaries.* Proceedings of the Microphytobenthos symposium, Amsterdam, The Netherlands, Royal Netherlands Academy of Arts and Sciences, 21–23, August, 2006; Kromkamp, J., Ed.; pp. 121–138.

60. Jesus, B.; Brotas, V.; Ribeiro, L.; Mendes, C.R.; Cartaxana, P.; Paterson, D.M. Adaptations of microphytobenthos assemblages to sediment type and tidal position. *Cont. Shelf Res.* **2009**, *29*, 1624–1634.

61. Cook, P.L.M.; Roy, H. Advective relief of CO_2 limitation in microphytobenthos in highly productive sandy sediments. *Limnol. Oceanogr.* **2006**, *51*, 1594–1601.

62. Kitidis, V.; Laverock, B.; McNeill, L.C.; Beesley, A.; Cummings, D.; Tait, K. Impact of ocean acidification on benthic and water column ammonia oxidation. *Geophys. Res. Lett.* **2011**, *38*, doi:10.1029/2011GL049095.

63. Dando, P.R.; Stüben, D.; Varnavas, S.P. Hydrothermalism in the Mediterranean Sea. *Progr. Oceanogr.* **1999**, *44*, 333–367.

64. Egge, J.K.; Aksnes, D.L. Silicate as a regulating nutrient in phytoplankton competition. *Mar. Ecol. Prog. Ser.* **1992**, *83*, 281–289.

65. Martin-Jézéquel, V.; Hildebrand, M.; Brzezinski, M.A. Silicon metabolism in diatoms: Implications for growth. *J. Phycol.* **2000**, *36*, 821–840.

66. Claquin, P.; Leynaert, A.; Sferratore, A.; Garnier, J.; Ragueneau, O. Physiological Ecology of Diatoms Along the Land-Sea Continuum. In *Land-Ocean Nutrient Fluxes: Silica Cycle*; Ittekot, V., Humborg, C., Garnier, J., Eds.; Island Press: Washington, DC, USA, 2006.

67. Thamatrakoln, K.; Hildebrand, M. Silicon uptake in diatoms revisited: A model for saturable and nonsaturable uptake kinetics and the role of silicon transporters. *Plant Physiol.* **2008**, *146*, 1397–1407.

68. Riebesell, U. Acid test for marine biodiversity. *Nature* **2008**, *454*, 46–47.

69. Gazeau, F.; Martin, S.; Hansson, L.; Gattuso, J.-P. Ocean acidification in the coastal zone. Available online: http://www.loicz.org/imperia/md/content/loicz/print/newsletter/Inprint_2011_3_online72.pdf (accessed on 24 November 2015)

70. Middelboe, A.L.; Hansen, P.J. High pH in shallow-water macroalgal habitats. *Mar. Ecol. Prog. Ser.* **2007**, *338*, 107–117.

71. Joint, I.; Doney, S.C.; Karl, D.M. Will ocean acidification affect marine microbes? *ISME J.* **2011**, *5*, 1–7.

72. Wooton, J.T.; Pfister, C.A.; Forester, J.D. Dyanamic patterns and ecological impacts of declining ocean pH in a high-resolution multi-year dataset. *Proc. Nat. Acad. Sci. USA* **2008**, *105*, 18848–18853.

73. Hoffmann, G.E.; Smith, J.E.; Johnson, K.S.; Send, U.; Levin, L.A.; Micheli, F.; Paytan, A.; Price, N.N.; Peterson, B.; Takeshita, Y.; *et al.* High-frequent dynamics of ocean pH: A multi-ecosystem comparison. *PLoS ONE* **2011**, *6*, doi:10.1371/journal.pone.0028983.

74. Breitbarth, E.; Bellerby, R.J.; Neill, C.C.; Ardelan, M.V.; Meyerhofer, M.; Zollner, E.; Croot, P.L.; Riebesell, U. Ocean acidification affects iron speciation during a coastal seawater mesocosm experiment. *Biogeosciences* **2010**, *7*, 1065–1073.

75. Richards, R.; Chaloupka, M.; Sanò, M.; Tomlinson, R. Modelling the effects of "coastal" acidification on copper speciation. *Ecol. Model.* **2011**, *22*, 3559–3567.

76. Millero, F.J.; Woosley, R.; DiTrolio, B.; Waters, J. Effect of ocean acidification on the speciation of metals in seawater. *Oceanography* **2009**, *22*, 72–85.

77. Dyson, K.E.; Bulling, M.T.; Solan, M.; Hernandez-Milian, G.; Raffaelli, D.G.; White, P.C.L.; Paterson, D.M. Influence of macrofaunal assemblages and environmental heterogeneity on microphytobenthic production in experimental systems. *Proc. R. Soc. B* **2007**, *274*, 2547–2554.

78. Cigliano, M.; Gambi, M.C.; Rodolfo-Metalpa. R.; Patti, F.P.; Hall-Spencer, J.M. Effects of ocean acidification on invertebrate settlement at volcanic CO_2 vents. *Mar. Biol.* **2010**, *157*, 2489–2502.

79. Arnold, T.; Mealey, C.; Leahey, H.; Miller, A.W.; Hall-Spencer, J.M.; Milazzo, M.; Maers, K. Ocean acidification and the loss of protective phenolics in seagrasses. *PLoS ONE* **2012**, *7*, e35107.

80. Kübler, J.E.; Johnston, A.M.; Raven, J.A. The effects of reduced and elevated CO_2 and O_2 on the seaweed *Lomentaria articulate*. *Plant Cell Environ.* **1991**, *22*, 1303–1310.

81. Porzio, L.; Buia, M.C.; Hall-Spencer, J.M. Effects of ocean acidification on macroalgal communities. *J. Exp. Mar. Biol. Ecol.* **2011**, *400*, 278–287.

82. Tribollet, A.; Atkinson, M.J.; Langdon, C. Effects of elevated pCO_2 on epilithic and endolithic metabolism on reef carbonates. *Glob. Chang. Biol.* **2006**, *12*, 2200–2208.

83. Badger, M.R.; Andrews, T.J.; Whitney, S.M.; Ludwig, M.; Yellowlees, D.C.; Leggat, W.; Price, G.D. The diversity and co-evolution of Rubisco, plastids, pyrenoids and chloroplast-based CCMs in the algae. *Can. J. Bot.* **1998**, *76*, 1052–1071.

84. Kaplan, A.; Reinhold, L. CO_2 concentrating mechanisms in photosynthetic microorganisms. *Annu. Rev. Plant Physiol. Plant Mol. Biol.* **1999**, *50*, 539–559.

85. Badger, M.R.; Price, G.D. CO_2 concentrating mechanisms in cyanobacteria: Molecular components, their diversity and evolution. *J. Exp. Bot.* **2002**, *54*, 609–622.

86. Carrasco, M.; Mercado, J.M.; Niel, F.X. Diversity of inorganic carbon acquisition mechanisms by intact microbial mats of *Microcoleus chthonoplastes* (*Cyanobacteriae, Oscillatoriaceae*). *Physiol. Plant.* **2008**, *133*, 49–58.

87. Hutchins, D.A.; Fu, F-X.; Zhang, Y.; Warner, M.E.; Feng, Y.; Portune, K.; Bernhardt, P.W.; Mullholland, M.R. CO_2 control of *Trichodesmium* N_2 fixation, photosynthesis, growth rates and elemental ratios: Implications for past, present and future ocean biogeochemistry *Limnol. Oceanogr.* **2007**, *52*, 1293–1304.

88. Levitan, O.; Rosenberg, G.; Setlik, I.; Setlikova, E.; Grigel, J.; Klepetar, J.; Prasil O.; Berman-Frank, I. Elevated CO_2 enhances nitrogen fixation and growth in the marine cyanobacterium *Trichodesmium*. *Glob. Chang. Biol.* **2007**, *13*, 531–538.

89. Kranz, S.A.; Sültemeyer, D.; Richter, K.U.; Rost, B. Carbon acquisition in *Trichodesmium*: Diurnal variation and effect of *p*CO_2. *Limnol. Oceanogr.* **2009**, *54*, 548–559.

90. Kletou, D.; Hall-Spencer, J.M. Threats to Ultraoligotrophic Marine Ecosystems. In *Marine Ecosystems*; Cruzado, A., Ed.; InTech—Open Access Publisher: Rijeka, Croatia, 2012.

91. Wu, Y.; Campbell, D.D.; Irwin, A.J.; Suggett, D.J.; Finkel, Z.V. Ocean acidification enhances the growth rate of larger diatoms. *Limnol. Oceanogr.* **2014**, *59*, 1027–1034.

92. Flynn, K.J.; Blackford, C.J.; Baird, M.E.; Raven, J.A.; Clark, D.R.; Beardall, J.; Brownlee, C.; Fabian, H.; Wheeler, G.L. Changes in pH at the exterior surface of plankton with ocean acidification. *Nat. Clim. Chang.* **2012**, *2*, 510–513.

93. Hervé, V.; Derr, J.; Douady, S.; Quinet, M.; Moisan, L.; Lopez, P.J. Multiparametric analyses reveal the pH-dependence of silicon biomineralization in diatoms. *PLoS ONE* **2012**, *7*, doi:10.1371/journal.pone.0046722.

94. Huang, R.; Boney, A.D. Growth interactions between littoral diatoms and juvenile marine algae. *J. Exp. Mar. Biol. Ecol.* **1984**, *81*, 21–45.

13

Soil Organic Carbon in Mangrove Ecosystems with Different Vegetation and Sedimentological Conditions

Naohiro Matsui [1,*], Wijarn Meepol [2] and Jirasak Chukwamdee [3]

[1]   Environment Department, The General Environmental Technos Co., Ltd., Osaka 541-0052, Japan

[2]   Ranong Mangrove Forest Research Center, Department of Marine and Coastal Resources, Tambon Ngao, Muang District, Ranong 85000, Thailand; E-Mail: wijarn.meepol@yahoo.com

[3]   Department of National Park, Wildlife and Plant Conservation, 61 Pholyothin Road, Ladyao, Chatuchak, Bangkok 10900, Thailand; E-Mail: j-chukwamdee@hotmail.com

*   Author to whom correspondence should be addressed: E-Mail: matui_naohiro@kanso.co.jp

Academic Editors: Joseph M. Smoak and Christian Joshua Sanders

Abstract: A large number of studies have been conducted on organic carbon (OC) variation in mangrove ecosystems. However, few have examined its relationship with soil quality and stratigraphic condition. Mangrove OC characteristics would be explicitly understood if those two parameters were taken into account. The aim of this study was to examine mangrove OC characteristics qualitatively and quantitatively after distinguishing mangrove OC from other OC. Geological survey revealed that the underground of a mangrove ecosystem was composed of three layers: a top layer of mangrove origin and two underlying sublayers of geologic origin. The underlying sublayers were formed from different materials, as shown by X-ray fluorescence analysis. Despite a large thickness exceeding 700 cm in contrast to the 100 cm thickness of the mangrove mud layer, the sublayers had much lower OC stock. Mangrove mud layer formation started from the time of mangrove colonization, which dated back to between 1330 and 1820 ^{14}C years BP, and OC stock in the mangrove mud layer was more than half of the total OC stock in the underground layers, which had been accumulating since 7200 ^{14}C years BP. pH and redox potential (Eh) of the surface soils varied depending on vegetation type. In the surface soils, pH correlated to C% ($r = -0.66$, $p < 0.01$). C/N ratios varied widely from 3.9 to 34.3, indicating that mangrove OC had various sources. The pH and Eh gradients were important factors affecting the OC stock and the mobility/uptake of chemical elements in the mangrove mud layer. Humic acids extracted from the mangrove mud layer had relatively high aliphatic contents, in contrast with the carboxylic acid rich sublayers, indicating that humification has not yet progressed in mangrove soil.

Keywords: mangrove; soil OC; redox potential; C/N ratio; humic acid; leaf analysis

---

1. Introduction

Mangrove ecosystems are recognized to have OC stocks equal to or higher than terrestrial tropical forests [1–5]. Mangroves are some of the most biogeochemically active regions on Earth and represent important carbon sinks in the biosphere [6–8]. However, mangrove ecosystems have been vastly exploited for anthropogenic uses; the rate of mangrove loss is 0.7%–7% annually, which is four times higher than the rate of rainforest loss [9–13].

To assess the carbon sink capacity of mangrove ecosystems, OC stock in mangrove ecosystems should be distinguished according to mangrove and non-mangrove origin to further clarify this ecosystem's carbon sink capacity. To this end, stratigraphic examination is adopted as it delineates the mangrove layer from the other layers and can be used in tandem with radioisotope methods to determine mangrove colonization time.

A considerable amount of data has been collected for OC stock as quantitative studies [14–17]. On the other hand, the source of OC has been studied as the qualitative determination of mangrove OC because OC source has a major impact on the overall carbon dynamics in intertidal mangrove ecosystems. C/N ratio is a useful parameter to predict the source of OC [18] reflecting levels of OC degradation, and X-ray fluorescence (XRF) analysis has been conducted to study the source of sediment deposited at different times [19]. Not only the source of OC, but also the quantity and the composition of the OC, may differ depending on the vegetation type [20]. Therefore, it is necessary to examine the spatial variation of OC in relation to vegetation type.

The nature of OC is another important qualitative feature because it may offer some insight into the stability of OC. Humic acid composition reflects the nature of OC as humic substances are chemically and biologically synthesized from litter and microorganism debris through microbial activity under the influence of the deposition environment. As humic acid composition is determined on the basis of the dynamic balance between OC input and decomposition [21], it is worth examining humic acid composition to predict the fate of OC.

This study has the following objectives: (1) to characterize OC of a mangrove ecosystem differentiating from non-mangrove OC; (2) to assess mangrove OC source in relation to vegetation type; and (3) to examine the nature of OC from the viewpoint of humic acid composition.

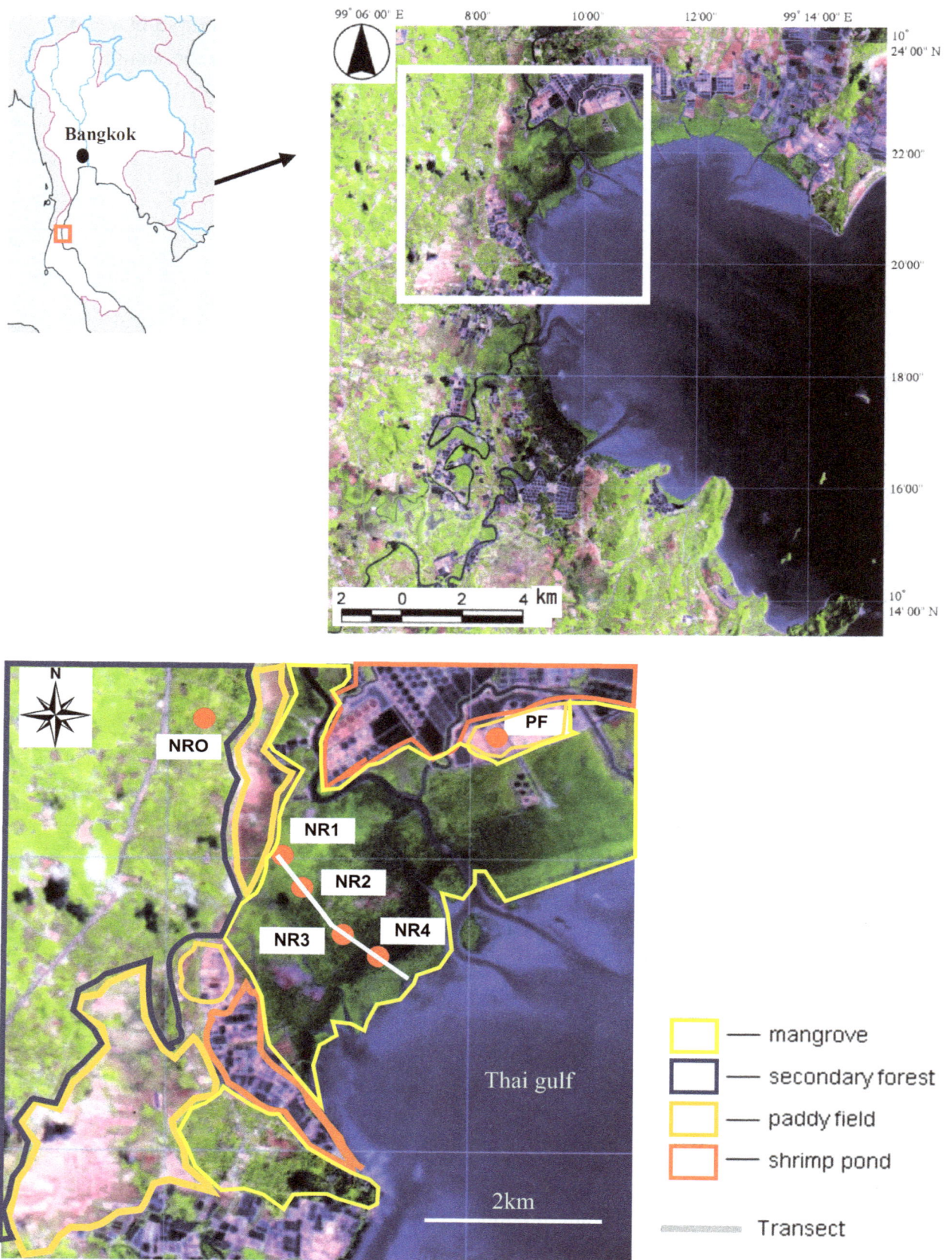

Figure 1. Location of study site is indicated by a white square in Landsat image. The transect was made by crossing different mangrove vegetation types. Boring survey was conducted in mangrove ecosystem (NR1–NR4), coconut plantation (NR0), and paddy field (PF).

2. Materials and Methods

2.1. Study Site

The study site was located in Tungka Bay, Chumphon, Thailand (Figure 1). The area has two monsoon seasons, which occur from December to April and from May to November, respectively. The mean monthly rainfall for the last 10 years (2005–2014) was 1883 mm. The rainy season was recorded during the months of October and November, whereas the low rainfall season was recorded from February to May, with February being the driest month. The average mean monthly temperature varied between 25.8 to 28.9 °C. The mean annual relative humidity was 81% (personal communication with Meteorological Office, Chumphon Province, Thailand, 2015). Mangrove forests at the study sites registered 19 species belonging to the families *Rhizophoraceae, Avicenniaceae, Sonneratiaceae, Meliaceae, Euphorbiaceae,* and *Combretaceae.*

Chumphon mangroves decreased from 8100 ha in 1961, to 3625 ha in 1986, and to 1818 ha in 1991 [22]. Considering that shrimp farm increased 2219 ha from 1987 to 1993 while mangrove area decreased by 1976 ha during the same period [23], loss of mangrove forest could be attributable to shrimp farming.

The study site was composed of coastal ecosystems (including mangroves) and terrestrial ecosystems, such as paddy field and secondary forest. Anthropogenic activities that were started in the 1960s, particularly shrimp aquaculture, had adversely affected mangroves in the study site. Those activities led to the continuous degradation and deforestation of mangroves, which resulted in significant soil carbon loss [24,25]. There were traces of human activity even in the midst of mangrove forests (Figure 1; NR3, NR4). In NR1 and NR2, which are located near Tungka village, human impact on the mangrove forests was enormous. Due to frequent exploitation, such as tree cutting since the 1980s, those areas were converted into *Acrostichum aureum* dominant fields.

2.2. Soil Sampling and Measurement

In order to characterize OC of a mangrove ecosystem, as well as to assess mangrove OC source and the nature of OC, soil samples were collected from surface and from belowground (Table 1). Sampling and analytical methods were decided *ad hoc* for surface soils along the transect (Figure 1) and for belowground samples via boring survey (Figure 2). A 2500 m long transect was made across different mangrove zones in August 2006 (Figure 1). Eighty-nine surface soil samples were collected from 0 to 5 cm soil depth with a 100 cc stainless steel cylinder (Daiki Rika Kogyo Co., Ltd., Akagidai, Japan) every 20 to 40 meters along the transect. Belowground samples were collected at four sampling points (NR1–NR4) in the mangrove areas along the transect, at a coconut plantation (NR0) and a paddy field (PF) in the terrestrial areas. Sampling was done with Soil Check Simplification Consortium (SCSC) in September 2007. SCSC is a boring machine equipped with an engine, and was developed for collecting undisturbed soil samples from large depths.

Table 1. Purpose of analysis, analytical items, places of soils/sediments sample collection and the methods of sample and analysis.

Purpose	Analytical items	Places of sample collection		Sample collection/ method
		Surface	**Belowground**	
Characterization of organic carbon (OC)	Three phase distribution	●	●	100 cc cylinder/Volumenometer
	Bulk density	●	●	100 cc cylinder/Cylindrical core method
	Redoxpotential (Eh)	●		Extract pore water *in situ*/Eh meter
	pH	●	●	Dilution/pH meter
	Electric conductivity (EC)	●	●	Extract pore water by centrifuge/Conductivity meter
	Particle size distribution	●	●	Bulk sample/Pipette method
	Leaf analysis	●		Leaf collection and digestion/ICP-MS
OC source	Total C and N	●	●	Bulk sample/NC analyzer
	Total chemical composition	●	●	2 mm sieved sample/X-ray fluorescence
	Radiocarbon dating		●	Shell, sediment/AMS, radiometric-standard method
OC nature	Humic acids	●	●	Soil, sediment/Fraction method

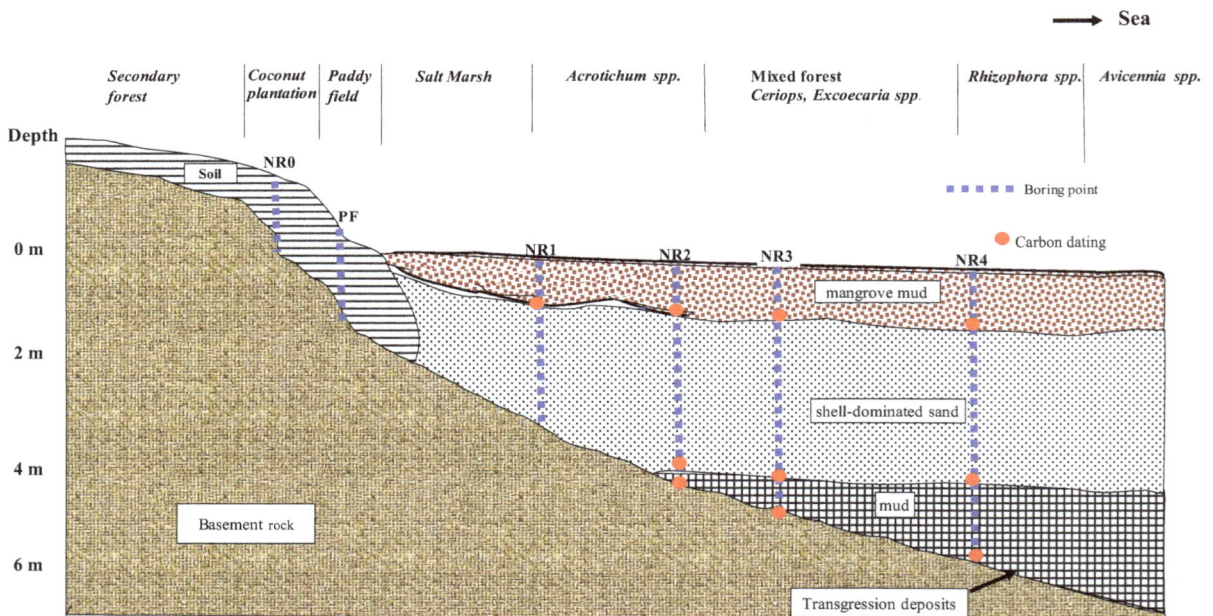

Figure 2. Geologic profile of study site. The mangrove area was composed of three layers: mangrove mud, shell-dominated sand, and mud layers.

Three-phase distribution and bulk density were determined from undisturbed soils removed with the same 100 cc stainless steel cylinder. For the three-phase distribution, solid and liquid phases of soil samples were measured with a volumenometer (DIK-1120, Daiki Rika Kogyo Co., Ltd., Saitama, Japan). After the wet weight was measured, the samples were dried at 105 °C for 3 days, and the weights of the soil samples were measured. Eh was measured *in situ* for extracted pore water, and pH of 1:5 dilutions of the soil samples was measured in the laboratory with a Horiba D-54 Multi-Parameter Water Quality Meter (Horiba Co., Ltd., Kyoto, Japan).

EC of 0.5 mL of pore water diluted 20 times with distilled water was measured with a conductivity meter (TOA Electronics, Hamburg, Germany, TOA CM305). The direct centrifugation drainage technique was applied to obtain pore water to minimize the risk of contamination [26]. Centrifugation was used to extract soil water at precise time intervals at the matric suction of 1500 kPa. Total carbon and nitrogen were measured by the combustion method at 800 °C using a sumigraph NC-800-13N (Sumitomo-Kagaku Co., Ltd., Tokyo, Japan).

Composite samples were prepared from five samples, each of which were collected at NR0 and PF, and one sample was collected from each of the four layers at NR1–NR4 for the measurement of particle size distribution and X-ray fluorescence analysis. Particle size distribution was determined by the pipette method [27]. For analysis, samples were passed through a 2 mm sieve after drying in an oven at 100–105 °C. Total chemical composition analysis was conducted according to the method of Ochi and Okashita [28] with a XRF analyzer VF320-A (Shimadzu Co., Ltd., Kyoto, Japan) using prepared glass beads. The dried samples were pulverized and homogenized in an agate mortar and used for the XRF bead analysis (Shimadzu Co., Ltd., Kyoto, Japan). Approximately 600 mg of pulverized sample was mixed with 3600 mg of lithium tetraborate ($Li_2B_4O_7$), pre-oxidized at 500 °C with NH_4NO_3, and fused to glass beads [29].

Statistical analysis of measured elements was carried out using Tukey's test (JMP 4.0, SAS Institute Inc., Cary, NC, USA) and differences at the $p < 0.05$ level were considered to be significant.

2.3. Radiocarbon Dating

Radiocarbon dating was performed on two types of samples collected from the four layers at NR1–NR4. One was shell, which was pre-treated by acid etch and then measured by accelerator mass spectrometry (AMS). The shell is an ideal sample because it is an autochthonous fossil that shows accurately the sedimentation time in the area where it is buried [30]. Only a small number of shells could be collected in the sampling. Nevertheless, radiocarbon dating was possible, as AMS required only a few grams of the sample. The other sample was organic sediment, which was pre-treated by acid wash and then measured by the radiometric-standard method, which is widely used in radiocarbon dating. Mangrove colonization time was estimated from the results of radiocarbon dating of the mangrove mud layers previously identified by a stratigraphic survey [4]. Sampling was done at the boundary of each layer in order to know the starting time of formation for each layer (Figure 2).

2.4. Leaf Measurement

Leaves were collected in June 2013 from different vegetation zones in mangrove ecosystems and terrestrial zones. Sampled plant species were sedge (*Cyperus microiria*), *Acrostichum* (mangrove associate), *Excoecaria, Ceriops, Bruguiera, Rhizophora, Sonneratia, Avicennia* which were in wetland, and secondary forest (species names were not identified), coconut tree, rice which were in the terrestrial zone. Fifty young leaves were collected from each type of vegetation, as it was reported that fifty was a reliable number for use as bioindicators of heavy metal contamination [31]. The leaves were immediately washed with distilled water to remove adhered matter, and dried. The dried leaves were pulverized by a mixer to make composite leaf samples. The composite leaf samples were digested by adding 5 mL of conc. HNO₃, and the mixture was left to stand overnight at room temperature for pre-digestion. Thereafter, another 5 mL of conc. HNO₃ was added, and the entire mixture was heated on a hot plate at 75 °C for 0.5 h, 130 °C for 0.5 h, and 200 °C for 2 h. The mixture was filtered through a membrane filter (pore size 0.45 μm), and the filtrate was used as the analysis solution for the determination of macro and micro elements, such as Na, Mg, Al, Si, P, S, Cl, K, Ca, Mn, and Fe, by an ICP-MS instrument (Model SPQ 8000A, Seiko Instruments, Chiba, Japan).

2.5. Humic Acid Determination

Humic acids were determined for the mangrove mud layer (NR1, NR2, NR3, NR4 0–5), the shell-dominated layer (NR4 305–310), and the mud layer (NR4 490–495). Six air-dried soil samples collected from each layer (four from the mangrove mud layer and one each from the shell-dominated layer and the mud layer) were used for the analysis. Humic acid components were separated following the fractionation method of Yonebayashi and Hattori [32]. Amberlite XAD-8 resin was pulverized and particles in the 50–200 μm size range were isolated. The sieved particles were washed with ethanol, acetonitrile, ethanol again, and packed into a column (20 cm × 1.8 cm i.d.). The column was conditioned with 0.1 M NaOH and then with the universal buffer adjusted to pH 3 with NaOH solution. Humic acid was dissolved in 0.1 M NaOH and treated with Amberlite IR-120 resin to transform it into the H⁺-saturated form, made to 2% solution, and loaded onto the column packed with XAD-8 resin. A pH-gradient solution was prepared by titrating 200 mL of 0.02 M universal buffer contained in an air-tight flask with 0.1 M NaOH using a peristaltic pump, and passed through the column at the flow rate of 1.5 mL·min⁻¹. The pH of the column eluate was measured with a pH meter. A water-ethanol gradient was generated by mixing 200 mL of distilled water contained in an air-tight flask with ethanol using a peristaltic pump. Elution was carried out at the flow rate of 1.5 mL·min⁻¹. The elution profile was determined by measuring the optical density at 400 nm after the eluate was alkalified to above pH 12 by the addition of 10 M NaOH. Stepwise elution was carried out with universal buffer solutions adjusted to pH 7 and pH 11, distilled water, and 50% ethanol. The elution profile was determined in the same way as that of pH-gradient chromatography. Each eluate was precipitated with sulfuric acid and dissolved in 0.1 M NaOH. As the humic acid fraction eluting at pH 7 was not precipitated by acidification, it was subjected to chromatography on a small XAD-8 column at pH 3 using NaOH solution as eluent. Each of the four eluates was dialyzed against distilled water and freeze-dried.

3. Results and Discussion

3.1. Soil/Sediment Properties

Mangrove soil had higher clay content (33.3%) than terrestrial soil (Table 2), and its clay content was almost twice that of PF soil (17.3%). Soil from PF that was converted from mangroves in the 1980s would have contained more clay but became sandy afterwards due to erosion. The shell-dominated layer had high sand content (83%), indicating that this layer was formed near the shore with the deposition of coarse materials. In contrast, the underlying mud layer (silt 23%, clay 35%) was formed offshore with the deposition of fine particles. The mangrove mud layer and the mud layer had almost the same clay content. However, XRF analysis showed that Mg content in the mangrove mud layer was lower than that of the mud layer (Table 3). This difference indicated that the two layers were formed from different materials. The mud layer was formed mainly from marine deposits whereas the mangrove mud layer formation was influenced by mangrove forest [33,34].

Table 2. Particle size distribution of soils/sediments collected from terrestrial and mangrove ecosystems. In the mangrove ecosystem, sublayers were also examined.

		Sand (%)	Silt (%)	Clay (%)
Terrestrial ecosystem	Coconut plantation	91.4	6.0	2.6
	Paddy	76.6	6.1	17.3
Mangrove	Mangrove mud	48.7	18.0	33.3
	Shell-dominated	83.0	6.0	11.0
	Mud	42.1	23.0	35.0

Table 3. XRF analysis of soils/sediments in terrestrial and mangrove ecosystems.

		Al (%)	Mg (%)	Si (%)	P (%)	K (%)	Ca (%)	Ti (%)	Mn (%)	Fe (%)
Terrestrial areas	Coconut	1.04	2.28	95.54	0.00	0.00	0.20	0.29	0.01	0.62
	Paddy field	6.32	0.20	90.75	0.01	0.80	0.06	0.34	0.01	1.52
Mangrove	Mangrove mud	10.23	0.89	79.41	0.03	1.94	1.28	0.70	0.04	5.49
	Shell-dominated	4.85	1.22	84.29	0.03	1.19	4.53	0.33	0.09	3.47
	Mud	13.64	1.85	73.07	0.06	2.17	2.46	0.81	0.08	5.88

As clay had high carbon capturing capacity [17,35], the high clay content in the mangrove mud layer was responsible for the high carbon sink capacity of the mangrove (Table 4). The mud layer had comparable clay content to the mangrove mud layer, but its carbon content was not high. It was likely that the mud layer received no fresh organic matter (OM) supply and maintained the same OM level that it had in the past.

Table 4. Physical and chemical characteristics of soils/sediments collected from terrestrial and mangrove ecosystems.

Land type and depth (cm)	Air	Liquid (%)	Solid	Porosity	Bulk density (g/mL)	pH	EC (dS/m)	Total-C (%)	Total-N (%)	CN
Secondary forest	24.31	31.87	43.82	56.18	1.12	6.61	2.61	1.244	0.079	16
Coconut plantation										
NR0 0–5	44.43	14.57	41.00	59.00	1.12		3.49	0.382	0.023	17
NR0 50–55	23.71	15.23	61.06	38.94	1.61		0.11	0.104	0.005	21
NR0 85–90	24.38	17.56	58.06	41.94	1.54		0.34	0.030	0.001	30
Paddy field										
PF 0803 0–5	8.36	36.44	55.20	44.80	1.44	4.91	0.48	0.515	0.035	15
PF 0803 10–15	5.59	33.65	60.76	39.24	1.57	4.78	0.23	0.636	0.041	16
PF 0803 20–25	5.30	29.76	64.94	35.06	1.68			0.647	0.026	25
PF 0803 30–35	6.44	29.29	64.27	35.73	1.68			0.270	0.009	30

Table 4. *Cont.*

PF 0803 40–45	8.16	31.23	60.61	39.39	1.59		0.25	0.233	0.006	39
PF 0803 50–55	9.98	30.95	59.07	40.93	1.57		0.41	0.177	0.005	35
Mangrove										
NR1 0–5	18.80	58.17	23.03	76.97	0.56		37.5	9.643	0.620	16
NR1 35–40	7.19	58.58	34.23	65.77	0.89		32.0	2.338	0.091	26
NR1 100–105	6.85	52.58	40.57	59.43	1.10	8.13	31.9	0.485	0.028	17
NR1 145–150	8.29	49.61	42.10	57.90	1.18	8.25	42.0	0.546	0.026	21
NR1 190–195	9.47	41.76	48.77	51.23	1.32	8.31	44.0	0.369	0.011	34
NR1 245–250	3.57	41.01	55.42	44.58	1.51	7.86	56.2	0.440	0.009	49
NR2 0–5	16.88	69.71	13.41	86.59	0.29		40.0	14.234	0.873	16
NR2 40–45	6.62	65.49	27.89	72.11	0.68	4.92	39.8	4.904	0.208	24
NR2 75–80	5.33	67.83	26.84	73.16	0.70	2.99	46.6	3.071	0.116	26
NR2 130–135	4.36	47.37	48.27	51.73	1.28		52.2	0.505	0.021	24
NR2 200–205	6.26	44.89	48.85	51.15	1.32		53.8	0.398	0.017	23
NR2 235–240	5.25	45.56	49.19	50.81	1.33		62.4	0.491	0.016	31
NR2 270–275	2.78	41.86	55.36	44.64	1.47	5.83	61.6	0.303	0.006	51
NR2 330–335	1.14	46.31	52.55	47.45	1.41	5.22	63.2	0.502	0.010	50
NR2 385–390	1.93	60.05	38.02	61.98	1.07	7.81	67.4	1.514	0.044	34
NR2 440–445	0.00	63.55	36.45	63.55	0.98	7.57	67.8	1.648	0.053	31
NR2 500–505	2.30	57.26	40.44	59.56	1.09	8.15	67.4	1.228	0.040	31
NR3 0–5	11.06	65.21	23.73	76.27	0.58		45.0	3.412	0.181	19
NR3 35–40	6.53	64.77	28.70	71.30	0.71	5.39	45.4	4.693	0.160	29
NR3 100–105	4.39	67.35	28.26	71.74	1.00		71.6	1.050	0.048	22
NR3 135–140	6.07	51.24	42.69	57.31	1.14			0.660	0.030	22
NR3 205–210	2.75	50.96	46.29	53.71	1.26		67.2	0.707	0.031	23
NR3 270–275	5.13	42.82	52.05	47.95	1.40	8.23	70.2	0.499	0.014	36
NR3 305–310	6.37	44.69	48.94	51.06	1.33		69.8	0.446	0.014	32
NR3 345–350	2.26	43.40	54.34	45.66	1.45	8.38	69.6	0.358	0.008	45
NR3 380–385	6.13	58.34	35.53	64.47	0.96	8.19	72.6	1.537	0.049	31
NR3 465–470	1.23	58.67	40.10	59.90	1.09		72.2	1.490	0.059	25
NR3 530–535	10.14	36.99	52.87	47.13	1.51	7.63	83.0	0.448	0.019	24
NR4 0–5	15.43	64.60	19.97	80.03	0.48	8.29	39.4	4.842	0.219	22
NR4 45–50	10.95	62.57	26.48	73.52	0.66	5.28	38.3	3.484	0.132	26
NR4 150–155	11.23	53.53	35.24	64.76	0.94	7.71	45.0	0.829	0.032	26
NR4 185–190	7.83	51.20	40.97	59.03	1.07		46.6	1.438	0.060	24
NR4 225–230	10.52	51.12	38.36	61.64	1.04		46.2	1.238	0.059	21
NR4 305–310	8.95	46.17	44.88	55.12	1.21		48.4	0.791	0.024	33
NR4 365–370	4.15	41.72	54.13	45.87	1.43		53.0	0.930	0.018	52
NR4 490–495	4.41	48.36	47.23	52.77	1.25		52.6	1.127	0.022	51
NR4 570–575	4.45	53.06	42.49	57.51	1.16		53.2	0.925	0.023	40
NR4 610–615	0.56	64.05	35.39	64.61	0.97	7.94	52.8	1.272	0.044	29

The low content of alkaline-earth metals such as K, Ca, and Mg meant that terrestrial soil was highly weathered (Table 3). In contrast, K, Ca, and Mg content was high in mangrove soil. This was because of not only the slow weathering due to water inundation, but also the continuous supply of those chemical elements from mangrove vegetation. PF soil had lower sand content and higher clay content than the coconut plantation soil. This would be related to land management, as PFs were artificially waterlogged for a certain period, and the waterlogged condition suppressed intensive weathering.

The mangrove soil was characterized by a high liquid content and a low solid content (Figure 3). The large proportion of pores that were filled with either water, gas, or air played a major role in the biogeochemical process in the mangrove ecosystem.

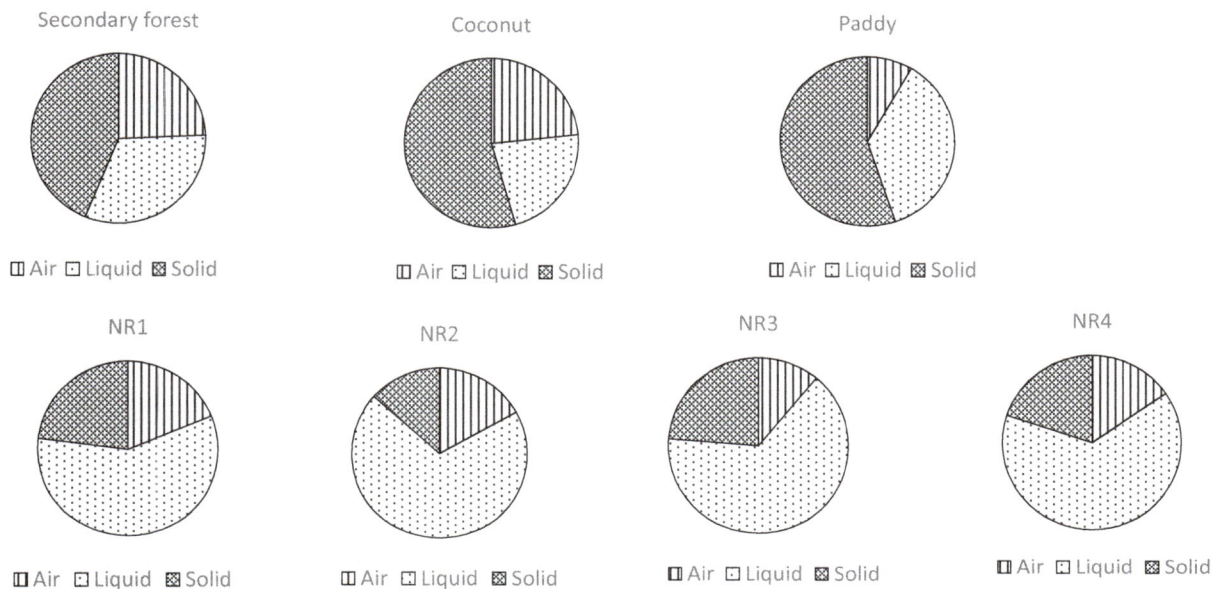

Figure 3. Three-phase distributions of soils collected at 5 cm soil depth from seven places. Top pie charts are for the terrestrial ecosystem (secondary forest, coconut plantation, and PF), and bottom ones are for the mangrove ecosystem.

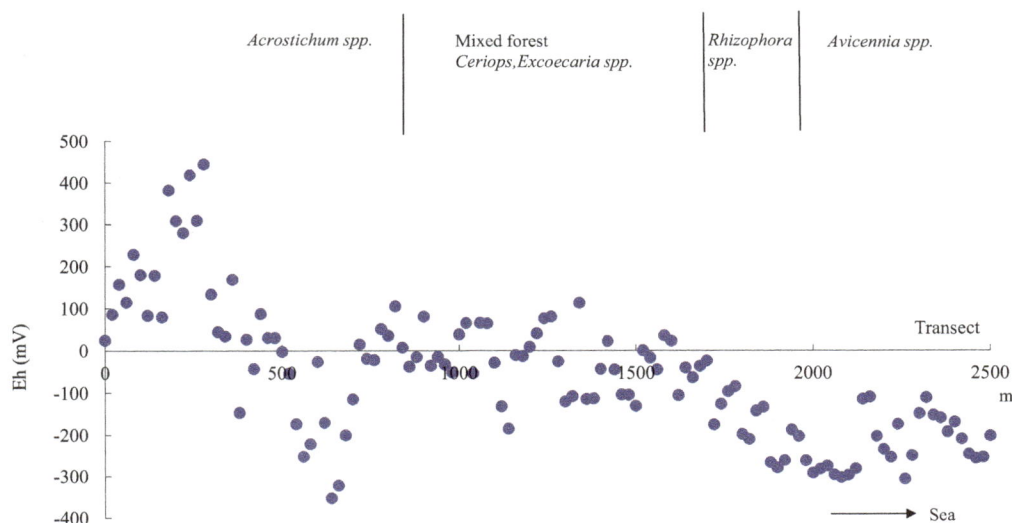

Figure 4. Redox potential (Eh) changes along the transect. Eh decreased towards the sea.

Eh varied markedly in the *Acrostichum* spp. zone (Figure 4). The degradation of OM, which was present at high concentrations in most wetland sediments, was partly responsible for the variation of sediment Eh [36]. The high variability of Eh was attributable to the variable redox state brought about by undulating microtopography, which created different conditions for the degradation of OM by location.

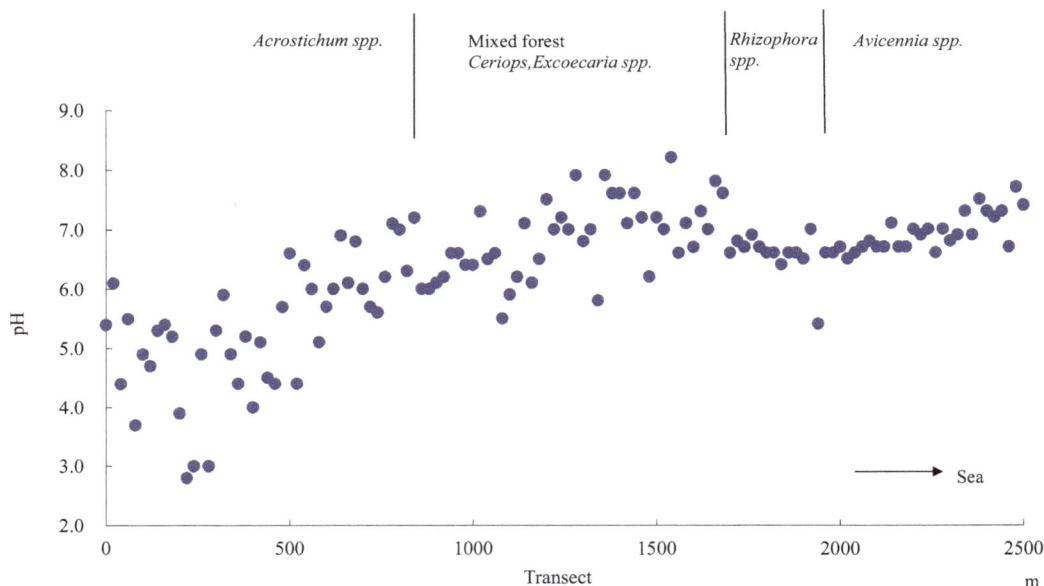

Figure 5. pH changes along the transect. The low pH in *Acrostichum* spp. zone is due to rapid OM decomposition.

Eh was rather low in the area around 600 m from the start of the transect. The rapid oxygen consumption by aerobic microorganisms, which was driven by the high OM content, generated a strong anaerobic condition. pH in the *Acrostichum* spp. zone was significantly different from that of the mangrove zone ($p < 0.01$); the mean pH in the *Acrostichum* spp. zone was 4.67 ($n = 19$) and that of the mangrove zone was 6.08 ($n = 42$). The pH values in the area at 200 m were lower than 3 as a result of the strong acidification (Figure 5). Human disturbance in the form of excavation, which led to soil oxidation, might have caused the generation of extremely acidic soil.

Eh increased slightly in the area around 2300 m in the *Avicennia* spp. zone. Nevertheless, no significant difference was found by statistical analysis. It was reported that Eh was significantly different between the *Avicennia* spp. zone and the *Rhizophora* spp. zone because of the difference in litter composition [37] and in OM decomposability [16]. As the major organic materials entering the coastal ecosystem were plant litter and root exudates, the significant differences in Eh among the areas are related to the differences in mangrove species.

3.2. OC in Surface Soils

Surface soil OC contents were higher in the mangrove areas than in the terrestrial areas (Table 4), implying that the mangrove ecosystem had higher capacity for OM production and storage. At NR2, a sampling point in the *Acrostichum* spp. zone, total carbon content was the highest at 14.2%. NR1, another sampling point in the *Acrostichum* spp. zone, also had high total carbon content, as shown by the mean OC values of 17.6% in the *Acrostichum* spp. zone ($n = 19$) and 5.1% in the mangrove area ($n = 42$) with a significance level of $p < 0.05$. It was speculated that *Acrostichum* spp. would have the capacity to produce a large amount of OM. Meanwhile, OC content in PF was low. As PF was originally a mangrove before the 1980s, carbon was lost by land use change. A significant amount of carbon might have been lost from the coastal ecosystem, considering that a large mangrove area was converted into shrimp ponds in the study area.

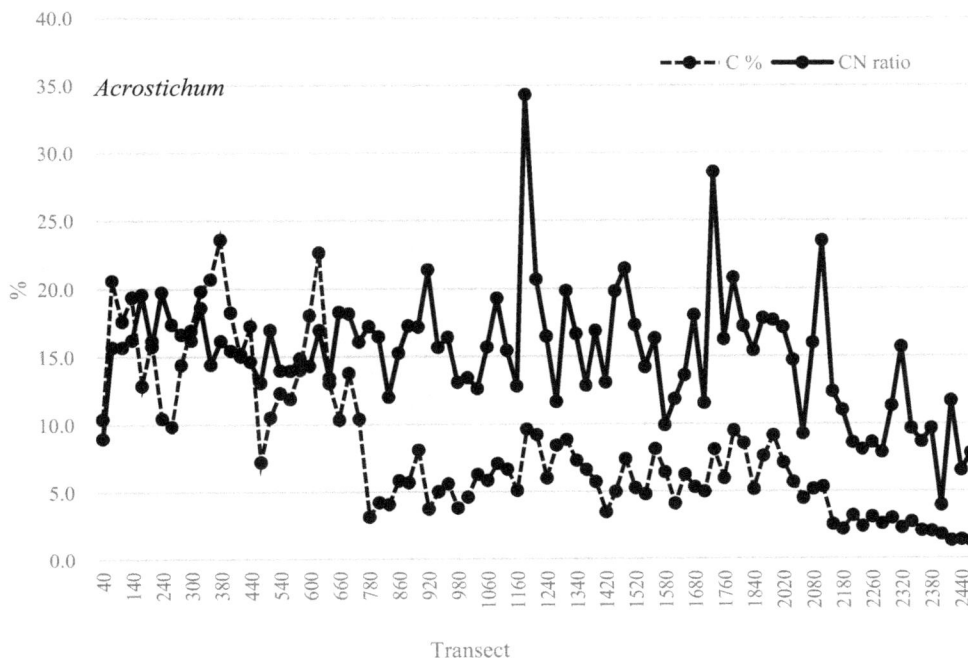

Figure 6. OC contents and C/N ratios in surface soils collected at 5 cm soil depth along the transect. Note that soil from *Acrostichum* spp. zone had high OC content and C/N ratio was low near the shoreline.

The C/N ratio of topsoil decreased seaward (Figure 6), but increased from the upper layer to the lower layer (Table 4). The C/N ratio was often used as an indicator of the source of OM in aquatic sediments. The C/N ratio in aquatic systems was governed by the mixing of terrestrial and autochthonous OM [38,39]. OM derived mainly from plankton had a C/N ratio of 6 to 9 [40,41], whereas OM derived from terrestrial vascular plants and their derivatives in sediments had a C/N ratio of 15 or higher [42,43]. Therefore, the low C/N ratio in *Avicennia* spp. zone could be influenced by carbon originating from biological sources, such as plankton or algae.

Correlation analysis of soil properties was conducted in 74 surface soil samples collected along the transect (Table 5). In the surface soil samples, pH was correlated to C% ($r = -0.66, p < 0.01$). Organic acids released by OM decomposition increased acidity. pH was also influenced by the reductive condition (Eh) ($r = -0.58, p < 0.01$). The strong correlation between carbon and nitrogen indicated that carbon was the major source of nitrogen. The majority of nitrogen in the soil samples existed in the organic form. After decomposition by microorganisms (mineralization), organic nitrogen was transformed into inorganic nitrogen in the form of NH_4 or NO_3, which are the available forms of nitrogen for utilization by plants. Mineralization provided much of the nitrogen needed to maintain high primary production in mangrove forests [44,45] and salt marshes [46].

Table 5. Correlation matrix of surface soil properties along the transect ($n = 74$).

	C%	N%	CN ratio	pH	Eh
N%	0.83 *				
CN ratio	0.34 **	−0.09			
pH	−0.66 *	−0.58 *	−0.13		
Eh	0.33 **	0.30 **	0.17	−0.58 *	
EC (ms/cm)	−0.37 **	−0.29 **	−0.20	0.35 **	−0.17

* and ** indicate 1% and 5% levels of significance, respectively.

The relatively weak correlation between carbon and C/N ratio meant that there were a number of carbon sources and different degrees of biological activity. Carbon source would primarily originate from vegetation or plankton, and its decomposability would be affected by both the chemical constituents in OM and the condition of the soil, *i.e.*, its pH and Eh.

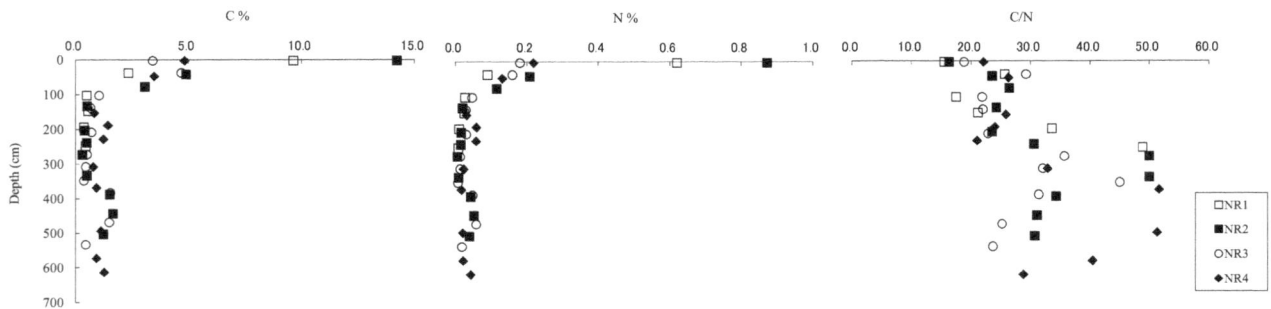

Figure 7. Belowground OC and N contents, and C/N ratios at different depths along the transect. Note that the mangrove mud layer had higher OC content and lower C/N ratio than the other layers.

3.3. Belowground OC and N

Belowground OC contents were higher in the mangrove mud layer than in the other layers (Figure 7), indicating that the mangrove mud layer had accumulated much OM provided by the mangrove forest. OC contents at NR1–NR4 in the shell-dominated layer and the mud layer were identical.

The belowground C/N ratios showed large variations (Figure 7), which were likely due to the difference in sources. The large variation in the shell-dominated layer was due to deposition that occurred nearshore, thereby accumulating different sources, either terrestrial or marine source.

3.4. Leaf Analysis

Chemical element uptake was significantly greater in plants grown under wetland conditions than in those grown under dryland conditions [47]. Compared with the terrestrial ecosystem, Na, P, Mg, and Cl showed greater accumulation in the flooded zone, including the mangrove area (Table 6, Figure 7). An increase of Na content in leaf was noted along the transect (Figure 8). This tendency corresponded to the salt tolerance of mangrove species. Na content in major mangrove species was higher than 15 g/kg, markedly contrasting that of terrestrial plants, which was less than 5 g/kg (Figure 7). The significantly high

Na content in sedge, the dominant plant species in salt marsh, could be due to its selective absorption capacity. Eh and pH gradients played an important role in the mobility and uptake of P and Mg. P became available when pH became alkaline. Therefore, mangrove plants absorbed more P than terrestrial plants.

Table 6. Macro and micro element concentrations (g/kg) determined by leaf analysis.

Vegetation	Na	Mg	Al	Si	P	S	Cl	K	Ca	Mn	Fe
SF 1	1.20	1.29	0.08	0.82	0.16	4.63	12.9	2.54	9.02	1.01	0.23
SF 2	0.15	8.03	0.05	0.77	0.23	7.03	5.4	3.97	10.53	1.70	0.22
Coconut	2.27	2.20	0.05	2.63	0.22	7.92	8.4	3.13	5.30	0.21	0.18
Rice	1.58	0.94	0.02	0.74	0.51	9.52	56.8	14.40	1.79	0.06	0.47
Sedge	29.09	1.18	0.01	0.22	0.23	16.44	125.0	17.21	2.02	0.03	0.14
Acrostichum	8.67	5.02	0.02	0.18	0.25	11.54	39.5	18.56	12.38	0.10	0.11
Excoecaria	2.60	3.02	0.04	0.41	0.32	11.43	45.7	9.85	1.70	0.19	0.25
Ceriops	8.54	9.79	0.06	0.06	0.23	29.14	54.0	6.29	14.08	0.23	0.17
Bruguiera	21.25	8.05	0.07	0.08	0.23	10.84	144.9	5.97	19.48	0.27	0.12
Rhizophora	16.38	7.37	0.34	0.21	0.38	4.61	0.0	8.87	9.54	0.46	0.36
Sonneratia	25.31	6.17	0.07	0.14	0.45	14.18	123.9	7.79	7.73	1.07	0.17
Avicennia	26.59	4.57	0.02	0.10	0.44	7.07	49.0	13.99	5.23	0.27	0.07

SF: Secondary forest.

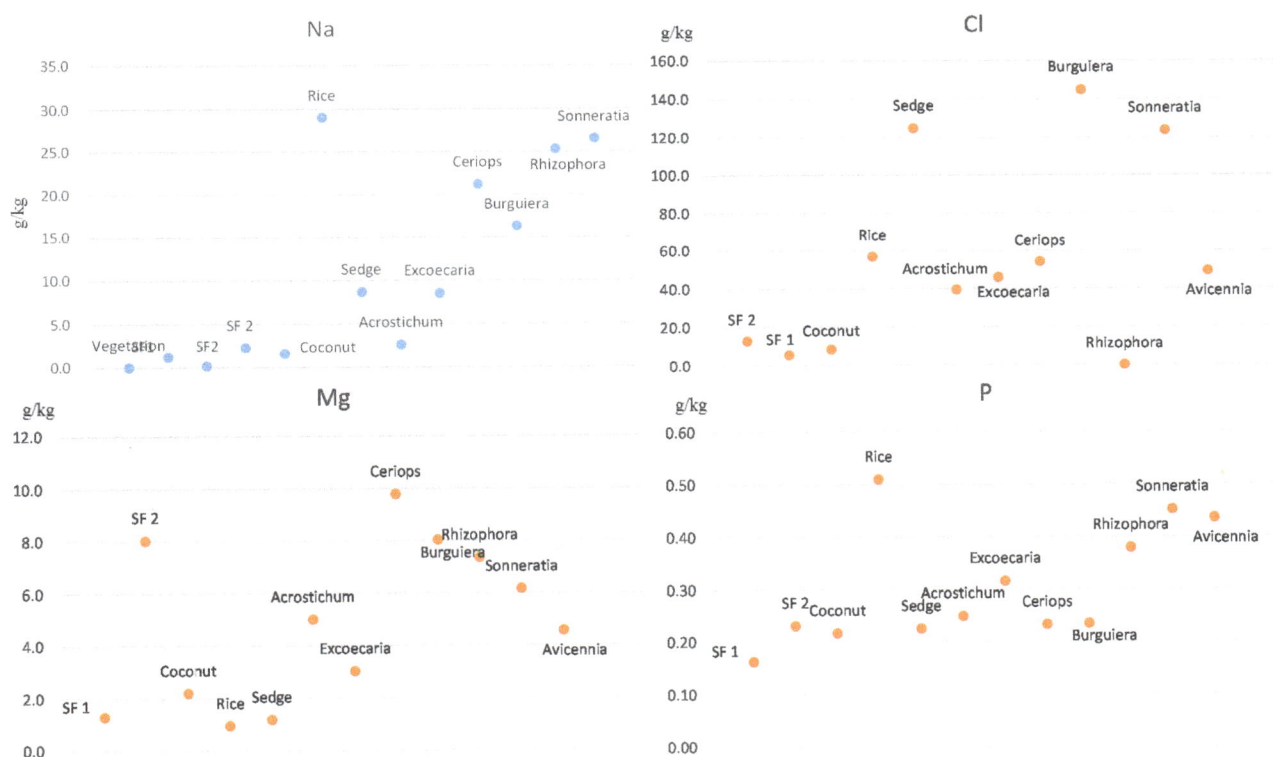

Figure 8. Leaf Na, P, Mg, and Cl contents in different types of vegetation along the transect. SF denotes secondary forest.

Significant differences were noted for particular elements among the three mangrove species. Fe content was high in *Rhizophora* leaf tissues, whereas Al, Mn, and S contents were high in *Avicennia*, *Sonneratia*, and *Ceriops*, respectively. Thus, mangrove species appeared to have uptake preference for elements, and this might indicate their possible use as ecological indicators for inorganic compound monitoring in mangrove ecosystems.

3.5. Radio Carbon Dating Analysis

The formation of the mud layer started around 6460–7030 ^{14}C years BP, as shown in the radiocarbon dating of mud layer bottom (NR2, 6462; NR3, 7540; NR4, 7030 ^{14}C years BP) (Table 7). As Fujimoto *et al.* [48] reported, the first regression occurred before 7200 ^{14}C years BP in the southwestern coast of Thailand; the mud layer could have been formed during the transgression period.

Table 7. Results of radiocarbon dating analysis.

Site	Layer	Depth *	Type of sample	Measured ^{14}C age (year BP)	δ^{13}C (‰)	Conventional ^{14}C age (year BP)
NR1	Mangrove mud	105–115	organic sediment	1820 ± 110	−25.00	1820 ± 110
NR2	Mangrove mud	100–104	organic sediment	1610 ± 70	−25.70	1600 ± 70
	Shell-dominated	400–404	organic sediment	6020 ± 80	−25.90	6000 ± 80
	Mud	530–540	shell	6070 ± 40	−1.40	6460 ± 40
NR3	Mangrove mud	105–112	organic sediment	1730 ± 100	−24.40	1730 ± 100
	Shell-dominated	385–393	organic sediment	6080 ± 80	−25.90	6070 ± 80
	Mud	535–542	organic sediment	7530 ± 40	−24.20	7540 ± 40
NR4	Mangrove mud	174–176	shell	1020 ± 40	−5.80	1330 ± 40
	Shell-dominated	387	shell	3400 ± 40	1.80	3840 ± 40
	Mud	700–710	organic sediment	7050 ± 40	−26.20	7030 ± 40

* Depth where sample was collected.

Figure 9. δ13C values and C/N values of soils and sediments in the layers.

The deposition of the organic-rich mangrove mud layer started around 1330–1820 ^{14}C years BP. The sea level started to change around 2200 ^{14}C years BP in the southwestern coast of Thailand [48]. The mangrove mud layer at the study site started to form after those periods.

A comparison of the carbon dating results obtained from different locations in the mangrove mud layer revealed that sedimentation was likely to start earlier inland than offshore. This was in agreement with the idea that the mangrove ecosystem was developing in the offshore direction. The distance between NR1 and NR4 was 2000 m, and their sedimentation times differed by 500 years. By computing those values, it was found that the mangrove would have moved 4 m seaward per year.

Bulk sediment δ13C and C/N values were the result of autochthonous inputs from wetland vegetation and allochthonous sources, such as algae and particulate organic matter [49–51]. δ13C (‰) in organic sediment was between −26‰ and −24‰ (Table 7), which resembled that of terrestrial carbon sources, whose range was between −33‰ and −25‰ [52], and of fresh water phytoplankton δ13C, whose range was between −30‰ and −25‰ [53]. However, δ13C (‰) in the mangrove mud layer varied greatly (Figure 9), which might indicate diverse source, *i.e.*, not only terrestrial but also marine source.

3.6. Humic Acid Determination

The type of humic acid is representative of the deposition environment. Carboxylic humic acid is composed of aromatic compounds that are most likely derived from lignin of vascular plants [54]. It is thought that, during humification, carboxylic acid content in OM increases [55]. Thus, carboxylic acid content can be used as an indicator of the degree of humification.

Carboxylic acid content in the mangrove mud layer ranged from 24% to 52%, and aliphatic content, from 34% to 66% (Table 8). Considering that carboxylic acid content and aliphatic content were generally 76%–95% and 5%–27%, respectively, in terrestrial ecosystems [56], the mangrove OM had not yet progressively humified. Mangrove soils were mostly reductive due to waterlogging, which retarded humification.

Table 8. Compositions (%) of humic acids at different sites and depths.

Site	Depth	Layer	Aliphatic	Phenolic	Carboxylic
NR1		Mangrove mud	38	10	52
NR2		Mangrove mud	66	10	24
NR3		Mangrove mud	36	12	52
NR4					
	0–5 cm	Mangrove mud	34	15	51
	305–310 cm	Shell-dominated	22	11	67
	490–495 cm	Mud	19	12	69

Humic acids at NR2 were characterized by a high aliphatic content, resembling the chemical characteristics of humic acids in sea-bottom or lake-bottom sediments. The high OC production of *A. aureum* (Figure 6) and the reductive condition (Figure 5) might have influenced the buildup of this aliphatic-rich humic composition.

Aliphatic-rich mangrove soils are susceptible to decomposition because they are prone to oxidation [57]. Under the aerobic condition, aliphatic compounds are easily degraded by microbial activity because the compounds possess a long aliphatic chain in their chemical structures [56]. Land use change, such as the conversion of mangrove into shrimp pond, may cause structural changes in mangrove humic substances. Carbon decomposition will be accelerated by the decrease of moisture regime and the increase of soil temperature. Therefore, it is necessary to study the fate of mangrove OCs in relation to land use change, with water/soil condition monitoring from the viewpoint of humic substances.

Carboxylic acid content was high, and aliphatic content was low in humic acids from the shell-dominated and mud layers at NR4. The increase of the carboxylic acid content and the decrease of the aliphatic content are related to structural changes caused by the humification process. In the event of dehydration, demethylation of the aliphatic components would proceed, increasing carboxylic acid components, which are weathering-resistant components, and hence advancing humification. Sediments in the sublayers, which partly included marine-originating deposits, were mostly transported from the land. Humification of terrestrial OM took place more rapidly because the terrestrial condition was normally drier and aerobic. OM with a high degree of humification was transported from the land and deposited as sublayers; therefore, humic acids in the sublayers had high carboxylic acid content but not aliphatic content. In PF, where water was stagnant in a pond, humification did not occur progressively due to the prolonged wet condition.

4. Conclusions

Surface soil OC content in mangrove stands varied greatly depending on the vegetation type along the 2500 m long transect. The large variations in $\delta 13C$ and the C/N ratio in surface soil indicated different sources, namely, autochthonous and allochthonous. The pH and Eh gradients were important factors affecting OC stock and the mobility/uptake of chemical elements. P and Mg content in the leaf increased with increasing pH. Mangrove species showed preference for uptake of particular elements. Eh was positively correlated with OC content, and OC content became high when Eh was positive in the *Acrostichum* spp. zone. The significant fluctuation of Eh in the *Acrostichum* spp. zone could be caused by human disturbance in the form of excavation. OC in the mangrove mud layer was characterized by high aliphatic content, indicating that the soils were not yet humified and were susceptible to decomposition.

Acknowledgments

This research was undertaken mostly as part of a Joint Research Project participated by the Department of Marine and Coastal Resources (DMCR), Thailand, Kansai Electric Power Co., Inc., Osaka, Japan, and The General Environmental Technos Co., Ltd. (KANSO), Osaka, Japan, from 1998 to 2007, and partly as an additional study conducted solely by KANSO from 2009 to 2013. The authors are deeply indebted to Emeritus Koyo Yonebayashi, Ishikawa Prefectural University, and Munehiro Ebato, National Institute of Livestock and Grassland Science, for their valuable guidance and advice both in laboratory work and in interpreting analyzed data from humic acid measurement. They also thank the members of Soil Science Laboratory, Kyoto University, for their valuable work in soil physicochemical and leaf analyses.

Author Contributions

Conceived and designed the study: N. Matsui. Perfomed the study: N. Matsui, W. Meepol, J. Chukwamdee. Collection of field data: W. Meepol, J. Chukwamdee. Analyzed the data and wrote the paper: N. Matsui.

Conflicts of Interest

The authors declare no conflict of interest.

References

1. Woodroffe, C.D.; Chappell, J.; Thom, B.G.; Wallensky, E. Depositional model of a macrotidal estuary and floodplain, South Alligator River, Northern Australia. *Sedimentolgy* **1989**, *36*, 737–756.
2. Fujimoto, K.; Miyagi, T. Development process of tidal-flat type mangrove habitats and their zonation in the Pacific Ocean. *Vegetatio* **1993**, *106*, 137–146.
3. Matsui, N. Estimated stocks of organic carbon in mangrove roots and sediments in Hinchinbrook Channel, Australia. *Mangrove Salt Marshes* **1998**, *2*, 199–204.
4. Matsui, N.; Yamatani, Y. Estimated total stocks of sedement carbon in relation to stratigraphy underlying the mangrove forests of Sawi Bay. *Phuket Mar. Biol. Cent. Spec. Publ.* **2000**, *22*, 15–25.
5. Mcleod, E.; Chmura, G.L.; Bouillon, S.; Salm, R.; Bjork, M.; Duarte, C.M.; Lovelock, C.E.; Schlesinger, W.H.; Silliman, B.R. A blueprint for blue carbon: Toward an improved understanding of the role of vegetated coastal habitats in sequestering CO_2. *Front. Ecol. Environ.* **2011**, *9*, 552–560.
6. Twillery, R.R.; Chen, R.H.; Hargis, T. Carbon sinks in mangroves and their implications to carbon budget of tropical coastal ecosystems. *Water Air Soil Pollut.* **1992**, *64*, 265–288.
7. Nellemann, C.; Corcoran, E.; Duarte, C.M.; Valdes, M.; de Young, L.; Fonseca, L.; Grimsditch, G. *Blue Carbon: The Role of Healthy Oceans in Binding Carbon—A Rapid Response Assessment*; GRID-Arendal (United Nations Environment Programme): Arendal, Norway, 2009.
8. Chmura, G.L.; Anisfeld, S.C.; Cahoon, D.R.; Lynch, J.C. Global carbon sequestration in tidal, saline wetland soils. *Glob. Biogeochem. Cycles* **2003**, *17*, 1111–1123.
9. Duarte, C.M.; Middelburg, J.; Caraco, N. Major role of marine vegetation on the oceanic carbon cycle. *Biogeosciences* **2005**, *2*, 1–8.
10. Bouillon, S.; Borges, A.V.; Castañeda-Moya, E.; Diele, K.; Dittmar, T.; Duke, N.C.; Kristensen, E.; Lee, S.Y.; Marchand, C.; Middelburg J.J.; *et al.* Mangrove production and carbon sinks: A revision of global budget estimates. *Glob. Biogeochem. Cycles* **2008**, *22*, 1–12.
11. Duarte, C.M.; Marbà, N.; Gacia, E.; Fourqurean, J.W.; Beggins, J.; Barrón, C.; Apostolaki, E.T. Seagrass community metabolism: Assessing the carbon sink capacity of seagrass meadows. *Glob. Biogeochem. Cycles* **2010**, *24*, 1–8.
12. Kennedy, H.; Beggins, J.; Duarte, C.M.; Fourqurean, J.W.; Holmer, M.; Marba, N.; Middelburg, J.J. Seagrass sediments as a global carbon sink: Isotopic constraints. *Glob. Biogeochem. Cycles* **2010**, *24*, 1–8.

13. United Nations Framework on Climate Change. Land-Use Change and Forestry. Available online: http://unfccc.int/methods_and_science/lulucf/items/3060.php (accessed on 12 June 2012).

14. Raza, M.; Zakaria, M.P.; Hashim, N.R. Spatial and Temporal Variation of Organic Carbon in Mangrove Sediment of Rembau-Linggi Estuary, Malaysia. *World Appl. Sci. J.* **2011**, *14*, 48–54.

15. Trevor, G.; Harifidy, R.R.; Lalao, R.; Garth, C.; Adia, B. Ecological Variability and Carbon Stock Estimates of Mangrove Ecosystems in Northwestern Madagascar. *Forests* **2014**, *5*, 177–205.

16. Lacerda, L.D.; Ittekkot, V.; Patchineelam, S.R. Biogeochemistry of mangrove soil organic matter: A comparison between Rhizophora and Avicennia soils in South-eastern Brazil. *Estuar. Coast. Shelf Sci.* **1995**, *40*, 713–720.

17. Diekow, J.; Mielniczuk, J.; Knicker, H.; Bayer, C.; Dick, D.P.; Kögel-Knabner, I. Carbon and nitrogen stocks in physical fractions of a subtropical Acrisol as influenced by long-term no-till cropping systems and N fertilization. *Plant Soil* **2005**, *268*, 319–328.

18. Gonneea, M.E.; Paytan, A.; Herrera-Silveira, J.A. Tracing organic matter sources and carbon burial in mangrove sediments over the past 160 years. *Estuar. Coast. Shelf Sci.* **2004**, *61*, 211–227.

19. Hennekam, R.; de Lange, G. X-ray fluorescence core scanning of wet marine sediments: Methods to improve quality and reproducibility of higher solution paleoenvironmental records. *Limnol. Oceanogr. Methods* **2012**, *10*, 991–1003.

20. Bouillon, S.; Rao, V.V.S.; Koedam, N.; Dehairs, F. Sources of organic carbon in mangrove sediments: variability and possible ecological implications. *Hydrobiologia* **2003**, *495*, 33–39.

21. Arlauskienė, A.; Maikštėnienė, S.; Šlepetienė, A. Effect of cover crops and straw on the humic substances in the clay loam Cambisol. *Agron. Res.* **2010**, *8*, 397–402.

22. Plathong, S.; Plathong, J. Past and Present Threats on mangrove ecosystem in peninsular Thailand. In *Coastal Biodiversity in Mangrove Ecosystems: Paper presented in UNU-INWEH-UNESCO International Training Course, held at Centre of Advanced Studies*; Annamalai University: Chidambaram, India, 2004; pp. 1–13.

23. Ratanasermpong, S.; Charuppat, T. Coastal zone environment management with emphasis on mangrove ecosystem, A case study of Ao-Sawi Thung Khla, Chumphon, Thailand. In Proceeding of the Asian Conference on Remote Sensing, Taipei, Taiwan, 4–8 December 2000.

24. Valiela, I.; Bowen, J.L.; York, J.K. Mangrove forests: One of the world's threatened major tropical environments. *BioScience* **2001**, *51*, 807–815.

25. Alongi, D.M. Present state and future of the world's mangrove forests. *Environ. Conserv.* **2002**, *29*, 331–349.

26. Tyler, G. Effects of sample pretreatment and sequential fractionation by centrifuge drainage on concentrations of minerals in a calcareous soil solution. *Geoderma* **2000**, *94*, 59–70.

27. Day, P.R. Particle fractionation and particle-size analysis. In *Methods of Soil Analysis. Part 1*; Black, C.A., Ed.; American Society of Agronomy, Inc.: Madison, WI, USA, 1965; pp. 545–567.

28. Ochi, H.; Okashita, H. Fluorescence analysis of new materials with fundamental parameter method—Analysis of nickel, cobalt and titanium base alloys. *Shimadzu Rev.* **1988**, *45*, 51–60.

29. Schnetger, B.; Brumsack, H.J.; Schale, H.; Hinrichs, J.; Dittert. L. Geochemical characteristics of deep-sea sediments from the Arabian Sea: A high-resolution study. *Deep Sea Res.* II **2000**, *47*, 2735–2768, doi:10.1016/S0967-0645(00)00047-3.

30. Tabuki, R. Paleoenvironment of the Plio-Pleistocene Daishaka Formation, Tsugaru basin, Northeast Japan. *Trans. Proc. Palaeontol. Soc. Jpn. New Ser* **1983**, *130*, 61–78.

31. Saenger, P.; McConchie, D. Heavy metals in mangroves: Methodology, monitoring and management. *Envis For. Bull.* **2004**, *4*, 52–62.

32. Yonebayashi, K.; Hattori, T. Chemical and biological studies on environmental humic acids. I. Composition of elemental and functional groups of humic acids. *Soil Sci. Plant Nutr.* **1988**, *34*, 524–571.

33. Hattori, T. Some properties of brackish sediments along the Chao Phraya river of Thailand. *Southeast Asian Stud.* **1972**, *9*, 522–532.

34. Furukawa, H. *Coastal Wetlands of Indonesia: Environment, Subsistence and Exploitation*; Kyoto University Press: Kyoto, Japan, 1994; p. 219.

35. Reis, C.E.S.; Dick, D.P.; Caldas, J.S.; Bayer, C. Carbon sequestration in clay and silt fractions of Brazilian soils under conventional and no-tillage systems. *Sci. Agric.* **2014**, *71*, 292–301.

36. Van Cappellen, P.; Gaillard, J.F. Biogeochemical dynamics in aquatic sediments. In *Reactive Transport in Porous Media, Reviews in Mineralogy*; Lichtner, P.C., Steefel, C.I., Oelkers, E.H., Eds.; Mineralogical Society of America: Washington, DC, USA, 1996; Volume 34, pp. 335–376.

37. Thibodeau, F.R.; Nickerson, N.H. Differential oxidation of mangrove substrate by Avicennia germinans and Rhizophora mangle. *Am. J. Bot.* **1988**, *73*, 512–516.

38. Thornton, S.F.; McManus, J. Application of organic carbon and nitrogen stable isotope and C/N ratios as source indicators of organic matter provenance in estuarine systems: Evidence form the Tay Estuary, Scotland. *Estuar. Coast. Shelf Sci.* **1994**, *38*, 219–233.

39. Meyers, P.A. Organic geochemical proxies of paleoceanographic, paleolimnologic, and paleoclimatic processes. *Org. Geochem.* **1997**, *27*, 213–250.

40. Prahl, F.G.; Bennett, J.T.; Carpenter, R. The early diagenesis of aliphatic hydrocarbons and organic matter in sedimentary particulates from Dabob Bay, Washington. *Geochim. Cosmochim. Acta* **1980**, *44*, 1967–1976.

41. Biggs, R.B.; Sharp, J.H.; Church, T.M.; Tramontano, J.M. Optical properties, suspended sediments, and chemistry associated with the turbidity maxima of the Delaware Estuary. *Can. J. Fish. Aquat. Sci.* **1983**, *40*, 172–179.

42. Hedges, J.I.; Clark, W.A.; Quay, P.D.; Ricihey, J.E.; Devol, A.H.; Santos, D.M. Compositions and fluxes of particulate organic material in the Amazon River. *Liimnol. Oceanogr.* **1986**, *31*, 717–738.

43. Orem, W.H.; Burnett, W.C.; Landing, W.M.; Lyons, W.B.; Showers, W. Jellyfish Lake, Palau: Early diagenesis of organic matter in sediments of an anoxic marine lake. *Liimnol. Oceanogr.* **1991**, *36*, 526–543.

44. Nedwell, D.B.; Blackburn, T.H.; Wiebe, W.J. Dynamic nature of the turnover of organic carbon, nitrogen and sulphur in the sediments of a Jamaican mangrove forest. *Mar. Ecol. Prog. Ser.* **1994**, *110*, 223–231.

45. Alongi, D.M.; Trott, L.A.; Wattayakorn, G.; Clough, B.F. Below-ground nitrogen cycling in relation to net canopy production in mangrove forests of southern Thailand. *Mar. Biol.* **2002**, *140*, 855–864.

46. Anderson, I.C.; Tobias, C.R.; Neikirk, B.B.; Wetzel, R.L. Development of a process-based nitrogen mass balance model for a Virginia (USA) *Spartina alterniora* salt marsh: Implications for net DIN flux. *Mar. Ecol. Prog. Ser.* **1997**, *159*, 13–27.

47. La Toya, T.K.; Donna, L.J.; Marinus, L.O. Multiple elements in Typha angustifolia rhizosphere and plants: Wetland *versus* dryland. *Environ. Exp. Bot.* **2011**, *72*, 232–241.

48. Fujimoto, K.; Imaya, A.; Tabuchi, R.; Kuramoto, S.; Utsugi, H.; Murofushi, T. Belowground carbon storage of Micronesian mangrove forests. *Ecol. Res.* **1999**, *14*, 409–413.

49. Fry, B.; Scalan, R.S.; Parker, L. Stable carbon isotope evidence for two sources of organic matter in coastal sediments: Seagrasses and plankton. *Geochim. Cosmochim. Acta* **1977**, *41*, 1875–1877.

50. Chmura, G.L.; Aharon, P. Stable carbon isotope signatures of sedimentary carbon in coastal wetlands as indicators of salinity regime. *J. Coast. Res.* **1995**, *11*, 124–135.

51. Middelburg, J.J.; Nieuwenhuize, J.; Lubberts, R.K.; van de Plassche, O. Organic carbon isotope systematics of coastal marshes. *Estuar. Coast. Shelf Sci.* **1997**, *45*, 681–687.

52. Lamb, A.L.; Wilson, G.P.; Leng, M.J. A review of coastal palaeoclimate and relative sea-level reconstructions using δ13C and C/N ratios in organic material. *Earth Sci. Rev.* **2006**, *75*, 29–57.

53. Anderson, T.F.; Arthur, M.A. Stable isotopes of oxygen and carbon and their application to sedimentologic and paleoenvironmental problems. In *Stable Isotopes in Sedimentary Geology*; Arthur, M.A., Anderson, T.F., Eds.; Society of Paleontologists and Mineralogists: Tulsa, OK, USA, 1983; p. 151.

54. Hatcher, P.G.; Maciel, G.E.; Dennis, L.W. Aliphatic structure of humic acids; a clue to their origin. *Org. Geochem.* **1981**, *3*, 43–48.

55. Day, X.Y.; Ping, C.L.; Candler, R.; Haumaier, L.; Zech, W. Characterization of soil organic matter fractions of Tundra soils in Artic Alaska by carbon-13 nuclear magnetic resonance spectroscopy. *Soil Sci. Soc. Am. J.* **2001**, *65*, 87–93.

56. Yonebayashi, K. Humic component distribution of humic acids as shown by adsorption chromatography using XAD-8 resin. In *Humic substances in the Global Environment and Implication on Human Health*; Elsevier Science: Amsterdam, The Netherlands, 1994; pp. 181–186.

57. Gonet, S.S.; Debska, B. Properties of humic acids developed during humification process of post-harvest plant residues. *Environ. Int.* **1998**, *24*, 603–608.

14

Tsunamigenic Earthquakes at Along-dip Double Segmentation and Along-strike Single Segmentation near Japan

Junji Koyama [1,2,†,*], Motohiro Tsuzuki [1,†] and Kiyoshi Yomogida [1,†]

[1] Division of Natural History Sciences, Graduate School of Science, Hokkaido University, N10 W8, Kita-ku, Sapporo, Hokkaido 0650810, Japan; E-Mails: 2007grad-340@mail.sci.hokudai.ac.jp (M.T.); yomo@mail.sci.hokudai.ac.jp (K.Y.)

[2] Hyotanjima Scholorship, Sapporo, Hokkaido 151-1854630, Japan

[†] These authors contributed equally to this work.

[*] Author to whom correspondence should be addressed; E-Mail: koyama@mail.sci.hokudai.ac.jp

Academic Editor: Valentin Heller

Abstract: A distinct difference of the earthquake activity in megathrust subduction zones is pointed out, concerning seismic segmentations in the vicinity of Japan—that is, the apparent distribution of earthquake hypocenters characterized by Along-dip Double Segmentation (ADDS) and Along-strike Single Segmentation (ASSS). ADDS is double aligned seismic-segmentation of trench-ward seismic segments along the Japan Trench and island-ward seismic segments along the Pacific coast of the Japan Islands. The 2011 Tohoku-oki megathrust earthquake of Mw9.0 occurred in ADDS. In the meantime, the subduction zone along the Nankai Trough, the western part of Japan, is the source region of a multiple rupture of seismic segments by the 1707 Houei earthquake, the greatest earthquake in the history of Japan. This subduction zone is narrow under the Japan Islands, which is composed of single aligned seismic-segmentation side by side along the Nankai Trough, which is typical of ASSS. Looking at the world seismicity, the 1960 and 2010 Chile megathrusts, for example, occurred in ASSS, whereas the 1952 Kamchatka and the 1964 Alaska megathrusts occurred in ADDS. These megathrusts in ADDS result from the rupture of strong asperity in the trench-ward seismic segments. Since the asperity of earthquakes in

ASSS is concentrated in the shallow part of subduction zones and the asperity of frequent earthquakes in ADDS is in deeper parts of the island-ward seismic segments than those of ASSS, there must be a difference in tsunami excitations due to earthquakes in ADDS and ASSS. An analysis was made in detail of tsunami and seismic excitations of earthquakes in the vicinity of Japan. Tsunami heights of ASSS earthquakes are about two times larger than those of ADDS earthquakes with the same value of seismic moment. The reason for this different tsunami excitation is also considered in relation to the seismic segmentations of ADDS and ASSS.

Keywords: ADDS; ASSS; tsunamigenic earthquake; tsunami magnitude

1. Introduction

Devastating tsunamis in the last two decades required a new paradigm of the earthquake occurrences and tsunami generations on a geologically extended time-span. The extraordinary tsunami in the Indian Ocean of 2004 was generated by the earthquake of more than 1000km in fault length, which includes historical fault ruptures of smaller scales along the Sumatra-Andaman subduction zone [1]. The 2011 Tohoku tsunami in Japan resulted from a fault rupture developed from the Japan Trench to the coastline of the Tohoku prefecture of Japan along the dip-direction of the Pacific plate as well as along the strike-direction of the Japan Trench. This wide-spread source region of the 2011 megathrust earthquake of Mw9.0 includes trench-ward seismic segments, and the segments have been believed to be an interface of aseismic segments [2,3] without any stress accumulation.

These tsunamis were due to megathrust earthquakes much larger than anticipated sizes in their respective seismological histories, although some geological evidence demonstrated past earthquakes and tsunamis larger than those in historical documents and instrumentally observed records. It has been pointed out recently that observations of their seismic activity in recent years revealed that the megathrust earthquakes occurred in seismic segments of either Along-strike Single Segmentation (ASSS) or Along-dip Double Segmentation (ADDS) [4–6]. The different seismic segmentations—ASSS and ADDS—have been identified, referring to their regional seismic-activity, focal mechanisms, rupture patterns, geometry of subduction zones, types of overriding plates and back-arc activity [5].

Previous studies discuss the seismic segmentation along the subduction zones as a more complex structure, being composed of aseismic-zone, unstable zone, conditionally stable zone, and stable sliding zone, e.g. [2]. The present ADDS/ASSS hypothesis would be one where the segmentation relates directly to observable seismic activity in the source regions of great (Mw~8) and megathrust earthquakes (Mw~9). The main purpose of this study is to discuss the tsunami excitation in relation to these segmentations so as to make a comprehensive understanding of source regions of megathrust earthquakes.

Previous studies showed that tsunami excitation in the Japan Sea is larger than that in the Pacific Sea side of Japan [7–9]. These studies attributed the difference to the dip angle of earthquake faults or to the rigidity of earthquake source-regions. In this study, a further discussion is made about the tsunami excitation referring to the recent understanding of earthquake sources in different segmentations of ADDS and ASSS in the vicinity of Japan.

2. Diversity of Megathrust Earthquakes in the World

Seismic activity off the Pacific coast of the Tohoku district, Japan has been extensively investigated from historical and instrumental records. It is characterized into regional seismic segments; there is a double aligned seismic-segmentation along the dip-direction of the subduction zone. The segments along the island-arc side of Japan frequently generated earthquakes as large as Mw8, but the segments along the Japan Trench had been considered to be aseismic. This double aligned seismic-segmentation is called Along-dip Double Segmentation (ADDS). The 2011 megathrust earthquake of Mw9.0 ruptured many segments along not only strike- but also dip-directions of the Japan Trench, covering an area of about 200×500 km^2, as shown in Figure 1.

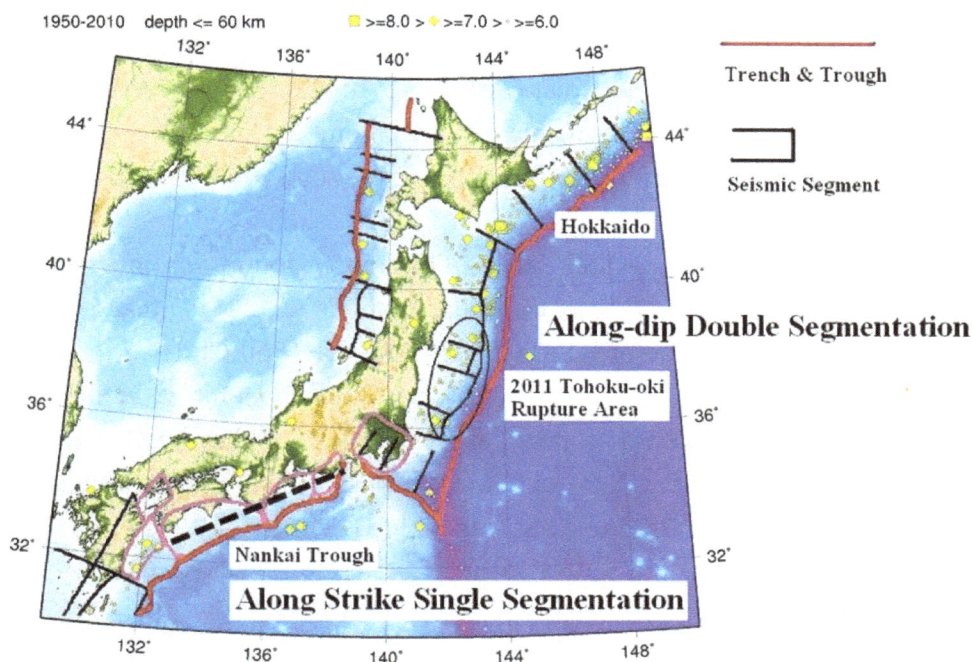

Figure 1. Seismic activity in Japan and in its vicinity. Seismic segments have been used in the official earthquake forecasting [10] of Evaluation of Major Subduction Zone Earthquakes by the Headquarters for Earthquake Research Promotion. Epicenters of earthquakes are plotted by yellow symbols from 1950 to 2010 with magnitudes larger than 5.9 and their focal depths shallower than 61km determined by the Japan Meteorological Agency [11]. Trench and trough near the Japan Islands are illustrated by red curves. The 2011 Tohoku-oki megathrust earthquake ruptured the area circled by a solid ellipse, where is Along-dip Double Segmentation (ADDS). Along-strike Single Segmentation (ASSS) can be found in the Nankai Trough, where little recent seismic activity has been observed. Such regions are often called seismic gaps [12]. The source extent of the 1707 Houei great earthquake in this segmentation is added by a broken line along the Nankai Trough to the original figure on [13].

The best-known megathrust earthquake in Japan is the 1707 Houei great earthquake along the Nankai Trough with three major segments in Figure 1. This historical event exhibits an interaction in the trench-axis direction among adjacent segments. The 1707 Houei great earthquake took place at a very different site from that of the 2011 megathrust event, where little seismic activity is observed in the single aligned seismic-segmentation along the axis of the Nankai Trough (Figure 1). The inactive seismicity in this region not only applies to the period analyzed in Figure 1 but also to the whole period from 1924 to the present, according to the Japan Meteorological Agency [11], except for the enhanced aftershock activity following the 1944 Tonankai earthquake Mw8.2 and the 1946 Nankaido earthquake Mw8.2, which successively occurred along the Nankai Trough. The Houei earthquake of multi-segment rupture is referred to an earthquake in Along-strike Single Segmentation (ASSS), which contrasts to the 2011 Tohoku-oki megathrust earthquake in ADDS.

The reason for a gigantic megathrust earthquake in ADDS to grow up to such the scale is due to the rupture of strongly-coupled asperity in the trench-ward segment with a longer recurrence time (e.g., millennial) after large earthquakes in the land-ward segments repeated with a shorter recurrence time (e.g., centennial). The activity in ASSS is characterized by almost 100% coupled areas of shallow subduction zones, which finally gives rise to a great earthquake. In other words, the difference between these two types of segmentations appears in a seismic gap [12] along the subduction zone in ASSS and in a doughnut pattern [14] of seismic activity in the subduction zone prior to a gigantic megathrust earthquake in ADDS.

These two types of segmentations (*i.e.*, ASSS and ADDS) can be found not only in the vicinity of Japan but also elsewhere in the world. The 1952 Kamchatka earthquake of Mw 9.0 and the 1964 Alaska earthquake of Mw9.2 are pointed out to be of ADDS type, while the 1960 Chile earthquake of Mw9.5 and the 2010 Maule earthquake of Mw8.8 are of ASSS [4,5]. Speaking about the 2004 Sumatra-Andaman earthquake of Mw9.3, the faulting process is peculiar. Judging from the detailed analysis on the seismicity around the source region of the 2004 earthquake [1], the 2004 event started in the ADDS region and extended into the Andaman-Nicobar ASSS region [13]. The latter region is a typical oblique subduction zone, and is quite similar to the source areas of great earthquakes of the 1957 Andreanof Mw8.6 and the 1965 Rat Island Mw8.7 in the Aleutian arc.

General descriptions of proposed ADDS and ASSS are summarized in Table 1, indicating the distinction from the previous seismic segmentations e.g., [2,3]. Seismic segments in ASSS are usually characterized by a narrow subduction zone from the oceanic trench to the island arc. The evidence has been presented, showing the asperity of the 2010 Maule earthquake in accordance with the strongly-coupled plate interface identified by GPS observation beforehand of the 2010 event [15]. Seismic segments in ADDS are, on the other hand, rather wide. The 2011 Tohoku-oki megathrust earthquake revealed that the strongly-coupled segments exist in the trench-ward of ADDS, where the millennial seismic-asperity of the event was observed [16,17]. This is also true in the 1964 Alaska earthquake [18], showing a very strong asperity in the Alaska Bay close to the Aleutian Trench. Large (not as large as gigantic) earthquakes occur repeatedly in the landward segments of ADDS with their foci in the deep part of the subduction zone. As a result, the asperity of repeated large earthquakes in ASSS is distributed in a shallow part of the subduction zone, while that of the repeated large earthquakes in ADDS stay in a rather deeper part. This would result in a different tsunami excitation between ADDS and ASSS earthquakes.

Figure 2 illustrates an image of earthquake cycles in ADDS and ASSS. An earthquake cycle in ASSS is commonly restricted in the shallow part of the subduction zone, where the contact of the plate interface is strong, forming the seismic gap all over the particular subduction zone. Successive rupture of multiple seismic segments along the trench direction results in a megathrust earthquake (metaphoric centennial event) in ASSS. On the other hand, an earthquake cycle in ADDS is characterized by repeated earthquakes as large as Mw8 and many smaller earthquakes are commonly found in the island-ward segments of the subduction zone, though the asperity in the trench-ward stays still, which eventually induces a megathrust earthquake (metaphoric millennial event) in ADDS. At the time of an ADDS megathrust event, the rupture extends in the dip-direction as well as in the strike-direction of the subduction zone.

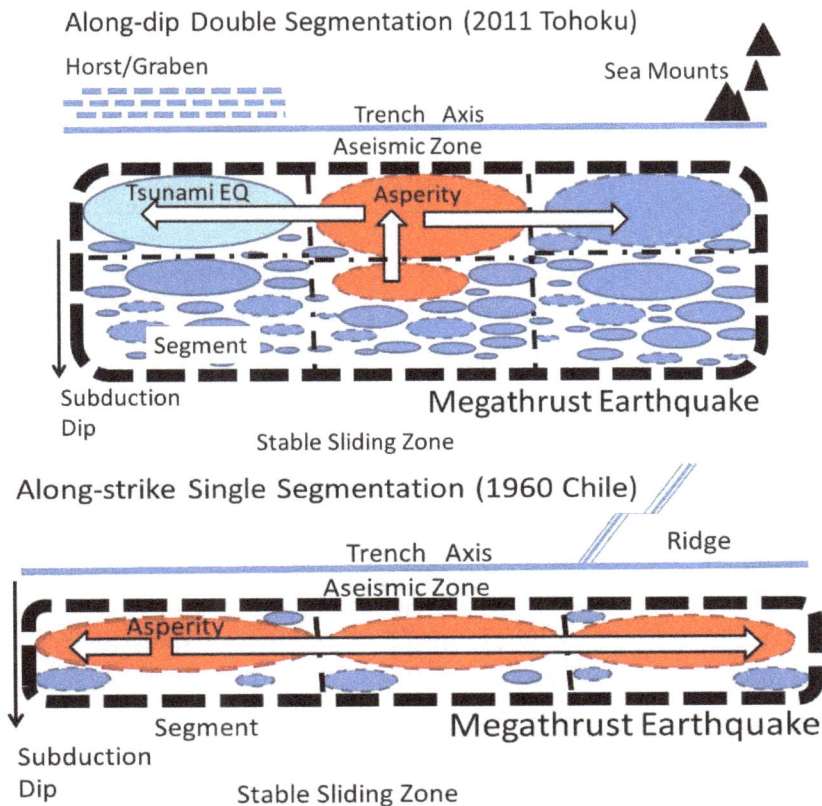

Figure 2. Schematic illustration of Along-dip Double Segmentation (ADDS) and Along- strike Single Segmentation (ASSS). ADDS is characterized by a double aligned seismic-segments along dip direction of subduction, one of which is a strongly-coupled asperity in the trench-ward segments, which eventually ruptures induced by an asperity in the island-ward segment(s) where repeated large earthquakes as large as Mw8 and many smaller-sized earthquakes (solid ellipses) occur. Rupturing all these segments results in a millennial megathrust earthquake (square by thick broken line) like the 2011 Tohoku event. ASSS shows a seismic gap of subduction interface along the trench axis in single aligned seismic-segments side by side. The asperity stays in a narrow part of the subduction interface in ASSS. Horst and Graben and seamounts are considered to be responsible for generating tsunami earthquakes [19–21]. As well as Horst and Graben, surface roughness such as seamounts and oceanic ridges plays an important role in blocking seismic segments (dotted broken lines) along the subduction zone [2].

Table 1. Characteristics of the seismic activity in Along-strike Single Segmentation (ASSS) and Along-dip Double Segmentation (ADDS).

	ASSS	ADDS
Alignment	Single Aligned	Double Aligned
Seismic Zone	Narrow	Wide
Width/Length	1:4	1:2
Interface Contact	Single Whole Contact	Island-ward/Trench-ward
Seismic Activity	Quiet Everywhere Seismic Gap	Active/Quiet Doughnut Pattern
Recurrence Time	A Few Hundred	A Few Hundred/A Thousand
Interface Contact	Whole Zone	Contact/Strong Contact
Previous Events	1960 Chile, 1707 Houei	1964 Alaska, 2011 Tohoku
Possible Region	Cascade, Canada	Hokkaido, Japan [21]

Although the segmentation of ADDS/ASSS is our proposed understanding on the seismic source region of megathrust/great earthquakes, we need to deepen our understanding of these segmentations in order to distinguish megathrust earthquakes in the future and also to reduce the disasters due to such different types of earthquakes and tsunamis. This study is an attempt to enlarge our knowledge on the tsunami excitations in relation to the different seismic segmentations.

3. Tsunami Magnitudes for Tsunamigenic Earthquakes near Japan

Tsunamis generated in the vicinity of Japan have been quantified from -1 to 4 using tsunami heights and disaster distribution near respective source regions [22,23]. This is called the Imamura-Iida scale of tsunamis, and it has been determined from historical documents and recorded evidence in Japan since 176 A.D. This is available at present, as listed in major tsunami catalogs, such as Historical Tsunami Data Base for the World Ocean [24]. This scale for tsunamis is similar to earthquake intensity, such as the Mercalli intensity scale. Local tsunami magnitude m is numerically defined [25,26], extending the Imamura-Iida scale. The definition of m is, using tsunami heights H (peak to trough in meter) recorded on tide-gauges and correction for tsunami travel distances Δ (km) as

$$m = 2.7 \log H + 2.7 \log \Delta - 4.3 \qquad (1)$$

Since this is the tsunami magnitude scale investigated well in Japan, at first we analyzed this local tsunami magnitude m comparing with seismic moment Mo of each earthquake that occurred near Japan. Another magnitude scale, Mt is the far-field tsunami magnitude [27], of which definition is

$$Mt = \log H + 9.1 + \Delta C \qquad (2)$$

where H is the maximum amplitude of tsunami heights in meter and ΔC is an empirical station correction or regionally-averaged correction. The correction term and the constant 9.1 are introduced so that Mt agrees with moment magnitude Mw [28] of each earthquake. Mt scale is extended to estimate tsunami magnitude Mt for local earthquakes near Japan [7] as

$$Mt = \log H + \log \Delta + 5.80 \qquad (3)$$

The same analysis is made to study the relation between Mt and Mo of tsunamigenic earthquakes near Japan, taking into account the seismic segmentations.

Historical tsunami heights until recent years were measured in tide gauge stations at best located in developed harbors, most of which face in the opposite direction to their respective open seas. Others are eyewitness records and traces of inundations and up-streams. All these data and their sources are listed on the worldwide tsunami catalogs [23,24]. Quantification of tsunami excitation and scaling relations of tsunamis to earthquake source parameters have been made based on those observations. After the recent installation of ocean-bottom pressure gauges, the observation makes it possible to study the tsunami generation free from the contaminations of detail oceanic bathymetry and coastal topography. Unfortunately or fortunately, only few records are available now e.g., [17]. Therefore, it is necessary to apply historical tsunami data to describe the general scaling relation of tsunami excitation to earthquakes and to study the generation mechanism of tsunamis.

In the above equations of (1) and (2), the dependence of tsunami magnitudes on tsunami heights are different. This is because local tsunami magnitude m is based on the energy of tsunami waves and far-field tsunami magnitude Mt on the amplitude of tsunami waves. It is true that these tsunami magnitudes are empirical parameters and needed to be rigorously quantified; however, the present study is interested only in their regional bias due to regional seismic segmentation. The analysis would be granted, since the discussion here is made in a relative manner based on the tsunami magnitudes, which would be free from the absolute uncertainty of tsunami magnitude determination.

4. Tsunami Magnitude and Seismic Moment of Tsunamigenic Earthquakes near Japan

Local tsunami magnitude m and seismic moment Mo of 59 tsunamigenic earthquakes in the vicinity of Japan since 1923 have been determined. Table 2 summarizes those data with their references. There is a variety of earthquakes occurring near Japan, which cannot be classified into ADDS nor ASSS, although tsunamis were excited and recorded. Some of them were intra-plate and outer-rise earthquakes of normal-fault type and the others are earthquakes near the shore of the Japan Islands of the strike-slip type [13]. Therefore, we classify them into ADDS, ASSS and NFSS (Normal Fault-Strike Slip). Figure 3 shows the locations of all the earthquakes in Table 2.

Figure 4 shows the least-squares regressions between m and Mo for ADDS (m_D; 22 events in Table 2) and ASSS (m_S; 10 events in Table 2).

$$m_D = 1.26 \log Mo - 24.4 \ (\pm 0.17)$$
$$m_S = 1.19 \log Mo - 22.3 \ (\pm 0.39)$$

(4)

The data of which m is smaller than 0 in Table 2 is not included in the regression analyses, because the tsunami height is smaller than 15cm at the distance of 100km for $m = -0.5$ and smaller than -0.5, which we considered large uncertainty in observations. The regression relation for the NFSS category is not calculated, because it is irrelevant in this study. However, it should be noted that the relationship looks similar to that for ADDS in general and deviates from ASSS.

The purpose of above analyses is not to introduce other empirical relationships but to quantify the difference in tsunami excitation due to the seismic segmentation. We find that m of ASSS is about 0.7 larger than that of ADDS for the earthquakes with Mo of 10^{20} Nm and 0.6 for 10^{21} Nm, which are within the 95% confidence interval. Considering the scatter of the data in Figure 4 and the basic data of tsunami heights to determine m, the difference of m would be about 1.0 at most.

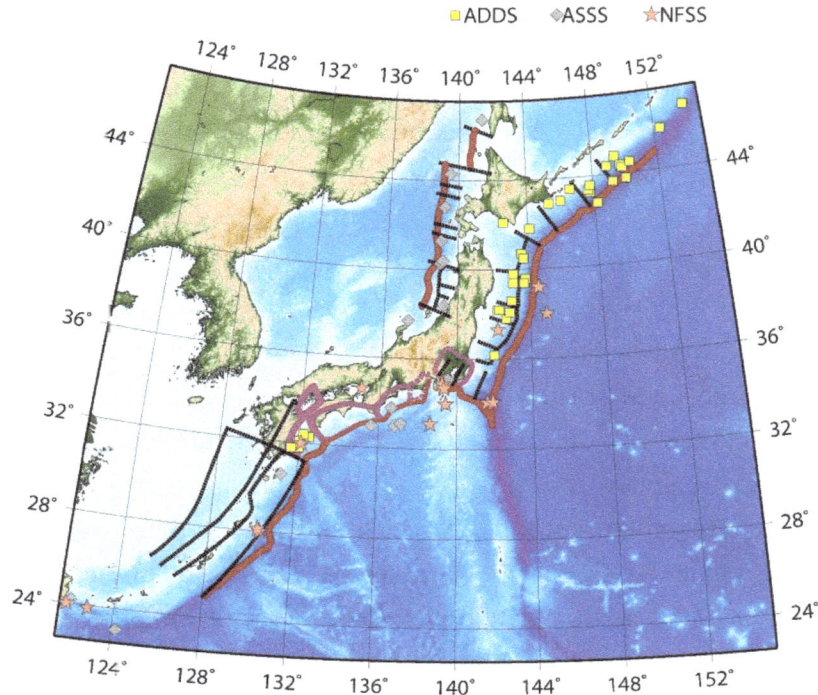

Figure 3. Epicenter locations of tsunamigenic earthquakes near Japan from 1923. Earth-quakes are categorized into ADDS, ASSS and Normal Fault/Strike Slip (NFSS), which are plotted by different symbols. Seismic segments in Figure 1 are also drawn.

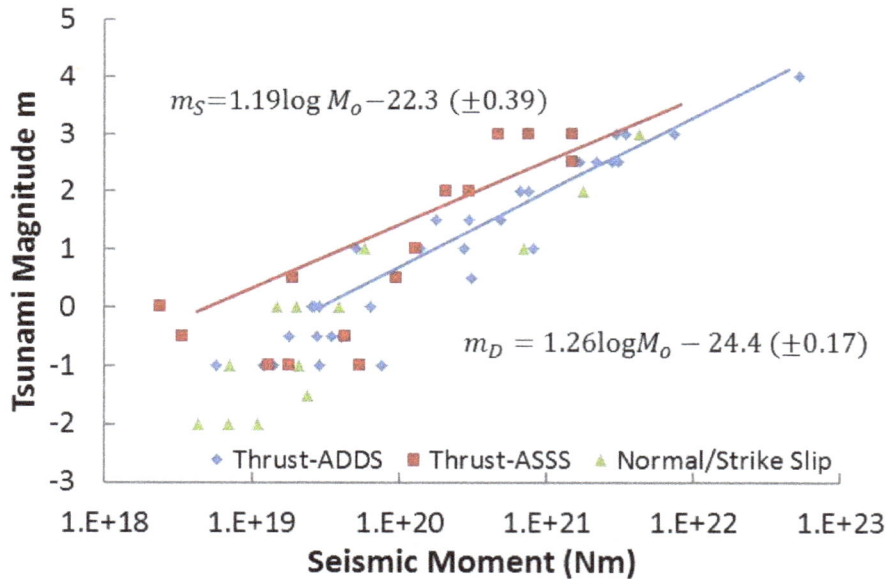

Figure 4. Tsunami magnitude m in relation to seismic moment Mo of corresponding earthquakes near Japan. Different symbols represent different category of earthquakes. Least squares regressions between m (m_D and m_S) and Mo for ADDS and ASSS earthquakes, respectively are derived.

Table 2. Tsunami magnitude and seismic moment of earthquakes near Japan since 1923.

Year	Month	Day	Location	Mw[#]	Mt[$]	m[%]	Mo (N·m)	Dip[+]	DipSlip[*]	Segment	Ref
1923	9	1	Kanto	7.9	8.0	2.0	7.60×10^{20}	0.56	0.17	ADDS	[29]
1933	3	3	Sanriku	8.4	8.3	3.0	4.30×10^{21}			NF	[30]
1938	11	5	Fukushima	7.8	7.6	1.0	7.00×10^{20}			NF	[31]
1940	8	2	W.Hokkaido	7.5	7.7	2.0	2.10×10^{20}	0.72	0.72	ASSS	[32]
1944	12	7	Tonankai	8.1	8.1	2.5	1.50×10^{21}	0.17	0.17	ASSS	[33]
1946	12	21	Nankaido	8.1	8.1	3.0	1.50×10^{21}	0.17	0.17	ASSS	[33]
1952	3	4	Tokachi-oki	8.1	8.2	2.5	1.70×10^{21}	0.34	0.33	ADDS	[34]
1963	10	13	Kurile	8.5	8.4	3.0	7.50×10^{21}	0.37	0.37	ADDS	[34]
1964	5	7	Oga-oki	7.0	7.1	−0.5	4.30×10^{19}			ASSS	[32]
1964	6	16	Niigata	7.6	7.9	2.0	3.00×10^{20}	0.94	0.94	ASSS	[35]
1968	4	1	Hyuganada	7.4	7.7	1.5	1.80×10^{20}	0.29	0.29	ADDS	[36]
1968	5	16	Tokachi-oki	8.2	8.2	2.5	2.80×10^{21}	0.34	0.21	ADDS	[37]
1968	6	12	Iwate-oki	7.1	7.4	1.0	5.10×10^{19}	0.50	0.24	ADDS	[38]
1969	8	12	Kurile	8.2	8.2	2.5	2.20×10^{21}	0.27	0.27	ADDS	[39]
1970	7	26	Hyuganada	7.0	7.1	−0.5	4.10×10^{19}			ADDS	[36]
1971	9	6	Sakhalin	7.3	7.5	0.5	9.50×10^{19}	0.63	0.62	ASSS	[32]
1973	6	17	Nemuro-oki	7.8	8.1	2.0	6.70×10^{20}	0.45	0.42	ADDS	[40]
1975	6	10	Kurile	7.0	7.9	1.5	3.00×10^{20}	0.22	0.22	ADDS	[41]
1978	1	14	Oshima	6.6	6.7	−2.0	1.10×10^{19}			SS	[42]
1978	6	12	Miyagi-oki	7.6	7.4	0.5	3.10×10^{20}	0.24	0.21	ADDS	[43,44]
1980	6	29	E. Izu	6.4	6.3	−2.0	7.00×10^{18}			SS	[43,44]
1982	3	21	Urakawa	6.9	7.1	0.0	2.60×10^{19}	0.47	0.47	ADDS	[43,44]
1982	7	23	Ibaraki-oki	7.0	7.0	−0.5	2.80×10^{19}			ADDS	[43,44]
1983	5	26	C.Nihonkai	7.9	8.1	3.0	7.60×10^{20}	0.45	0.45	ASSS	[43,44]
1983	6	21	W. Aomori	7.0	7.3	0.5	1.90×10^{19}	0.68	0.68	ASSS	[43,44]
1984	3	24	Etorof-oki	7.1	7.1	0.0	6.40×10^{19}	0.29	0.28	ADDS	[43,44]
1984	8	7	Hyuganada	6.9	6.9	−1.0	2.90×10^{19}			ADDS	[43,44]
1984	9	19	Boso-oki	6.8	7.3	0.0	2.00×10^{19}			NF	[43,44]
1986	11	15	Taiwan-oki	7.3	7.6	1.0	1.30×10^{20}	0.54	0.54	ASSS	[43,44]
1989	10	29	Sanriku-oki	6.9	6.8	−1.0	5.80×10^{18}			ADDS	[43,44]
1989	11	2	Sanriku-oki	7.2	7.5	1.0	1.40×10^{20}	0.23	0.23	ADDS	[43,44]
1990	2	20	N. Oshima	6.2	6.5	−2.0	4.30×10^{18}			SS	[43,44]
1990	9	24	Tokai-oki	6.5	6.8	−1.0	7.10×10^{18}			SS	[43,44]
1991	12	22	Uruppu Isl	7.5	7.5	1.0	2.80×10^{20}	0.28	0.27	ADDS	[43,44]
1992	7	18	Sanriku-oki	6.8	7.2	0.0	2.70×10^{19}	0.19	0.18	ADDS	[43,44]
1993	2	7	Noto Pen.	6.6	6.7	−0.5	3.40×10^{18}			ASSS	[43,44]
1993	7	12	SW Hokkaido	7.7	8.1	3.0	4.70×10^{20}	0.57	0.57	ASSS	[43,44]
1994	10	4	E Hokkaido	8.1	8.2	3.0	3.00×10^{21}	0.66	0.27	ADDS	[43,44]
1994	12	28	Sanriku-oki	7.7	7.7	1.5	4.90×10^{20}	0.21	0.19	ADDS	[43,44]
1995	1	17	S Hyougo	6.8	6.4	−1.5	2.40×10^{19}			SS	[43,44]
1995	10	18	Kikaijima	6.9	7.6	1.0	5.90×10^{19}			NF	[43,44]
1995	10	19	Kikaijima	6.7	7.3	0.0	1.50×10^{19}			NF	[43,44]
1995	12	4	Etorof-oki	7.6	7.6	1.0	8.20×10^{20}	0.21	0.21	ADDS	[43,44]
1996	10	19	Hyuganada	6.6	6.9	−1.0	1.40×10^{19}			ADDS	[43,44]

Table 2. *Cont.*

1996	12	3	Hyuganada	6.7	6.7	−1.0	1.20×10^{19}			ADDS	[43,44]
2001	12	18	Yonagunijima	6.8	6.8	−1.0	2.10×10^{19}			NF	[43,44]
2002	3	31	E Taiwan	7.1	7.2	−1.0	5.40×10^{19}			ASSS	[43,44]
2003	9	26	Tokachi-oki	8.1	8.1	2.5	3.10×10^{21}	0.19	0.14	ADDS	[43,44]
2003	10	31	Fukushima-oki	6.8	7.0	−0.5	3.50×10^{19}			ADDS	[43,44]
2005	8	16	Miyagi-oki	7.1	7.0	−1.0	7.60×10^{19}			ADDS	[43,44]
2005	11	15	Sanriku-oki	6.9	7.3	0.0	3.90×10^{19}			NF	[43,44]
2006	11	15	Kurile	7.9	8.2	3.0	3.50×10^{21}	0.26	0.26	ADDS	[43,44]
2007	1	13	Kurile	8.2		2.0	1.80×10^{21}			NF	[43,44]
2007	3	25	Noto Pen.	6.9		−1.0	1.30×10^{19}			ASSS	[43,44]
2007	8	2	Sakhalin	6.2		0.0	2.40×10^{18}	0.67	0.67	ASSS	[43,44]
2008	7	19	Fukushima-oki	6.9		0.0	2.90×10^{19}	0.28	0.28	ADDS	[43,44]
2008	9	11	Tokachi-oki	7.1		−0.5	1.80×10^{19}			ADDS	[43,44]
2009	10	30	Amamioshima	6.8		−1.0	1.80×10^{19}			ASSS	[43,44]
2011	3	11	Tohoku-oki	9.0	9.1	4.0	5.31×10^{22}	0.17	0.17	ADDS	[43,44]

$Mw^{\#}$; Moment magnitude [45], $Mt^{\$}$; Tsunami magnitude [46], $m^{\%}$; Tsunami magnitude [8,25,26]; $^{+}$Effect of fault dip; sin δ, *Effect of fault dip and slip; sinδsinλ.

Since the increase of one unit in m indicates the increase by the factor of 2.24 in tsunami heights [25,26], the above difference of 1.0 means the tsunami heights of ASSS earthquakes are about two times larger than those of ADDS earthquakes with the same seismic moment.

Figure 5 shows the relations between Mt and Mo of earthquakes in ADDS (Mt_D) and ASSS (Mt_S):

$$Mt_D = 1/1.5 \log Mo − 6.04 \pm 0.03$$
$$Mt_S = 1/1.5 \log Mo − 5.86 \pm 0.04 \tag{5}$$

where the coefficient of 1/1.5 of the relations is from the definition of Mw [28] with the assumption of $Mt = Mw$;

$$Mw = 1/1.5 \log Mo − 6.06 \tag{6}$$

Although there seems to be a little difference in plots between ADDS and ASSS in Figure 5, the difference is within the estimation uncertainty. It is about 0.2 in Mt from Equation (5), which would suggest that tsunami heights by ASSS earthquakes are larger by a factor of about 1.8 than those by ADDS earthquakes with the same value of seismic moment. All the discussions here suggest that the tsunami excitations by ASSS earthquakes are larger, by more or less about two times, than those by ADDS earthquakes near Japan.

5. Discussion

The difference in tsunami excitation is obtained by earthquakes in ADDS and ASSS near Japan, as shown in Figures 4 and 5. The difference is by a factor of about two. Similar emphasis has been made for tsunamigenic earthquakes in the Japan Sea through the analyses including those of tsunamis with smaller tsunami magnitude [7–9]. Generally speaking, the larger dip angles and dip components of dislocations on the faults of earthquakes in the Japan Sea give rise to the larger tsunami excitation. This is the major reason for the difference stated in the previous studies, in addition to the effect of a soft ocean-bottom structure. However, the previous emphasis should not be restricted to tsunamis in the Japan Sea, since ASSS earthquakes outside of the Japan Sea show larger tsunami excitation similar to those in the Japan Sea (Table 3). Therefore, the difference may not be attributed to the local ocean-bottom structures nor to faulting parameters but to the global characteristics of seismic segmentation or to the geological activity which makes such the surface structures.

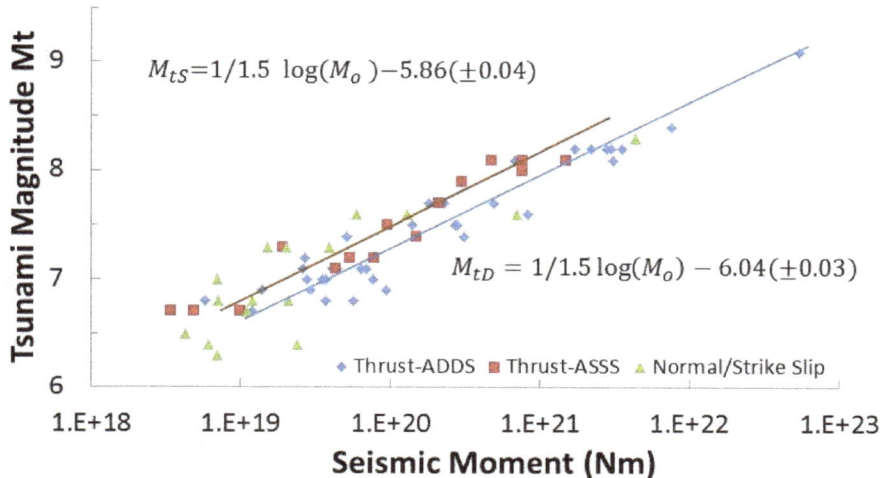

Figure 5. Tsunami magnitude Mt [27] in relation to seismic moment Mo of each earth-quake near Japan. Others are the same as those in Figure 4.

Table 3. Effect of dip angles and slip angles of dislocations on faults resposible for ocean-bottom vertical deformation. Average values are calculated from Table 2.

	Dip Angle ($\sin\delta$)	Sigma %	Dip and Slip Angle ($\sin\delta\sin\lambda$)	Sigma %
ADDS	0.32	0.13	0.26	0.08
ASSS	0.55	0.24	0.55	0.24
ASSSexJapan *	0.44	0.25	0.44	0.25

% Standard deviation, * Earthquakes classified ASSS but not in the Japan Sea.

Tsunami height is basically responsible for the ocean-bottom vertical deformation due to earthquake faults [47], so that the difference obtained indicates larger ocean-bottom vertical deformation for ASSS earthquakes than that for ADDS earthquakes of the same seismic moment. There would be many reasons to lead to such the difference as follows;

(A) Since the focal depth of ASSS earthquakes is in general shallower than those of ADDS earthquakes, the dislocation by ASSS earthquakes on particular underground faults produces larger effect on ocean-bottom deformation.

(B) Since the dip angle of faults by ASSS earthquakes is larger than those by ADDS earthquakes, the focal mechanism affects larger ocean-bottom deformation.

(C) Slip angle on the faults specifies the partition of along dip and along strike components of the dislocation on a fault. Focal mechanism of ASSS earthquakes provides more along dip components than that of ADDS earthquakes.

(D) Due to three dimensional structures of ocean-bottom topography, such as trench and continental shelf, tsunami excitations become large in ASSS.

(E) Focusing effect of tsunamis due to ocean-bottom topography near ASSS earthquake sources results in larger tsunamis.

(F) Soft skin layer in the ocean bottom enhances the vertical deformation, therefore, the soft sediment in the Japan Sea side would enhance the tsunami excitation there.

The reason (A) is the result of the character of ASSS asperities and/or coupled zones existing in the shallower depth of plate interfaces than that of ADDS asperities, as described in Section 2 and Figure 2. The reason (B) is related to (F), that is, dip angles of earthquake faults in the Japan Sea are generally larger than those of the subduction zone earthquakes in the Pacific Ocean side (Table 3). However, ASSS earthquakes in this study include earthquakes in the Nankai Trough and the Okinawa Trough along the Philippine Sea Plate subduction as well as earthquakes in the west coast of Sakhalin. It is also apparent from the result in Table 3 that the effect of dip and slip angles of faults classified into ASSS earthquakes excluding the events in the Japan Sea is consistent with that of ASSS earthquakes. Smaller tsunamis ($m \leq -0.5$) which were not included in the present analysis do show the similar tendency of the larger tsunami excitations in ASSS compared to those in ADDS (Figure 4). Furthermore, a theoretical study on the deformation due to earthquake faults gives that the difference in the maximum deformation on the surface between dip angles with 30 and 60 degrees is within about 30% [48–50]. That is too small to comprehend the observed difference in Equation (4) between the Japan Sea and the Pacific Sea side of Japan. For (B) and (C), dip angles and slip angles of faults in ADDS and ASSS are evaluated and listed in Table 2. It is found that there are larger values for the earthquakes in the Japan Sea side than those in the Pacific Ocean side, but these earthquakes are not restricted in the Japan Sea as is mentioned in (B).

Regarding reason (D), ASSS earthquakes include those occurring in the Japan Sea side, in the back-arc basin without apparent trenches, as well as along the trench with developed continental shelves in the Nankai Trough. They generate larger tsunamis than those by ADDS earthquakes elsewhere. This eliminates the possibility of (D) for the different tsunami excitation. Speaking about (E), the results from both the local and far-field tsunami magnitude give consistent results in Equations (4) and (5). Therefore, the effect of local ocean-bottom structures alone is not the reason. Considering the reason (F) in the above, we should remind readers that tsunami earthquakes, earthquakes with slow-slip characteristics along the Sanriku Coast of the Tohoku, have occurred, where the source region of the 2011 megathrust earthquake located. The subduction zone shows a developed Horst and Graben structure [19–21], which is formed by sedimentation and related to the generation of tsunami earthquakes. The source region is not similar to the sedimentary layer in the Japan Sea.

6. Conclusions

The relationship has been derived between moment magnitude Mw of earthquake sources and local tsunami magnitude m [25,26] for the tsunamigenic earthquakes near Japan, and it turns out to be similar to the relation between Mt and Mw [27]. These two tsunami magnitude scales suggest that the tsunami excitations in ASSS are about two times larger than those of ADDS near Japan. Observations of tsunamis are of special importance in measuring the strength of historical large earthquakes before the dawn of instrumental seismometry in the 19th century. Previous studies pointed out that there are larger tsunami

...s in the Japan Sea compared to those in the Pacific Ocean side of Japan; however, it was found ...s study that the larger excitation of tsunamis is not only restricted for earthquakes in the Japan Sea ...ut also for those in the extended seismic region of Along-strike Single Segmentation (ASSS). The reason inducing this difference has been discussed in detail and the most probable reason is suggested that the asperity of earthquakes in ASSS spreads over the whole shallow area of the subduction interface. The rupture of the shallow asperity causes large ocean-bottom deformation. On the other hand, an asperity in the deeper seismic segments of the subduction interface of ADDS exists as well as an everlasting asperity in the shallower seismic segments, which form the double aligned seismic segments along the dip direction of the subduction. In the deeper part, repeated large earthquakes occur. A rupture of such asperity induces a smaller amount of the ocean-bottom deformation compared to that by ASSS events. The asperity in the shallower part of ADDS ruptures only in the case of a megathrust event as a millennial event. This rupture process has been observed in the 2011 Tohoku-oki megathrust and the 1964 great Alaska earthquake, and the other candidate for this type of a gigantic tsunami is the 17th century tsunami in Hokkaido, Japan [21]. The global importance of different tsunami excitation due to the seismic segmentation of ADDS or ASSS is stressed and is now open to future studies.

It is necessary to take into account the different excitation of tsunamis due to the seismic segmentation pointed out in this study and to consider the effect of dip and slip angles to retrieve the information on the source size of historical earthquakes. This is also important in estimating the strength of tsunamis for future earthquakes judging from so-called seismic gaps and/or doughnut patterns as well as from geodetic deformations by GPS observations.

Acknowledgments

A.R. Gusman kindly discussed with us on the tsunami excitation, showing many of his synthetic calculations to us. We thank K.T. Ginboi for his thorough review of the galley proof of the manuscript.

Author Contributions

All co-authors contributed to the present study. J. Koyama performed the data analysis and prepared the manuscript. M. Tsuzuki performed the data collection and analysis. K. Yomogida took part in discussions regarding the analyses, reviewed and edited the manuscript.

Conflicts of Interest

The authors declare no conflict of interest.

References

1. Lay, T.; Kanamori, H.; Ammon, C.J.; Nettles, M.; Ward, S.N.; Aster, R.; Beck, S.L.; Bilek, S.L.; Brudzinski, M.R.; Butler, R.; *et al.* The great Sumatra-Andaman earthquake of 26 December, 2004. *Science* **2005**, *308*, 1127–1138.
2. Bilek, S.L. Invited review paper: Seismicity along the South American subduction zone: Review of large earthquakes, tsunamis, and subduction zone complexity. *Tectonophysics* **2010**, *495*, 2–14.

3. Peng, Z.; Gomber, J. An integrated perspective of the continuum betwee slow-slip phenomena. *Nature Geosci.* **2010**, *3*, 599–607, doi:10.1038/NGEO940

4. Yomogida, K.; Yoshizawa, K.; Koyama, J.; Tsuzuki, M. Along-dip segmentation the Pacific coast of Tohoku earthquake and comparison with other megathrust eart *Planets Space* **2011**, *63*, 697–701.

5. Koyama, J.; Yoshizawa, K.; Yomogida, K.; Tsuzuki, M. Variability of megathrust earthquakes in the world revealed by the 2011 Tohoku-oki earthquake. *Earth Planets Space* **2012**, *64*, 1189–1198.

6. Kopp, H. The control of subduction zone structural complexity and geometry on margin segmentation and seismicity. *Tectonophysics* **2013**, *589*, 1–16.

7. Abe, K. Quantification of major earthquake tsunamis of the Japan Sea. *Phys. Earth Planet. Inter.* **1985**, *38*, 214–223.

8. Hatori, T. Characteristics of tsunami magnitude near Japan. *Rep. Tsunami Eng.* **1996**, *13*, 17–26.

9. Watanabe, H. Regional variation of the formulae for the determination of tsunami magnitude in and around Japan. *J. Seism. Soc. Japan* **1995**, *48*, 271–280.

10. Available online: http://www.j-shis.bosai.go.jp/map/?lang=en (accessed on 21 September 2015).

11. Available online: http://www.jma.go.jp/jma/en/Activities/earthquake.html (accessed on 21 September 2015).

12. Kelleher, J.; Savino, J. Distribution of seismicity before large strike slip and thrust-type earthquakes. *J. Geophys. Res.* **1975**, *80*, 260–271.

13. Koyama, J.; Tsuzuki, M. Activity of significant earthquakes before and after megathrust earthquakes in the world. *J. Seism. Soc. Jpn.* **2014**, *66*, 83–95.

14. Mogi, K. Seismicity in western Japan and long-term earthquake forecasting. *Earthquake Prediction*, Maurice Ewing Series, 4th ed.; Simpson, D.W., Richards, P.G., Eds.; AGU: Washington, DC, USA, 1981; pp. 43–51.

15. Moreno, M.; Rosenau, M.; Oncken, O. Maule earthquake slip correlates with pre-seismic locking of Andean subduction zone. *Nature* **2010**, *467*, 198–202.

16. Kohketsu, K.; Yokota, Y.; Nishimura, N.; Yagi, Y.; Miyazaki, S.I.; Satake, K.; Fujii, Y.; Miyake, H.; Sakai, S.; Yamanakae, Y.; *et al.* A unified source model for the 2011 Tohoku earthquake. *Earth Planet. Sci. Let.* **2011**, *310*, 480–487.

17. Maeda, T.; Furumura, T.; Sakai, S.; Shinohara, M. Significant tsunami observed at ocean-bottom pressure gauges during the 2011 off the Pacific coast of Tohoku earthquake. *Earth Planet Space* **2011**, *63*, 803–808.

18. Ruff, L.; Kanamori, H. The rupture process and asperity distribution of three great earthquakes from long-period diffracted P-waves. *Phys. Earth Planet. Inter.* **1983**, *31*, 202–230.

19. Tanioka, Y.; Ruff, L.; Satake, K. What controls the lateral variation of large earthquake occurrence along the Japan trench. *The Island Arc* **1998**, *6*, 261–266.

20. Bell, R.; Holden, C.; Power, W.; Wang, X.; Downes, G. Hikurangi margin tsunami earthquake generated by slow seismic rupture over a subducted seamount. *Earth Planet. Sci. Let.* **2014**,*397*, 1–9.

21. Ioki, K.; Tanioka, Y. Re-evaluated fault model of the 17th century great earthquake off Hokkaido using tsunami deposit data. *Earth Planet. Sci. Lett.* **2015**, submitted.

22. Imamura, M. Chronological table of tsunamis in Japan. *J. Seism. Soc. Japan* **1949**, *2*, 23–28.

, K. Magnitude and energy of earthquakes associated by tsunami, and tsunami energy. *J. Earth Sci. Nagoya Univ.* **1958**, *6*, 101–112.

24. Available online: http://www.ngdc.noaa.gov/nndc/struts/form?t=101650&s=70&d=7 (accessed on 21 September 2015).

25. Hatori, T. Relation between tsunami magnitude and wave energy. *Bull. Earthq. Res. Inst. Tokyo Univ.* **1979**, *54*, 531–541.

26. Hatori, T. Classification of tsunami magnitude scale. *Bull. Earthq. Res. Inst. Tokyo Univ.* **1986**, *61*, 473–515.

27. Abe, K. Size of great earthquakes of 1837–1974 inferred from tsunami data. *J. Geophys. Res.* **1979**, *84*, 1561–1568.

28. Hanks, T.; Kanamori, H. A moment magnitude scale. *J. Geophys. Res.* **1979**, *84*, 2348–2350.

29. Kanamori, H. Faulting of the great Kanto earthquake of 1923 as revealed by seismological data. *Bull. Earthq. Res. Inst. Tokyo Univ.* **1971**, *49*, 13–18.

30. Kanamori, H. Seismological evidence for a lithospheric normal faulting—The Sanriku earthquake of 1933. *Phys. Earth Planet. Int.* **1971**, *4*, 289–300.

31. Abe, K. Tectonic implications of the large Shioya-oki earthquakes of 1938. *Tectonophysics* **1977**, *41*, 269–289.

32. Fukao, Y.; Furumoto, M. Mechanism of large earthquakes along the eastern margin of the Japan Sea. *Tectonophysics* **1975**, *25*, 247–266.

33. Kanamori, H. Tectonic implications of the 1944 Tonankai and the 1946 Nankaido earthquakes. *Phys. Earth Planet. Int.* **1972**, *5*, 129–139.

34. Kanamori, H. Synthesis of long-period surface waves and its application to earthquake source studies—Kurile Islands earthquake of October 13, 1963. *J. Geophys. Res.* **1970**, *75*, 5011–5027.

35. Aki, K. Generation and propagation of G waves from the Niigata earthquake of June 16, 1964, Part 1. A statistical analysis. *Bull. Earthq. Res. Inst. Tokyo Univ.* **1966**, *44*, 33–72.

36. Shiono, K.; Mikumo, T.; Ishikawa, Y. Tectonics of the Kyushu-Ryukyu arc as evidenced from seismicity and focal mechanism of shallow to intermediate-depth earthquakes. *J. Phys. Earth* **1980**, *28*, 17–43.

37. Kanamori, H. Focal mechanism of the Tokachi-Oki earthquake of May 16, 1968: Contortion of the lithosphere at a junction of two trenches. *Tectonophysics* **1971**, *12*, 1–13.

38. Yoshioka, N.; Abe, K. Focal mechanism of the Iwate-oki earthquake of June 12, 1968. *J. Phys. Earth* **1976**, *24*, 251–262.

39. Abe, K. Tsunami and mechanism of great earthquakes. *Phys. Earth Planet. Int.* **1973**, *7*, 143–153.

40. Shimazaki, K. Nemuro-oki earthquake of June 17, 1973: A lithospheric rebound at the upper half of the interface. *Phys. Earth Planet. Int.* **1974**, *9*, 315–327.

41. Takemura, M.; Koyama, J.; Suzuki, Z. Source process of the 1974 and 1975 earthquakes in Kurile Islands in special relation to the difference in excitation of tsunami. *Tohoku Geophys. J.* **1977**, *24*, 113–132.

42. Shimazaki, K.; Sommerville, P. Static and dynamic parameters of the Izu-Oshima, Japan, earthquake of June 14, 1978. *Bull. Seism. Soc. Am.* **1979**, *69*, 1343–1378.

43. Available online: http://earthquake.usgs.gov/earthquakes/eqarchives/sopar/ (accessed on 21 September 2015).

44. Available online: http://www.globalcmt.org/CMTsearch.html (accessed on 21 Se~~

45. Available online: http://earthquake.usgs.gov/earthquakes/world/historical_cou~. on 21 September 2015).

46. Available online: http://wwweic.eri.u-tokyo. ac.jp/tsunamiMt.html (accessed on 21 Se~

47. Gusman, A.R.; Tanioka, Y. W phase inversion and tsunami inundation modeling for ~~ami early warning: Case study for the 2011 Tohoku event. *Pure Appl. Geophys.* **2013**, *170*, doi:10.1007/s00024-013-0680-z.

48. Matsu'ura, M.; Sato, R. Displacement fields due to the faults. *J. Seism. Soc. Jpn.* **1975**, *28*, 429–434.

49. Himematsu, Y. Hokkaido Univ. prepared numerical calculations on the displacements fields based on Okada, Y. *Bull. Seism. Soc. Am.* **1992**, *82*, 1018–1040.

50. Available online: http:/www.bosai.go.jp/study/application/dc3d/DC3Dhtml_J.html (accessed on 21 September 2015).

ermissions

The contributors of this book come from diverse backgrounds, making this book a truly international effort. This book will bring forth new frontiers with its revolutionizing research information and detailed analysis of the nascent developments around the world.

We would like to thank all the contributing authors for lending their expertise to make the book truly unique. They have played a crucial role in the development of this book. Without their invaluable contributions this book wouldn't have been possible. They have made vital efforts to compile up to date information on the varied aspects of this subject to make this book a valuable addition to the collection of many professionals and students.

This book was conceptualized with the vision of imparting up-to-date information and advanced data in this field. To ensure the same, a matchless editorial board was set up. Every individual on the board went through rigorous rounds of assessment to prove their worth. After which they invested a large part of their time researching and compiling the most relevant data for our readers.

The editorial board has been involved in producing this book since its inception. They have spent rigorous hours researching and exploring the diverse topics which have resulted in the successful publishing of this book. They have passed on their knowledge of decades through this book. To expedite this challenging task, the publisher supported the team at every step. A small team of assistant editors was also appointed to further simplify the editing procedure and attain best results for the readers.

Apart from the editorial board, the designing team has also invested a significant amount of their time in understanding the subject and creating the most relevant covers. They scrutinized every image to scout for the most suitable representation of the subject and create an appropriate cover for the book.

The publishing team has been an ardent support to the editorial, designing and production team. Their endless efforts to recruit the best for this project, has resulted in the accomplishment of this book. They are a veteran in the field of academics and their pool of knowledge is as vast as their experience in printing. Their expertise and guidance has proved useful at every step. Their uncompromising quality standards have made this book an exceptional effort. Their encouragement from time to time has been an inspiration for everyone.

The publisher and the editorial board hope that this book will prove to be a valuable piece of knowledge for researchers, students, practitioners and scholars across the globe.

List of Contributors

Su Yean Teh
School of Mathematical Sciences, Universiti Sains Malaysia, Penang 11800, Malaysia

Michael Turtora
U.S. Geological Survey, Caribbean-Florida Water Science Center, 4446 Pet Lane, Suite #108, Lutz, FL 33559-630, USA

Donald L. DeAngelis
U.S. Geological Survey, Southeast Ecological Science Center, Coral Gables, FL 33124, USA

Jiang Jiang
Jiang Jiang, Forestry College of Nanjing Forestry University, Key Laboratory of soil and water conservation and Ecological Restoration, Nanjing Forestry University, Nanjing 210037, China

Leonard Pearlstine
Leonard Pearlstine, Everglades National Park, South Florida Natural Resources Center, 950 N Krome Ave, Homestead, FL 33030, USA

Thomas J. Smith III
U.S. Geological Survey, 600 Fourth Street South, St. Petersburg, FL 33701, USA

Hock Lye Koh
Hock Lye Koh, Sunway University Business School, Jalan Universiti, Bandar Sunway, Selangor 47500, Malaysia

Nusrat Jahan, Jason Fawcett, Alexander M. McPherson, Katherine N. Robertson and Jason A. C. Clyburne
Atlantic Centre for Green Chemistry, Departments of Chemistry and Environmental Science, Saint Mary's University, Halifax, NS B3H 3C3, Canada

Thomas L. King
Centre for Offshore Oil, Gas and Energy Research, Bedford Institute of Oceanography, Dartmouth, NS B2Y 4A2, Canada

Ulrike Werner-Zwanziger
Department of Chemistry and Institute for Research in Materials, Dalhousie University, Halifax, NS B3H 4J3, Canada

Hanh Nguyen-Kim
Institute of Research for Development (IRD), National Center for Scientific Research (CNRS), UMR MARBEC, Montpellier 34095 cedex, France
Institute of Oceanography (IO), Vietnam Academy of Science and Technology (VAST), Nha Trang 650000, Vietnam

Thierry Bouvier and Corinne Bouvier
Institute of Research for Development (IRD), National Center for Scientific Research (CNRS), UMR MARBEC, Montpellier 34095 cedex, France

Van Ngoc Bui
Institute of Biotechnology (IBT), Vietnam Academy of Science and Technology (VAST),Hanoi, 100000, Vietnam

Huong Le-Lan
Institute of Oceanography (IO), Vietnam Academy of Science and Technology (VAST), Nha Trang 650000, Vietnam

Yvan Bettarel
Institute of Research for Development (IRD), National Center for Scientific Research (CNRS), UMR MARBEC, Montpellier 34095 cedex, France
IRD–Van Phuc Diplomatic Compound, Bldg 2G, Appt 202, 298 Kim Ma, Ba Dinh, Hanoi 100000, Vietnam

Fabrice Teletchea
Research Unit Animal and Functionalities of Animal Products (URAFPA), University of Lorraine–INRA, 2 Avenue de la Forêt de Haye, BP 172, 54505 Vandoeuvre-lès-Nancy, France

Climent Molins, Pau Trubat and Alexis Campos
Department of Civil and Environmental Engineering, Universitat Politècnica de Catalunya–BarcelonaTech, Jordi Girona 1-3, Campus Nord C1-206, 08018 Barcelona, Spain

Xavi Gironella
Department of Civil and Environmental Engineering, Universitat Politècnica de Catalunya–BarcelonaTech, Jordi Girona 1-3, Campus Nord D1- 111A, 08018 Barcelona, Spain

Gloria Biern, Gemma Monyarch and Carme Fuster
Unitat de Biologia Cel•lular i Genètica Mèdica, Facultat de Medicina, Universitat Autònoma de Barcelona (UAB), 08193-Bellaterra, Barcelona, Spain

Jesús Giraldo
Unitat de Bioestadística i Institut de Neurociències, Facultat de Medicina, Universitat Autònoma de Barcelona (UAB), 08193-Bellaterra, Barcelona, Spain

Jan-Paul Zock and Ana Espinosa
Centre de Recerca en Epidemiologia Ambiental (CREAL), 08003-Barcelona, Spain
Universitat Pompeu Fabra, 08002-Barcelona, Spain
CIBER Epidemiología y Salud Pública (CIBERESP), 28029-Madrid, Spain

...odríguez-Trigo
...icio de Neumología, Hospital Clínico San Carlos,
28040-Madrid, Spain
Facultad de Medicina, Universidad Complutense,
28040-Madrid, Spain
CIBER Enfermedades Respiratorias (CIBERES), Bunyola,
07004-Mallorca, Spain

Federico Gómez and Joan-Albert Barberà
CIBER Enfermedades Respiratorias (CIBERES), Bunyola,
07004-Mallorca, Spain
Departamento de Medicina Respiratòria, Hospital Clínic-
Institut d'Investigacions Biomèdiques
August Pi I Sunyer (IDIBAPS), 08036-Barcelona, Spain

Francisco Pozo-Rodríguez
CIBER Enfermedades Respiratorias (CIBERES), Bunyola,
07004-Mallorca, Spain
Departamento de Medicina Respiratoria, Unidad
Epidemiologia Clínica, Hospital 12 de Octubre,
28047-Madrid, Spain

Gerrit Lohmann and Martin Butzin
Alfred-Wegener-Institut Helmholtz Zentrum für Polar-
und Meeresforschung, Bussestr. 24, 27570 Bremerhaven,
Germany
MARUM, Center for Marine Environmental Sciences,
University of Bremen, P.O. Box 330440, 28334 Bremen,
Germany

Torsten Bickert
MARUM, Center for Marine Environmental Sciences,
University of Bremen, P.O. Box 330440, 28334 Bremen,
Germany

**Tessa Gordelier David Parish, Philipp R. Thies and
Lars Johanning**
University of Exeter, Treliever Road, Penryn, Cornwall
TR10 9FE, UK

Aggeliki Barberopoulou
National Observatory of Athens, Institute of Geodynamics,
Lofos Nymphon, 11810 Athens, Greece
AIR Worldwide, 131 Dartmouth Street, Boston, MA
02116, USA

Mark R. Legg
Legg Geophysical, 16541 Gothard St # 107, Huntington
Beach, CA 92647, USA

Edison Gica
Pacific Marine Environmental Laboratory, Center for
Tsunami Research, National Oceanic and Atmospheric
Administration, 7600 Sand Point Way NE, Seattle, WA
98115, USA
Joint Institute for the Study of the Atmosphere and
Ocean, University of Washington, 3737 Brooklyn Ave
NE, Box 355672, Seattle, WA 98195-5672, USA

**Sam D. Weller, Philipp R. Thies, Tessa Gordelier and
Lars Johanning**
University of Exeter, Penryn Campus, Cornwall, TR10
9FE, UK

**Tonya Cross Hansel, Howard J. Osofsky, Joy D. Osofsky
and Anthony Speier**
Department of Psychiatry, Louisiana State University
Health Sciences Center, 1542 Tulane Avenue, New
Orleans, LA 70433, USA

Vivienne R. Johnson and Jason M. Hall-Spencer
Marine Biology and Ecology Research Centre, School of
Marine Science and Engineering, Plymouth University,
Plymouth PL4 8AA, UK

Colin Brownlee
The Marine Biological Association of the United Kingdom,
Citadel Hill, Plymouth PL1 2PB, UK

Marco Milazzo
Department of Earth and Marine Sciences, University of
Palermo, I-90123 Palermo, Italy

Naohiro Matsui
Environment Department, The General Environmental
Technos Co., Ltd., Osaka 541-0052, Japan

Wijarn Meepol
Ranong Mangrove Forest Research Center, Department
of Marine and Coastal Resources, Tambon Ngao, Muang
District, Ranong 85000, Thailand

Jirasak Chukwamdee
Department of National Park, Wildlife and Plant
Conservation, 61 Pholyothin Road, Ladyao, Chatuchak,
Bangkok 10900, Thailand

Junji Koyama
Division of Natural History Sciences, Graduate School of
Science, Hokkaido University, N10 W8, Kita-ku, Sapporo,
Hokkaido 0650810, Japan
Hyotanjima Scholorship, Sapporo, Hokkaido 151-1854630,
Japan

Motohiro Tsuzuki and Kiyoshi Yomogida
Division of Natural History Sciences, Graduate School of
Science, Hokkaido University, N10 W8, Kita-ku, Sapporo,
Hokkaido 0650810, Japan